KU-733-167

Structural Engineer's Pocket Book

Second edition

Fiona Cobb

ELSEVIER

AMSTERDAM • BOSTON • HEIDELBERG • LONDON • NEW YORK • OXFORD
PARIS • SAN DIEGO • SAN FRANCISCO • SINGAPORE • SYDNEY • TOKYO
Butterworth-Heinemann is an imprint of Elsevier

Butterworth-Heinemann is an imprint of Elsevier
Linacre House, Jordan Hill, Oxford OX2 8DP, UK
30 Corporate Drive, Suite 400, Burlington, MA 01803, USA

First edition 2004
Reprinted 2004 (twice), 2005, 2006
Second edition 2009

British Library Cataloguing in Publication Data
A catalogue record for this book is available from the British Library

Library of Congress Cataloging-in-Publication Data
A catalog record for this book is available from the Library of Congress

ISBN: 978-0-7506-8686-0

For information on all Butterworth-Heinemann publications
visit our website at books.elsevier.com

Printed and bound in *Great Britain*

09 10 10 9 8 7 6 5

Working together to grow
libraries in developing countries

www.elsevier.com | www.bookaid.org | www.sabre.org

ELSEVIER   BOOK AID
            International   Sabre Foundation

Contents

Preface to Second Edition

When the *Structural Engineer's Pocket Book* was first conceived, I had no idea how popular and widely used it would become. Thank you to all those who took the time to write to me with suggestions. I have tried to include as many as I can, but as the popularity of the book is founded on a delicate balance between size, content and cover price, I have been unable to include everything asked of me. Many readers will notice that references to Eurocodes are very limited. The main reason being that the book is not intended as a text book and is primarily for use in scheme design (whose sizes do not vary significantly from those determined using British Standards). However Eurocode data will be included in future editions once the codes (and supporting documents) are complete, the codes have completed industry testing and are more widely used.

As well as generally updating the British Standards revised since 2002, the main additions to the second edition are: a new chapter on sustainability, addition of BS 8500, revised 2007 Corus steel section tables (including 20 new limited release UB and UC sections) and a summary of Eurocode principles and load factors.

Once again, I should say that I would be interested to receive any comments, corrections or suggestions on the content of the book by email at sepb@inmyopinion.co.uk.

Fiona Cobb

Preface to First Edition

As a student or graduate engineer it is difficult to source basic design data. Having been unable to find a compact book containing this information, I decided to compile my own after seeing a pocket book for architects. I realized that a *Structural Engineer's Pocket Book* might be useful for other engineers and construction industry professionals. My aim has been to gather useful facts and figures for use in preliminary design in the office, on site or in the IStructE Part 3 exam, based on UK conventions.

The book is not intended as a textbook; there are no worked examples and the information is not prescriptive. Design methods from British Standards have been included and summarized, but obviously these are not the only way of proving structural adequacy. Preliminary sizing and shortcuts are intended to give the engineer a 'feel' for the structure before beginning design calculations. All of the data should be used in context, using engineering judgement and current good practice. Where no reference is given, the information has been compiled from several different sources.

Despite my best efforts, there may be some errors and omissions. I would be interested to receive any comments, corrections or suggestions on the content of the book by email at sepb@inmyopinion.co.uk. Obviously, it has been difficult to decide what information can be included and still keep the book a compact size. Therefore any proposals for additional material should be accompanied by a proposal for an omission of roughly the same size – the reader should then appreciate the many dilemmas that I have had during the preparation of the book! If there is an opportunity for a second edition, I will attempt to accommodate any suggestions which are sent to me and I hope that you find the *Structural Engineer's Pocket Book* useful.

Fiona Cobb

Acknowledgements

Thanks to the following people and organizations:

Price & Myers for giving me varied and interesting work, without which this book would not have been possible! Paul Batty, David Derby, Sarah Fawcus, Step Haiselden, Simon Jewell, Chris Morrisey, Mark Peldmanis, Sam Price, Helen Remordina, Harry Stocks and Paul Toplis for their comments and help reviewing chapters. Colin Ferguson, Derek Fordyce, Phil Gee, Alex Hollingsworth, Paul Johnson, Deri Jones, Robert Myers, Dave Rayment and Andy Toohey for their help, ideas, support, advice and/or inspiration at various points in the preparation of the book. Renata Corbani, Rebecca Rue and Sarah Hunt at Elsevier. The technical and marketing representatives of the organizations mentioned in the book. Last but not least, thanks to Jim Cobb, Elaine Cobb, Iain Chapman for his support and the loan of his computer and Jean Cobb for her help with typing and proof reading.

Additional help on the second edition:

Lanh Te, Prashant Kapoor, Meike Borchers and Dave Cheshire.

Text and illustration credits

Permission to reproduce extracts from the British Standards is granted by BSI. British Standards can be obtained in PDF or hard copy formats from the BSI online shop: www.bsigroup.com/Shop or by contacting BSL Customer Services for hard copies only: Tel: +44 (0)20 8996 9001, Email cservices@bsigroup.com

Figures 2.1, 3.1, 3.4, 5.3 reproduced under the terms of the Click-Use Licence.

1
General Information

Metric system

The most universal system of measurement is the International System of Units, referred to as SI, which is an absolute system of measurement based upon the fundamental quantities of mass, length and time, independent of where the measurements are made. This means that while mass remains constant, the unit of force (newton) will vary with location. The acceleration due to gravity on earth is $9.81\,m/s^2$.

The system uses the following basic units:

Length	**m**	metre
Time	**s**	second
Luminous intensity	**cd**	candela
Quantity/substance	**mol**	mole $(6.02 \times 10^{23}$ particles of substance (Avogadro's number))
Mass	**kg**	kilogram
Temperature	**K**	kelvin ($0°C = 273°K$)
Unit of plane angle	**rad**	radian

The most commonly used prefixes in engineering are:

giga	**G**	1 000 000 000	1×10^9
mega	**M**	1 000 000	1×10^6
kilo	**k**	1000	1×10^3
centi	**c**	0.01	1×10^{-2}
milli	**m**	0.001	1×10^{-3}
micro	**μ**	0.000001	1×10^{-6}
nano	**N**	0.000000001	1×10^{-9}

The base units and the prefixes listed above, imply a system of supplementary units which forms the convention for noting SI measurements, such as the pascal for measuring pressure where $1\,Pa = 1\,N/m^2$ and $1\,MPa = 1\,N/mm^2$.

Typical metric units for UK structural engineering

Mass of material	kg
Density of material	kg/m^3
Bulk density	kN/m^3
Weight/force/point load	kN
Bending moment	kNm
Load per unit length	kN/m
Distributed load	kN/m^2
Wind loading	kN/m^2
Earth pressure	kN/m^2
Stress	N/mm^2
Modulus of elasticity	kN/mm^2
Deflection	mm
Span or height	m
Floor area	m^2
Volume of material	m^3
Reinforcement spacing	mm
Reinforcement area	mm^2 or mm^2/m
Section dimensions	mm
Moment of inertia	cm^4 or mm^4
Section modulus	cm^3 or mm^3
Section area	cm^2 or mm^2
Radius of gyration	cm or mm

Imperial units

In the British Imperial System the unit of force (pound) is defined as the weight of a certain mass which remains constant, independent of the gravitational force. This is the opposite of the assumptions used in the metric system where it is the mass of a body which remains constant. The acceleration due to gravity is 32.2 ft/s^2, but this is rarely needed. While on the surface it appears that the UK building industry is using metric units, the majority of structural elements are produced to traditional Imperial dimensions which are simply quoted in metric.

The standard units are:

Length

1 mile	= 1760 yards
1 furlong	= 220 yards
1 yard (yd)	= 3 feet
1 foot (ft)	= 12 inches
1 inch (in)	= 1/12 foot

Area

1 sq. mile	= 640 acres
1 acre	= 4840 sq. yd
1 sq. yd	= 9 sq. ft
1 sq. ft	= 144 sq. in
1 sq. in	= 1/144 sq. ft

Weight

1 ton	= 2240 pounds
1 hundredweight (cwt)	= 112 pounds
1 stone	= 14 pounds
1 pound (lb)	= 16 ounces
1 ounce	= 1/16 pound

Capacity

1 bushel	= 8 gallons
1 gallon	= 4 quarts
1 quart	= 2 pints
1 pint	= 1/2 quart
1 fl. oz	= 1/20 pint

Volume

1 cubic yard	= 27 cubic feet
1 cubic foot	= 1/27 cubic yards
1 cubic inch	= 1/1728 cubic feet

Nautical measure

1 nautical mile	= 6080 feet
1 cable	= 600 feet
1 fathom	= 6 feet

Conversion factors

Given the dual use of SI and British Imperial Units in the UK construction industry, quick and easy conversion between the two systems is essential. A selection of useful conversion factors are:

Mass	1 kg	= 2.205 lb	1 lb	= 0.4536 kg
	1 tonne	= 0.9842 tons	1 ton	= 1.016 tonnes
Length	1 mm	= 0.03937 in	1 in	= 25.4 mm
	1 m	= 3.281 ft	1 ft	= 0.3048 m
	1 m	= 1.094 yd	1 yd	= 0.9144 m
Area	1 mm^2	= 0.00153 in^2	1 in^2	= 645.2 mm^2
	1 m^2	= 10.764 ft^2	1 ft^2	= 0.0929 m^2
	1 m^2	= 1.196 yd^2	1 yd^2	= 0.8361 m^2
Volume	1 mm^3	= 0.000061 in^3	1 in^3	= 16390 mm^3
	1 m^3	= 35.32 ft^3	1 ft^3	= 0.0283 m^3
	1 m^3	= 1.308 yd^3	1 yd^3	= 0.7646 m^3
Density	1 kg/m^3	= 0.06242 lb/ft^3	1 lb/ft^3	= 16.02 kg/m^3
	1 tonne/m^3	= 0.7524 ton/yd^3	1 ton/yd^3	= 1.329 tonne/m^3
Force	1 N	= 0.2248 lbf	1 lbf	= 4.448 N
	1 kN	= 0.1004 tonf	1 tonf	= 9.964 kN
Stress and pressure	1 N/mm^2	= 145 lbf/in^2	1 lbf/in^2	= 0.0068 N/mm^2
	1 N/mm^2	= 0.0647 tonf/in^2	1 tonf/in^2	= 15.44 N/mm^2
	1 N/m^2	= 0.0208 lbf/ft^2	1 lbf/ft^2	= 47.88 N/m^2
	1 kN/m^2	= 0.0093 tonf/ft^2	1 tonf/ft^2	= 107.3 kN/m^2
Line loading	1 kN/m	= 68.53 lbf/ft	1 lbf/ft	= 0.0146 kN/m
	1 kN/m	= 0.03059 tonf/ft	1 tonf/ft	= 32.69 kN/m
Moment	1 Nm	= 0.7376 lbf ft	1 lbf ft	= 1.356 Nm
Modulus of elasticity	1 N/mm^2	= 145 lbf/in^2	1 lbf/in^2	= 6.8×10^{-3} N/mm^2
	1 kN/mm^2	= 145032 lbf/in^2	1 lbf/in^2	= 6.8×10^{-6} kN/mm^2
Section modulus	1 mm^3	= 61.01×10^{-6} in^3	1 in^3	= 16390 mm^3
	1 cm^3	= 61.01×10^{-3} in^3	1 in^3	= 16.39 cm^3
Second moment of area	1 mm^4	= 2.403×10^{-6} in^4	1 in^4	= 416200 mm^4
	1 cm^4	= 2.403×10^{-2} in^4	1 in^4	= 41.62 cm^4
Temperature	x°C	= $[(1.8x + 32)]$°F	y°F	= $[(y - 32)/1.8]$°C

NOTES:
1. 1 tonne = 1000 kg = 10 kN.
2. 1 ha = 10000 m^2.

Measurement of angles

There are two systems for the measurement of angles commonly used in the UK.

English system

The English or sexagesimal system which is universal:

1 right angle = 90° (degrees)
1° (degree)   = 60′ (minutes)
1′ (minute)   = 60″ (seconds)

International system

Commonly used for the measurement of plane angles in mechanics and mathematics, the radian is a constant angular measurement equal to the angle subtended at the centre of any circle, by an arc equal in length to the radius of the circle.

π radians   = 180° (degrees)

1 radian $= \dfrac{180°}{\pi} = \dfrac{180°}{3.1416} = 57°\ 17′\ 44″$

Equivalent angles in degrees and radians and trigonometric ratios

Angle θ in radians	0	$\dfrac{\pi}{6}$	$\dfrac{\pi}{4}$	$\dfrac{\pi}{3}$	$\dfrac{\pi}{2}$
Angle θ in degrees	0°	30°	45°	60°	90°
$\sin \theta$	0	$\dfrac{1}{2}$	$\dfrac{1}{\sqrt{2}}$	$\dfrac{\sqrt{3}}{2}$	1
$\cos \theta$	1	$\dfrac{\sqrt{3}}{2}$	$\dfrac{1}{\sqrt{2}}$	$\dfrac{1}{2}$	0
$\tan \theta$	0	$\dfrac{1}{\sqrt{3}}$	1	$\sqrt{3}$	∞

Construction documentation and procurement

Construction documentation

The members of the design team each produce drawings, specifications and schedules which explain their designs to the contractor. The drawings set out in visual form how the design is to look and how it is to be put together. The specification describes the design requirements for the materials and workmanship, and additional schedules set out sizes and co-ordination information not already covered in the drawings or specification. The quantity surveyor uses all of these documents to prepare bills of quantities, which are used to help break down the cost of the work. The drawings, specifications, schedules and bills of quantities form the tender documentation. 'Tender' is when the bills and design information are sent out to contractors for their proposed prices and construction programmes. 'Procurement' simply means the method by which the contractor is to be chosen and employed, and how the building contract is managed.

Certain design responsibilities can be delegated to contractors and subcontractors (generally for items which are not particularly special or complex, e.g. precast concrete stairs or concealed steelwork connections etc.) using a Contractor Design Portion (CDP) within the specifications. The CDP process reduces the engineer's control over the design, and therefore it is generally quicker and easier to use CDPs only for concealed/straightforward structural elements. They are generally unsuitable for anything new or different (when there is perhaps something morally dubious about trying to pass off design responsibility anyway).

With the decline of traditional contracts, many quantity surveyors are becoming confused about the differences between CDP and preliminaries requirements – particularly in relation to temporary works. Although temporary works should be allowed for in the design of the permanent works, its design and detailing is included as the contractor's responsibility in the contract preliminaries (normally NBS clause A36/320). Temporary works should not be included as a CDP as it is not the designer's responsibility to delegate. If it is mistakenly included, the designer (and hence the client) takes on additional responsibilities regarding the feasibility and co-ordination of the temporary works with the permanent works.

Traditional procurement

Once the design is complete, tender documentation is prepared and sent out to the selected contractors (three to six depending on how large the project is) who are normally only given a month to absorb all the information and return a price for the work. Typically, a main contractor manages the work on site and has no labour of his own. The main contractor gets prices for the work from subcontractors and adds profit and preliminaries before returning the tenders to the design team. The client has the option to choose any of the tenderers, but the selection in the UK is normally on the basis of the lowest price. The client will be in contract with the main contractor, who in turn is in contract with the subcontractors. The architect normally acts as the contract administrator for the client. The tender process is sometimes split to overlap part of the design phase with a first stage tender and to achieve a quicker start on site than with a conventional tender process.

Construction management

Towards the end of the design process, the client employs a management contractor to oversee the construction. The management contractor takes the tender documentation, splits the information into packages and chooses trade contractors (a different name for a subcontractor) to tender for the work. The main differences between construction management and traditional procurement are that the design team can choose which trade contractors are asked to price and the trade contractors are directly contracted to the client. While this type of contractual arrangement can work well for straightforward buildings it is not ideal for refurbishment or very complex jobs where it is not easy to split the job into simple 'trade packages'.

Design and Build

This procurement route is preferred by clients who want cost security and it is generally used for projects which have economy, rather than quality of design, as the key requirement. There are two versions of Design and Build. This first is for the design team to work for the client up to the tender stage, before being 'novated' to work for the main contractor. (A variant of this is a fixed sum contract where the design team remain employed by the client, but the cost of the work is fixed.) The second method is when the client tenders the project to a number of consortia on an outline description and specification. A consortium is typically led by a main contractor who has employed a design team. This typically means that the main contractor has much more control over the construction details than with other procurement routes.

Partnering

Partnering is difficult to define, and can take many different forms, but often means that the contractor is paid to be included as a member of the design team, where the client has set a realistic programme and budget for the size and quality of the building required. Partnering generally works best for teams who have worked together before, where the team members are all selected on the basis of recommendation and past performance. Ideally the contractor can bring his experience in co-ordinating and programming construction operations to advise the rest of the team on choice of materials and construction methods. Normally detailing advice can be more difficult as main contractors tend to rely on their subcontractors for the fine detail. The actual contractual arrangement can be as any of those previously mentioned and sometimes the main contractor will share the risk of costs increases with the client on the basis that they can take a share of any cost savings.

Drawing conventions

Drawing conventions provide a common language so that those working in the construction industry can read the technical content of the drawings. It is important for everyone to use the same drawing conventions, to ensure clear communication. Construction industry drawing conventions are covered by BS EN ISO 7519 which takes over from the withdrawn BS 1192 and BS 308.

A drawing can be put to its best use if the projections/views are carefully chosen to show the most information with the maximum clarity. Most views in construction drawings are drawn orthographically (drawings in two dimensions), but isometric (30°) and axonometric (45°) projections should not be forgotten when dealing with complicated details. Typically drawings are split into: location, assembly and component. These might be contained on only one drawing for a small job. Drawing issue sheets should log issue dates, drawing revisions and reasons for the issue.

Appropriate scales need to be picked for the different type of drawings:

Location/site plans – Used to show site plans, site levels, roads layouts, etc. Typical scales: 1:200, 1:500 and up to 1:2500 if the project demands.

General arrangement (GA) – Typically plans, sections and elevations set out as orthographic projections (i.e. views on a plane surface). The practical minimum for tender or construction drawings is usually 1:50, but 1:20 can also be used for more complicated plans and sections.

Details – Used to show the construction details referenced on the plans to show how individual elements or assemblies fit together. Typical scales: 1:20, 1:10, 1:5, 1:2 or 1:1.

Structural drawings should contain enough dimensional and level information to allow detailing and construction of the structure.

For small jobs or early in the design process, 'wobbly line' hand drawings can be used to illustrate designs to the design team and the contractor. The illustrations in this book show the type of freehand scale drawings which can be done using different line thicknesses and without using a ruler. These sorts of sketches can be quicker to produce and easier to understand than computer drawn information, especially in the preliminary stages of design.

Line thicknesses

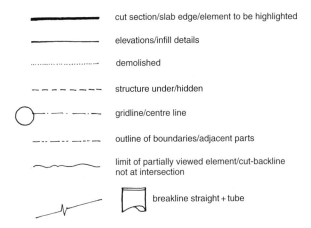

cut section/slab edge/element to be highlighted

elevations/infill details

demolished

structure under/hidden

gridline/centre line

outline of boundaries/adjacent parts

limit of partially viewed element/cut-backline not at intersection

breakline straight + tube

Hatching

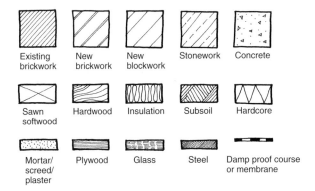

Existing brickwork

New brickwork

New blockwork

Stonework

Concrete

Sawn softwood

Hardwood

Insulation

Subsoil

Hardcore

Mortar/ screed/ plaster

Plywood

Glass

Steel

Damp proof course or membrane

Steps, ramps and slopes

Stairs

Ramp

Landscape slope

Slope/pitch

Arrow indicates 'up'

Common arrangement of work sections

The Common Arrangement of Work Sections for Building Work (CAWS) is intended to provide a standard for the production of specifications and bills of quantities for building projects, so that the work can be divided up more easily for costing and for distribution to subcontractors. The full document is very extensive, with sections to cover all aspects of the building work including: the contract, structure, fittings, finishes, landscaping and mechanical and electrical services. The following sections are extracts from CAWS to summarize the sections most commonly used by structural engineers:

A Preliminaries/ general conditions	A1	The project generally	A2	The contract
	A3	Employer's requirements	A4	Contractor's general costs
C Existing site/ buildings/ services	C1	Demolition	C2	Alteration – composite items
	C3	Alteration – support	C4	Repairing/renovating concrete/masonry
	C5	Repairing/renovating metal/timber		
D Groundwork	D1	Investigation/ stabilization/ dewatering	D2	Excavation/filling
	D3	Piling	D4	Diaphragm walling
	D5	Underpinning		
E In situ concrete/ large precast concrete	E1	In situ concrete	E2	Formwork
	E3	Reinforcement	E4	In situ concrete sundries
	E5	Precast concrete large units	E6	Composite construction
F Masonry	F1	Brick/block walling	F2	Stone walling
	F3	Masonry accessories		
G Structural/ carcassing in metal or timber	G1	Structural/carcassing metal	G2	Structural/carcassing timber
	G3	Metal/timber decking		
R Disposal systems	R1	Drainage	R2	Sewerage

There is a very long list of further subheadings which can be used to cover sections in more detail (e.g. F10 is specifically for Brick/block walling). However, the list is too extensive to be included here.

Source: CPIC (1998).

Summary of ACE conditions of engagement

The Association of Consulting Engineers (ACE) represents the consulting sector of the engineering profession in the UK. The ACE Conditions of Engagement, Agreement B(1), (2004) is used where the engineer is appointed directly to the client and works with an architect who is the lead consultant or contract administrator. A summary of the Normal Services from Agreement B(1) is given below with references to the lettered work stages (A–L) defined by the Royal Institute of British Architects (RIBA).

Feasibility		
Work Stage A	Appraisal	Identification of client requirements and development constraints by the Lead Consultant, with an initial appraisal to allow the client to decide whether to proceed and to select the probable procurement method.
Stage B	Strategic briefing	Confirmation of key requirements and constraints for or by the client, including any topographical, historical or contamination constraints on the proposals. Consider the effect of public utilities and transport links for construction and post construction periods on the project. Prepare a site investigation desk study and if necessary bring the full site investigation forward from Stage C. Identify the Project Brief, establish design team working relationships and lines of communication and discuss with the client any requirements for site staff or resident engineer. Collaborate on the design with the design team and prepare a stage report if requested by the client or lead consultant.
Pre-construction phase		
Stage C	Outline proposals	Visit the site and study any reports available regarding the site. Advise the client on the need and extent of site investigations, arrange quotes and proceed when quotes are approved by the client. Advise the client of any topographical or dimensional surveys that are required. Consult with any local or other authorities about matters of principle and consider alternative outline solutions for the proposed scheme. Provide advice, sketches, reports or outline specifications to enable the Lead Consultant to prepare his outline proposals and assist the preparation of a Cost Plan. Prepare a report and, if required, present to the client.
Stage D	Detailed proposals	Develop the design of the detailed proposals with the design team for submission of the Planning Application by the Lead Consultant. Prepare drawings, specifications, calculations and descriptions in order to assist the preparation of a Cost Plan. Prepare a report and, if required, present to the client.

Summary of ACE conditions of engagement – continued

Pre-construction phase – continued		
Stage E	Final proposals	Develop and co-ordinate all elements of the project in the overall scheme with the design team, and prepare calculations, drawings, schedules and specifications as required for presentation to the client. Agree a programme for the design and construction of the Works with the client and the design team.
Stage F	Production information	Develop the design with the design team and prepare drawings, calculations, schedules and specifications for the Tender Documentation and for Building Regulations Approval. Prepare any further drawings and schedules necessary to enable Contractors to carry out the Works, excluding drawings and designs for temporary works, formwork, and shop fabrication details (reinforcement details are not always included as part of the normal services). Produce a Designer's Risk Assessment in line with Health & Safety CDM Regulations. Advise the Lead Consultant on any special tender or contract conditions.
Stage G	Tender documents	Assist the Lead Consultant in identifying and evaluating potential contractors and/or specialists for the construction of the project. Assist the selection of contractors for the tender lists, assemble Tender Documentation and issue it to the selected tenderers. On return of
Stage H	Tender action	tenders, advise on the relative merits of the contractors proposals, programmes and tenders.
Construction phase		
Stage J	Mobilization	Assist the Client and Lead Consultant in letting the building contract, appointing the contractor and arranging site hand over to the contractor. Issue construction information to the contractor and provide further information to the contractor as and when reasonably required. Comment on detailed designs, fabrication drawings, bar bending schedules and specifications submitted by the Contractors, for general dimensions, structural adequacy and conformity with the design. Advise on the need for inspections or tests arising during the construction phase and the appointment and duties of Site Staff.
Stage K	Construction to practical completion	Assist the Lead Consultant in examining proposals, but not including alternative designs for the Works, submitted by the Contractor. Attend relevant site meetings and make other periodic visits to the site as appropriate to the stage of construction. Advise the Lead Consultant on certificates for payment to Contractors. Check that work is being executed generally to the control documents and with good engineering practice. Inspect the construction on completion and, in conjunction with any Site Staff, record any defects. On completion, deliver one copy of each of the final structural drawings to the planning supervisor or client. Perform work or advise the Client in connection with any claim in connection with the structural works.
Stage L	After practical completion	Assist the Lead Consultant with any administration of the building contract after practical completion. Make any final inspections in order to help the Lead Consultant settle the final account.

Source: ACE (2004).

2
Statutory Authorities and Permissions

Planning

Planning regulations control individuals' freedom to alter their property in an attempt to protect the environment in UK towns, cities and countryside, in the public interest. Different regulations and systems of control apply in the different UK regions. Planning permission is not always required, and in such cases the planning department will issue a Lawful Development Certificate on request and for a fee.

England and Wales

The main legislation which sets out the planning framework in England and Wales is the Town and Country Planning Act 1990. The government's statements of planning policy may be found in White Papers, Planning Policy Guidance Notes (PPGs), Mineral Policy Guidance Notes (MPGs), Regional Policy Guidance Notes (RPGs), departmental circulars and ministerial statements published by the Department for Communities and Local Government (DCLG).

Scotland

The First Minister for Scotland is responsible for the planning framework. The main planning legislation in Scotland is the Town and Country Planning Act (Scotland) 1997 and the Planning (Listed Buildings and Conservation Areas) (Scotland) Act 1997. The legislation is supplemented by the Scottish Government who publish National Planning Policy Guidelines (NPPGs) which set out the Scottish policy on land use and other issues. In addition, a series of Planning Advice Notes (PANs) give guidance on how best to deal with matters such as local planning, rural housing design and improving small towns and town centres.

Northern Ireland

The Planning (NI) Order 1991 could be said to be the most significant of the many different Acts which make up the primary and subordinate planning legislation in Northern Ireland. As in the other UK regions, the Northern Ireland Executive publishes policy guidelines called Planning Policy Statements (PPSs) which set out the regional policies to be implemented by the local authority.

Building regulations and standards

Building regulations have been around since Roman times and are now used to ensure reasonable standards of construction, health and safety, energy efficiency and access for the disabled. Building control requirements, and their systems of control, are different for the different UK regions.

The legislation is typically set out under a Statutory Instrument, empowered by an Act of Parliament. In addition, the legislation is further explained by the different regions in explanatory booklets, which also describe the minimum standards 'deemed to satisfy' the regulations. The 'deemed to satisfy' solutions do not preclude designers from producing alternative solutions provided that they can be supported by calculations and details to satisfy the local authority who implement the regulations. Building control fees vary around the country but are generally calculated on a scale in relation to the cost of the work.

England and Wales

England and Wales has had building regulations since about 1189 when the first version of a London Building Act was issued. Today the relevant legislation is the Building Act 1984 and the Statutory Instrument Building Regulations 2000. The Approved Documents published by the DCLG are the guide to the minimum requirements of the regulations.

Applications may be made as 'full plans' submissions well before work starts, or for small elements of work as a 'building notice' 48 hours before work starts. Completion certificates demonstrating Building Regulations Approval can be obtained on request. Third parties can become approved inspectors and provide building control services.

Approved documents (as amended)

A   Structure
       A1 Loading
       A2 Ground Movement
       A3 Disproportionate Collapse
B   Fire Safety (volumes 1 and 2)
C   Site Preparation and Resistance to Moisture
D   Toxic Substances
E   Resistance to the Passage of Sound
F   Ventilation
G   Hygiene
H   Drainage and Waste Disposal
J   Combustion Appliances and Fuel Storage Systems
K   Protection from Falling, Collision and Impact
L   Conservation of Fuel and Power
       L1A New Dwellings
       L1B Existing Dwellings
       L2A New Buildings (other than dwellings)
       L2B Existing Buildings (other than dwellings)
M   Access to and Use of Buildings
N   Glazing
P   Electrical Safety – Dwellings
Regulation 7 Materials and Workmanship

Scotland

Building standards have been in existence in Scotland since around 1119 with the establishment of the system of Royal Burghs. The three principal documents which currently govern building control are the Building (Scotland) Act 2003 and the Technical Standards 1990 – the explanatory guide to the regulations published by the Scottish Government.

Applications for all building and demolition work must be made to the local authority, who assess the proposals for compliance with the technical standards, before issuing a building warrant, which is valid for five years. For simple works a warrant may not be required, but the regulations still apply. Unlike the other regions in the UK, work may only start on site once a warrant has been obtained. Buildings may only be occupied at the end of the construction period once the local authority have issued a completion certificate. Building control departments typically will only assess very simple structural proposals and for more complicated work, qualified engineers must 'self-certify' their proposals, overseen by the Scottish Building Standards Agency (SBSA). Technical handbooks are to be updated annually and are free to download from the SBSA's website.

Technical handbooks (domestic or non-domestic)

0 General
1 Structure
2 Fire
3 Environment
4 Safety
5 Noise
6 Energy
Appendix A Defined Terms
Appendix B List of Standards and other Publications

Northern Ireland

The main legislation, policy and guidelines in Northern Ireland are the Building Regulations (Northern Ireland) Order 1979 as amended by the Planning and Building Regulations (Northern Ireland) (Amendment) Order 1990; the Building Regulations (NI) 2000 and the technical booklets – which describe the minimum requirements of the regulations published by the Northern Ireland Executive.

Building regulations in Northern Ireland are the responsibility of the Department of Finance and Personnel and are implemented by the district councils. Until recently the regulations operated on strict prescriptive laws, but the system is now very similar to the system in England and Wales. Applicants must demonstrate compliance with the 'deemed to satisfy' requirements. Applications may be made as a 'full plans' submission well before work starts, or as a 'building notice' for domestic houses just before work starts. Builders must issue stage notices for local authority site inspections. Copies of the stage notices should be kept with the certificate of completion by the building owner.

Technical booklets

A   Interpretation and General
B   Materials and Workmanship
C   Preparation of Sites and Resistance to Moisture
D   Structure
E   Fire Safety
F   Conservation of Fuel and Power
         F1 Dwellings
         F2 Buildings other than dwellings
G   Sound
         G1 Sound (Conversions)
H   Stairs, Ramps, Guarding and Protection from Impact
J   Solid Waste in Buildings
K   Ventilation
L   Combustion Appliances and Fuel Storage Systems
N   Drainage
P   Sanitary Appliances and Unvented Hot Water Storage Systems
R   Access to and Use of Buildings
V   Glazing

Listed buildings

In the UK, buildings of 'special architectural or historic interest' can be listed to ensure that their features are considered before any alterations are agreed to the exterior or interior. Buildings may be listed because of their association with an important architect, person or event or because they are a good example of design, building type, construction or use of material. Listed building consent must be obtained from the local authority before any work is carried out on a listed building. In addition, there may be special conditions attached to ecclesiastical, or old ecclesiastical, buildings or land by the local diocese or the Home Office.

England and Wales

English Heritage (EH) in England and CADW in Wales work for the government to identify buildings of 'special architectural or historic interest'. All buildings built before 1700 (and most buildings between 1700 and 1840) with a significant number of original features will be listed. A building normally must be over 30 years old to be eligible for listing. There are three grades: I, II* and II, and there are approximately 500 000 buildings listed in England, with about 13 000 in Wales. Grades I and II* are eligible for grants from EH for urgent major repairs and residential listed buildings may be VAT zero rated for approved alterations.

Scotland

Historic Scotland maintains the lists and schedules for the Scottish Government. All buildings before 1840 of substantially unimpaired character can be listed. There are over 40 000 listed buildings divided into three grades: A, B and C. Grade A is used for buildings of national or international importance or little altered examples of a particular period, style or building type, while a Grade C building would be of local importance or be a significantly altered example of a particular period, style or building type.

Northern Ireland

The Environment and Heritage Service (EHS) within the Northern Ireland Executive has carried out a survey of all the building stock in the region and keeps the Northern Ireland Buildings Database. Buildings must be at least 30 years old to be listed and there are currently about 8500 listed buildings. There are three grades of listing: A, B+ and B (with two further classifications B1 and B2) which have similar qualifications to the other UK regions.

Conservation areas

Local authorities have a duty to designate conservation areas in any area of 'special architectural or historic interest' where the character or appearance of the area is worth preserving or enhancing. There are around 8500 conservation areas in England and Wales, 600 in Scotland and 30 in Northern Ireland. The character of an area does not just come from buildings and so the road and path layouts, greens and trees, paving and building materials and public and private spaces are protected. Conservation area consent is required from the local authority before work starts to ensure any alterations do not detract from the area's appearance.

Tree preservation orders

Local authorities have specific powers to protect trees by making Tree Protection Orders (TPOs). Special provisions also apply to trees in conservation areas. A TPO makes it an offence to cut down, lop, top, uproot, wilfully damage or destroy the protected tree without the local planning authority's permission. All of the UK regions operate similar guidelines with slightly different notice periods and penalties.

The owner remains responsible for the tree(s), their condition and any damage they may cause, but only the planning authority can give permission to work on them. Arboriculturalists (who can give advice on work which needs to be carried out on trees) and contractors (who are qualified to work on trees) should be registered with the Arboricultural Association. In some cases (including if the tree is dangerous) no permission is required, but notice (about 5 days (or 6 weeks in a conservation area) depending on the UK region) must be given to the planning authority. When it is agreed that a tree can be removed, this is normally on the condition that a similar tree is planted as a replacement. Permission is generally not required to cut down or work on trees with a trunk less than 75 mm diameter (measured at 1.5 m above ground level) or 100 mm diameter if thinning to help the growth of other trees. Fines of up to £20 000 can be levied if work is carried out without permission.

Archaeology and ancient monuments

Archaeology in Scotland, England and Wales is protected by the Ancient Monuments and Archaeology Areas Act 1979, while the Historic Monuments and Archaeology Objects (NI) Order 1995 applies in Northern Ireland.

Archaeology in the UK can represent every period from the camps of hunter gatherers 10 000 years ago to the remains of twentieth century industrial and military activities. Sites include places of worship, settlements, defences, burial grounds, farms, fields and sites of industry. Archaeology in rural areas tends to be very close to the ground surface, but in urban areas, deep layers of deposits were built up as buildings were demolished and new buildings were put directly on the debris. These deposits, often called 'medieval fill', are an average of 5 m deep in places like the City of London and York.

Historic or ancient monuments are structures which are of national importance. Typically monuments are in private ownership but are not occupied buildings. Scheduled monument consent is required for alterations and investigations from the regional heritage bodies: English Heritage, Historic Scotland, CADW in Wales and EHS in Northern Ireland.

Each of the UK regions operates very similar guidelines in relation to archaeology, but through different frameworks and legislation. The regional heritage bodies develop the policies which are implemented by the local authorities. These policies are set out in PPG 16 for England and Wales, NPPG 18 for Scotland and PPS 6 for Northern Ireland. These guidance notes are intended to ensure that:

1. Archaeology is a material consideration for a developer seeking planning permission.
2. Archaeology strategy is included in the urban development plan by the local planning authority.
3. Archaeology is preserved, where possible, in situ.
4. The developer pays for the archaeological investigations, excavations and reporting.
5. The process of assessment, evaluation and mitigation is a requirement of planning permission.
6. The roles of the different types of archaeologists in the processes of assessment, evaluation and mitigation are clearly defined.

Where 'areas of archaeological interest' have been identified by the local authorities, the regional heritage bodies act as curators (English Heritage, Historic Scotland, CADW in Wales and EHS in Northern Ireland). Any developments within an area of archaeological interest will have archaeological conditions attached to the planning permission to ensure that the following process is put into action:

1. Early consultation between the developers and curators so that the impact of the development on the archaeology (or vice versa) can be discussed and the developer can get an idea of the restrictions which might be applied to the site, the construction process and the development itself.
2. Desk study of the site by an archaeologist.
3. Field evaluation by archaeologists using field walking, trial pits, boreholes and/or geophysical prospecting to support the desk study.
4. Negotiation between the site curators and the developer's design team to agree the extent of archaeological mitigation. The developer must submit plans for approval by the curators.
5. Mitigation – either preservation of archaeology in situ or excavation of areas to be disturbed by development. The archaeologists may have either a watching brief over the excavations carried out by the developer (where they monitor construction work for finds) or on significant sites, carry out their own excavations.
6. Post-excavation work to catalogue and report on the archaeology, either store or display the findings.

Generally the preliminary and field studies are carried out by private consultants and contractors employed by the developers to advise the local authority planning department. In some areas advice can also be obtained from a regional archaeologist. In Northern Ireland, special licences are required for every excavation which must be undertaken by a qualified archaeologist. In Scotland, England and Wales, the archaeological contractors or consultants have a 'watching brief'.

Field evaluations can often be carried out using geotechnical trial pits with the excavations being done by the contractor or the archaeologist depending on the importance of the site. If an interesting find is made in a geotechnical trial pit and the archaeologists would like to keep the pit open for inspection by, say, the curators, the developer does not have to comply if there would be inconvenience to the developer or building users, or for health and safety reasons.

Engineers should ensure for the field excavation and mitigation stages that the archaeologists record **all** the features in the excavations up to this century's interventions as these records can be very useful to the design team. Positions of old concrete footings could have as much of an impact on proposed foundation positions as archaeological features!

Party Wall etc. Act

The Party Wall etc. Act 1996 came into force in 1997 throughout England and Wales. In 2008 there is no equivalent legislation in Northern Ireland. In Scotland, The Tenements (Scotland) Act 2004 applies, but this seems to relate more to management and maintenance of shared building assets.

Different sections of the Party Wall Act apply, depending on whether you propose to carry out work to an existing wall or structure shared with another property; build a freestanding wall or the wall of a building astride a boundary with a neighbouring property, and/or excavate within 3 m of a neighbouring building or structure. Work can fall within several sections of the Act at one time. A building Owner must notify his neighbours and agree the terms of a Party Wall Award before starting any work.

The Act refers to two different types of Party Structure: 'Party Wall' and 'Party Fence Wall'. Party Walls are loosely defined as a wall on, astride or adjacent to a boundary enclosed by building on one or both sides. Party Fence Walls are walls astride a boundary but not part of a building; it does not include things like timber fences. A Party Structure is a wide term which can sometimes include floors or partitions.

The Notice periods and sections 1, 2 and 6 of the Act are most commonly used, and are described below.

Notice periods and conditions

In order to exercise rights over the Party Structures, the Act says that the Owner must give Notice to Adjoining Owners; the building Owner must not cause unnecessary inconvenience, must provide compensation for any damage and must provide temporary protection for buildings and property where necessary. The Owner and the Adjoining Owner in the Act are defined as anyone with an interest greater than a tenancy from year to year. Therefore this can include shorthold tenants, long leaseholders and freeholders for any one property.

A building Owner, or surveyor acting on his behalf, must send a Notice in advance of the start of the work. Different Notice periods apply to different sections of the Act, but work can start within the Notice period with the written agreement of the Adjoining Owner. A Notice is only valid for one year from the date that it is served and must include the Owner's name and address, the building's address (if different); a clear statement that the Notice is under the provisions of the Act (stating the relevant sections); full details of the proposed work (including plans where appropriate) and the proposed start date for the work.

The Notice can be served by post, in person or fixed to the adjoining property in a 'conspicuous part of the premises'. Once the Notice has been served, the Adjoining Owner can consent in writing to the work or issue a counter Notice setting out any additional work he would like to carry out. The Owner must respond to a counter Notice within 14 days. If the Owner has approached the Adjoining Owners and discussed the work with them, the terms of a Party Wall Award may have already been agreed in writing before a Notice is served.

If a Notice is served and the Adjoining Owner does not respond within 14 days, a dispute is said to have arisen. If the Adjoining Owner refuses to discuss terms or appoint a surveyor to act on his behalf, the Owner can appoint a surveyor to act on behalf of the Adjoining Owner. If the Owners discuss, but cannot agree terms they can jointly appoint a surveyor (or they can each appoint one) to draw up the Party Wall Award. If two surveyors cannot agree, a nominated Third Surveyor can be called to act impartially. In complex cases, this can often take over a year to resolve and in such cases the Notice period can run out, meaning that the process must begin again by serving another Notice. In all cases, the surveyors are appointed to consider the rights of the Owner over the wall and not to act as advocates in the negotiation of compensation! The building Owner covers the costs associated with all of the surveyors and experts asked about the work.

When the terms have been agreed, the Party Wall Award should include a description (in drawings and/or writing) of what, when and how work is to be carried out; a record of the condition of the adjoining Owner's property before work starts; arrangements to allow access for surveyors to inspect while the works are going on and say who will pay for the cost of the works (if repairs are to be carried out as a shared cost or if the adjoining Owner has served a counter Notice and is to pay for those works). Either Owner has 14 days to appeal to the County Court against an Award if an Owner believes that the person who has drafted the Award has acted beyond their powers.

An Adjoining Owner can ask the owner for a 'bond'. The bond money becomes the property of the Adjoining Owner (until the work has been completed in accordance with the Award) to ensure that funds are available to pay for the completion of the works in case the Owner does not complete the works.

The Owner must give 14 days' Notice if his representatives are to access the Adjoining Owner's property to carry out or inspect the works. It is an offence to refuse entry or obstruct someone who is entitled to enter the premises under the Act if the offender knows that the person is entitled to be there. If the adjoining property is empty, the Owner's workmen and own surveyor or architect may enter the premises if they are accompanied by a police officer.

Section 1 – new building on a boundary line

Notice must be served to build on or astride a boundary line, but there is no right to build astride if your neighbour objects. You can build foundations on the neighbouring land if the wall line is immediately adjacent to the boundary, subject to supervision. The Notice is required at least **1 month** before the proposed start date.

Section 2 – work on existing party walls

The most commonly used rights over existing Party Walls include cutting into the wall to insert a DPC or support a new beam bearing; raising, underpinning, demolishing and/or rebuilding the Party Wall and/or providing protection by putting a flashing from the higher over the lower wall. Minor works such as fixing shelving, fitting electrical sockets or replastering are considered to be too trivial to be covered in the Act.

A building Owner, or Party Wall Surveyor acting on the Owner's behalf must send a Notice at least **2 months** in advance of the start of the work.

Section 6 – excavation near neighbouring buildings

Notice must be served at least **1 month** before an Owner intends to excavate or construct a foundation for a new building or structure within 3 m of an adjoining Owner's building where that work will go deeper than the adjacent Owner's foundations, or within 6 m of an adjoining Owner's building where that work will cut a line projecting out at 45° from the bottom of that building's foundations. This can affect neighbours who are not immediately adjacent. The Notice must state whether the Owner plans to strengthen or safeguard the foundations of the Adjoining Owner. Adjoining Owners must agree specifically in writing to the use of 'special foundations' – these include reinforced concrete foundations. After work has been completed, the Adjoining Owner may request particulars of the work, including plans and sections.

Source: DETR (1997).

CDM

The Construction Design & Management (CDM) Regulations 2007 were developed to assign responsibilities for health and safety to the client, design team and principal contractor. The Approved Code of Practice is published by the Health and Safety Executive for guidance to the Regulations.

The client is required to appoint a CDM Co-ordinator (CDMC) who has overall responsibility for co-ordinating health and safety aspects of the design and planning stages of a project. The duties of the CDMC can theoretically be carried out by any of the traditional design team professionals. The CDMC must ensure that the designers avoid, minimize or control health and safety risks for the construction and maintenance of the project, as well as ensuring that the contractor is competent to carry out the work and briefing the client on health and safety issues during the works.

The CDMC prepares the pre-construction information for inclusion in the tender documents which should include project-relevant health and safety information gathered from the client and designers. This should highlight any unusual aspects of the project (also highlighted on the drawings) that a competent contractor would not be expected to know. This document is taken on by the successful principal contractor and developed into the construction phase health and safety plan by the addition of the contractor's health and safety policy, risk assessments and method statements as requested by the designers. The health and safety plan is intended to provide a focus for the management and prevention of health and safety risks as the construction proceeds.

The health and safety file is generally compiled at the end of the project by the contractor and the CDMC who collect the design information relevant to the life of the building. The CDMC must ensure that the file is compiled and passed to the client or the people who will use, operate, maintain and/or demolish the project. A good health and safety file will be a relatively compact maintenance manual including information to alert those who will be owners, or operators of the new structure, to the risks which must be managed when the structure and associated plant is maintained, repaired, renovated or demolished. After handover the client is responsible for keeping the file up to date.

Full CDM regulation provisions apply to projects over 30 days or involve 500 person days of construction work, but not to projects with domestic (i.e. owner-occupier) clients.

3
Design Data

Design data checklist

The following design data checklist is a useful reminder of all of the limiting criteria which should be considered when selecting an appropriate structural form:

- Description/building use
- Client brief and requirements
- Site constraints
- Loadings
- Structural form: load transfer, stability and robustness
- Materials
- Movement joints
- Durability
- Fire resistance
- Performance criteria: deflection, vibration, etc.
- Temporary works and construction issues
- Soil conditions, foundations and ground slab
- Miscellaneous issues

Structural form, stability and robustness

Structural form

It is worth trying to remember the different structural forms when developing a scheme design. A particular structural form might fit the vision for the form of the building. Force or moment diagrams might suggest a building shape. The following diagrams of structural form are intended as useful reminders:

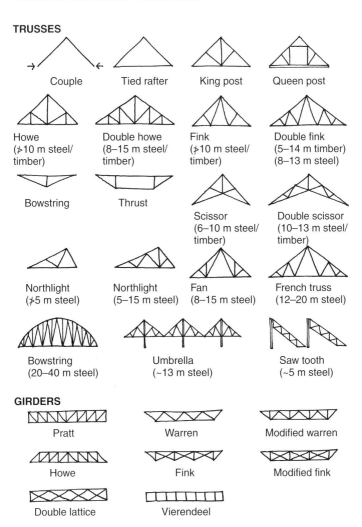

TRUSSES

Couple   Tied rafter   King post   Queen post

Howe
(⊁10 m steel/
timber)

Double howe
(8–15 m steel/
timber)

Fink
(⊁10 m steel/
timber)

Double fink
(5–14 m timber)
(8–13 m steel)

Bowstring   Thrust

Scissor
(6–10 m steel/
timber)

Double scissor
(10–13 m steel/
timber)

Northlight
(⊁5 m steel)

Northlight
(5–15 m steel)

Fan
(8–15 m steel)

French truss
(12–20 m steel)

Bowstring
(20–40 m steel)

Umbrella
(~13 m steel)

Saw tooth
(~5 m steel)

GIRDERS

Pratt   Warren   Modified warren

Howe   Fink   Modified fink

Double lattice   Vierendeel

PORTAL FRAMES

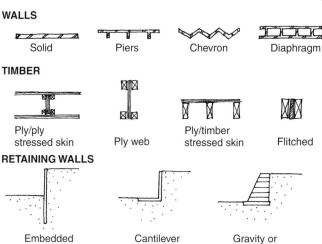

All fixed   2 pin   2 pin mansard   3 pin

ARCHES

Thrust   Tied   3 pin

SUSPENSION

Cable stay   Suspension   Closed suspension

WALLS

Solid   Piers   Chevron   Diaphragm

TIMBER

Ply/ply
stressed skin   Ply web   Ply/timber
stressed skin   Flitched

RETAINING WALLS

Embedded   Cantilever   Gravity or
reinforced earth

Stability

Stability of a structure must be achieved in two orthogonal directions. Circular structures should also be checked for rotational failure. The positions of movement and/or acoustic joints should be considered and each part of the structure should be designed to be independently stable and robust. Lateral loads can be transferred across the structure and/or down to the foundations by using any of the following methods:

- Cross bracing which carries the lateral forces as axial load in diagonal members.
- Diaphragm action of floors or walls which carry the forces by panel/plate/shear action.
- Frame action with 'fixed' connections between members and 'pinned' connections at the supports.
- Vertical cantilever columns with 'fixed' connections at the foundations.
- Buttressing with diaphragm, chevron or fin walls.

Stability members must be located on the plan so that their shear centre is aligned with the resultant of the overturning forces. If an eccentricity cannot be avoided, the stability members should be designed to resist the resulting torsion across the plan.

Robustness and disproportionate collapse

All structural elements should be effectively tied together in each of the two orthogonal directions, both horizontally and vertically. This is generally achieved by specifying connections in steel buildings as being of certain minimum size, by ensuring that reinforced concrete junctions contain a minimum area of steel bars and by using steel straps to connect walls and floors in masonry structures. It is important to consider robustness requirements early in the design process.

The 2004 revision of the Building Regulations made substantial alterations to part A3. The requirements of the regulations and various material codes of practice are summarized in the following table.

Disproportionate collapse requirements with British Standard clause references

Building class	Building type and occupancy	Building regulations requirements	BS 5268 – Timber	BS 5628 – Masonry	BS 5950 – Steel	BS 8110 – Concrete
1	Houses not exceeding 4 storeys. Agricultural buildings. Buildings into which people rarely go or come close to.	Basic Requirements	Structures should be constructed so that no collapse should be disproportionate to the cause and reduce the risk of localized damage spreading – but that permanent deformation of members/connections is acceptable.			C.1.2.2.2: effective horizontal ties AND designed resistance to notional lateral load of 1.5% design dead load.
2A	5 storey single occupancy house. Hotels not exceeding 4 storeys. Flats, apartments and other residential buildings not exceeding 4 storeys. Offices not exceeding 4 storeys. Industrial buildings not exceeding 3 storeys. Retailing premises not exceeding 3 storeys and less than 2000m² at each storey.	Option 1: Effective anchorage of suspended floors to walls	C.1.6.3.2: Figure M.3 or details in BS 5628-1 Annex D.	As for Class 1 plus C.1.33.4: details in BS 5628-1 Annex D or BS 8103-1.	Generally N/A but C.1.2.4.5.2: bearing details of precast concrete units to conform to C.1.5.2.3 of BS 8110-1.	C.1.5.2.3: precast bearings not less than 90mm or half load bearing wall/leaf thickness.
		Option 2: Provision of horizontal ties	C.1.6.3.3 and Figure M.1.	As for Class 1 plus C.1.33.4 and Table 12.	C.1.2.1.1.1 and C.1.2.4.5.2.	As Class 1 plus C.1.2.2.2.2 and C.1.3.12.3.6.
2B	Hotels, flats, apartments and other residential buildings greater than 4 storeys but not exceeding 15 storeys. Educational buildings greater than 1 storey but less than 15 storeys. Retailing premises greater than 3 storeys but less than 15 storeys. Hospitals not exceeding 3 storeys. Offices greater than 4 storeys but less than 15 storeys. All buildings to which members of the public are admitted which contain floor areas exceeding 2000m² but less than 5000m² at each storey. Car parking not exceeding 6 storeys.	Option 1: Provision of horizontal and vertical ties	As Class 2A Option 2 plus C.1.6.3.4.	As Class 2A Option 2 plus C.1.33.5 and Table 13.	As Class 2A Option 2 plus C.1.2.4.5.3.	As Class 2A Option 2 plus C.1.2.2.2.2 and C.1.3.12.3.7.
		Option 2: Check notional removal of load bearing elements	C.1.6.3.5	As for Class 1 plus Table 11: 'without collapse' rather than limited areas.	C.1.2.4.5.3 if Class 2B Option 1 cannot be satisfied.	C.1.2.6.3 of BS 8110-2.
			Check notional removal of load bearing elements such that for removal of any element the building remains stable and that the area of floor at any storey at risk of collapse is less than the lesser of 70m² or 15% of the floor area of that storey. The nominal length of load bearing wall should be the distance between vertical lateral restraints (not exceeding 2.25H for reinforced concrete walls or internal walls of masonry, timber or steel stud). If catenary action is assumed allowance should be made for the necessary horizontal reactions.			
		Option 3: Key element design	C.1.6.3.6.	As for Class 1 plus C.1.33.2.	C.1.2.4.5.4 if Class 2B Options 1 and 2 cannot be satisfied.	C.1.2.6.2 of BS 8110-2.
			Design of key elements to be capable of withstanding 34 kN/m² applied one direction at a time to the member and attached components subject to the limitations of their strength and connections, such accidental loading should be considered to act simultaneously with full dead loading and 1/3 of all normal wind/imposed loadings unless permanent storage loads etc. Where relevant, partial load factors of 1.05 or 0.9 should be applied for overturning and restoring loads respectively. Elements providing stability to key elements should be designed as key elements themselves.			
3	All buildings defined above as Class 2A and 2B that exceed the limits of area or number of storeys. Grandstands accommodating more than 5000 spectators. Buildings containing hazardous substances and/or processes.	Systematic risk assessment of the building should be undertaken taking into account all the normal hazards that may be reasonably forseen, together with any abnormal hazards.	Lack of clear guidance.	Lack of clear guidance.	C.1.2.4.5.1: Class 2B required as a minimum.	Class 2B required as a minimum.

NOTES:
1. Refer to the detailed British Standard clauses for full details of design and detailing requirements.
2. Where provided, horizontal and vertical ties should be safeguarded against damage and corrosion.
3. Key elements may be present in any class of structure and should be designed accordingly.
4. The construction details required by Class 2B can make buildings with load bearing walls difficult to justify economically.
5. In Class 2B and 3 buildings, precast concrete elements not acting as ties should be effectively anchored (C.1.5.1.8.3), such anchorage being capable of carrying the dead weight of the member.

Source: Adapted from Table 11, Part A3 Approved Document A, HMSO.

Structural movement joints

Joints should be provided to control temperature, moisture, acoustic and ground move-ments. Movement joints can be difficult to waterproof and detail and therefore should be kept to a minimum. The positions of movement joints should be considered for their effect on the overall stability of the structure.

Primary movement joints

Primary movement joints are required to prevent cracking where buildings (or parts of buildings) are large, where a building spans different ground conditions, changes height considerably or where the shape suggests a point of natural weakness. Without detailed calculation, joints should be detailed to permit 15–25 mm movement. Advice on joint spacing for different building types can be variable and conflicting. The following figures are some approximate guidelines based on the building type:

Concrete	25 m (e.g. for roofs with large thermal differentials)–50 m c/c.
Steel industrial buildings	100 m typical–150 m maximum c/c.
Steel commercial buildings	50 m typical–100 m maximum c/c.
Masonry	40 m–50 m c/c.

Secondary movement joints

Secondary movement joints are used to divide structural elements into smaller elements to deal with the local effects of temperature and moisture content. Typical joint spacings are:

Clay bricks	Up to 12 m c/c on plan (6 m from corners) and 9 m vertically or every three storeys if the building is greater than 12 m or four storeys tall (in cement mortar).
Concrete blocks	3 m–7 m c/c (in cement mortar).
Hardstanding	70 m c/c.
Steel roof sheeting	20 m c/c down the slope, no limit along the slope.

Fire resistance periods for structural elements

Fire resistance of structure is required to maintain structural integrity to allow time for the building to be evacuated. Generally, roofs do not require protection. Architects typically specify fire protection in consultation with the engineer.

Building types		Minimum period of fire resistance minutes					
		Basement[7] storey including floor over		Ground or upper storey			
		Depth of a lowest basement		Height of top floor above ground, in a building or separated part of a building			
		>10m	<10m	>5m	<18m	<30m	>30m
Residential flats and maisonettes		90	60	30[1]	60[2]	90[2]	120[2]
Residential houses		n/a	30[1]	30[1]	60[3]	n/a	n/a
Institutional residential[4]		90	60	30[1]	60	90	120[5]
Office	not sprinklered	90	60	30[1]	60	90	X
	sprinklered	60	60	30[1]	30[1]	90	120[5]
Shops & commercial	not sprinklered	90	60	60	60	90	X
	sprinklered	60	60	30[1]	60	60	120[5]
Assembly & recreation	not sprinklered	90	60	60	60	90	X
	sprinklered	60	60	30[1]	60	60	120[5]
Industrial	not sprinklered	120	90	60	90	120	X
	sprinklered	90	60	30[1]	60	90	120[5]
Storage and other non-residential	not sprinklered	120	90	60	90	120	X
	sprinklered	90	60	30[1]	60	90	120[5]
Car park for light vehicles	open sided	n/a	n/a	15[1]	15[1,8]	15[1,8]	60
	all others	90	60	30[1]	60	90	120[5]

NOTES:

X   Not permitted

1. Increased to 60 minutes for compartment walls with other fire compartments or 30 minutes for elements protecting a means of escape.
2. Reduced to 30 minutes for a floor in a maisonette not contributing to the support of the building.
3. To be 30 minutes in the case of three storey houses and 60 minutes for compartment walls separating buildings.
4. NHS hospitals should have a minimum of 60 minutes.
5. Reduced to 90 minutes for non-structural elements.
6. Should comply with Building Regulations: B3 section 12.
7. The uppermost floor over basements should meet provision for ground and upper floors if higher.
8. Fire engineered steel elements with certain H_p/A ratios are deemed to satisfy. See Table A2, Approved Document B for full details.

Source: Building Regulations Approved Document B (2007).

Typical building tolerances

SPACE BETWEEN WALLS

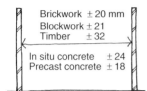

Brickwork ± 20 mm
Blockwork ± 21
Timber   ± 32

In situ concrete   ± 24
Precast concrete ± 18

SPACE BETWEEN COLUMNS

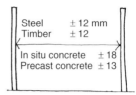

Steel   ± 12 mm
Timber   ± 12

In situ concrete   ± 18
Precast concrete ± 13

WALL VERTICALITY

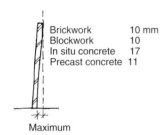

Brickwork   10 mm
Blockwork   10
In situ concrete   17
Precast concrete 11

Maximum

COLUMN VERTICALITY

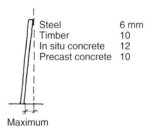

Steel   6 mm
Timber   10
In situ concrete   12
Precast concrete   10

Maximum

VERTICAL POSITION OF BEAMS

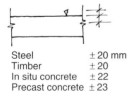

Steel   ± 20 mm
Timber   ± 20
In situ concrete   ± 22
Precast concrete ± 23

VERTICAL POSITION OF FLOORS

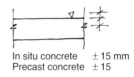

In situ concrete   ± 15 mm
Precast concrete ± 15

PLAN POSITION

Brickwork   ± 10 mm
Steel   ± 10
Timber   ± 10
In situ concrete   ± 12
Precast concrete   ± 10

FLATNESS OF FLOORS

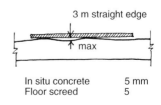

3 m straight edge

max

In situ concrete   5 mm
Floor screed   5

Source: BS 5606: 1990.

Historical use of building materials

Masonry and timber

	Georgian including William IV		Victorian	Edwardian	Inter Wars	Post Wars
	1714	1800	1837	1901	1919	1945

MASONRY

Bonding timbers	————————————————
Non hydraulic lime mortar	———————————— – – – –
Mathematical tiles	84 ———————— 30s – – –
Hydraulic lime mortar	50s – – – – 96 ——— 90s – – – – – – – – .
Clinker concrete blocks	– – – . ———— 60s
Cavity walls	00s – – – – – – – – – 50 ——— 10 – – –
Pressed bricks	51
Flettons	70s
Concrete bricks	– – – – 20
Dense concrete blocks	– – – 50s
Sand line bricks	20s – – – –
Stretcher bond	20s
Mild steel cavity wall ties	– – – – – – – – – 40s
Galvanised steel cavity wall ties	60s ———— 45s ——— 80s
Stainless steel cavity wall ties	– – – – – – – – – – – 65 — 80s
Aerated concrete blocks	53 — 60s – – –

TIMBER

Trussed timber girders	33 ———— 50 – – – – 92
King + queen post trusses	50 ———————— 50s
Wrought iron flitched beams	10s ———— 70 – – – – .
Belfast trusses	60 ———— 40s
Trussed rafters	50s
Ply stressed skin panels	60s
Mild steel flitched beams	– – – – – 40s

Source: Richardson, C. (2000).

Concrete and steel

	Georgian including William IV	Victorian	Edwardian	Inter Wars	Post Wars
Timeline	1714 — 1800	1837	1901	1919	1945
CONCRETE					
Limecrete/Roman cement	96	80s			
Jack arch floors	96	62			
Portland cement	24	51		30	
Filler joists			70s	30s	
Clinker concrete		80		30	
RC framed buildings		54	97		
RC shells + arches				20s	
Hollow pot slabs				25	80
Flat slabs			00s	31	
Lightweight concrete				32 50	
Precast concrete floors					50
Composite metal deck slabs					52 64
Woodwool permanent shutters					69 90s
Waffle/coffered slabs					60s
Composite steel + concrete floors with shear keys					70s
CAST IRON (CI) + WROUGHT IRON (WI)					
CI columns	70s 92			30s	
CI beams	96	65			
WI rods + flats		10s	80		
WI roof trusses		37			
WI built up beams		40			
WI rolled sections		50s			
'Cast steel' columns			90s 10s		
MILD STEEL					
Plates + rods		80			
Riveted sections		90s		60	
Hot rolled sections		83			
Roof trusses		90s			
Steel framed buildings			96		
Welds				55	
Castellated beams				38	
High strength friction grip bolts (HSFG)					50
Hollow sections					60
STAINLESS STEEL Bolts, straps, lintels, shelf angles, etc.				13	70s

Source: Richardson, C. (2000).

Typical weights of building materials

Material	Description	Thickness/ quantity of unit	Unit load kN/m^2	Bulk density kN/m^3
Aggregate				16
Aluminium	Cast alloy			27
	Longstrip roofing	0.8 mm	0.022	
Aluminium bronze				76
Asphalt	Roofing – 2 layers	25 mm	0.58	
	Paving			21
Ballast	see Gravel			
Balsa wood				1
Bituminous felt roofing	3 layers and vapour barrier		0.11	
Bitumen				11–13
Blockboard	Sheet	18 mm	0.11	
Blockwork	Lightweight – dense			10–20
Books	On shelves			7
	Bulk			8–11
Brass	Cast			85
Brickwork	Blue			24
	Engineering			22
	Fletton			18
	London stock			19
	Sand lime			21
Bronze	Cast			83
Cast stone				23
Cement				15
Concrete	Aerated			10
	Lightweight aggregate			18
	Normal reinforced			24
Coal	Loose lump			9
Chalk				22
Chipboard				7
Chippings	Flat roof finish	1 layer	0.05	
Clay	Undisturbed			19
Copper	Cast			87
	Longstrip roofing	0.6 mm	0.05	

Typical weights of building materials – continued

Material	Description	Thickness/ quantity of unit	Unit load kN/m^2	Bulk density kN/m^3
Cork	Granulated			1
Double decker bus	see Vehicles			
Elephants	Adult group		3.2	
Felt	Roofing underlay		0.015	6
	Insulating	50 mm	0.05	
Glass	Crushed/refuse			16
	Clear float			25
Glass wool	Quilt	100 mm	0.01	
Gold				194
Gravel	Loose			16
	Undisturbed			21
Hardboard				6–8
Hardcore				19
Hardwood	Greenheart			10
	Oak			8
	Iroko, teak			7
	Mahogany			6
Hollow clay pot slabs	Including ribs and mortar but excluding topping	300 mm thick overall		12
		100 mm thick overall		15
Iron	Cast			72
	Wrought			77
Ivory				19
Lead	Cast			114
	Sheet	1.8 mm	0.21	
	Sheet	3.2 mm	0.36	
Lime	Hydrate (bags)			6
	Lump/quick (powder)			10
	Mortar (putty)			18
Linoleum	Sheet	3.2 mm	0.05	
Macadam	Paving			21
Magnesium	Alloys			18
MDF	Sheet			8
Mercury				136
Mortar				17–18
Mud				17–20
Partitions	Plastered brick	102 + 2 × 13 mm	2.6	21
	Medium dense plastered block	100 + 2 × 13 mm	2.0	16
	Plaster board on timber stud	100 + 2 × 13 mm	0.35	3

Patent glazing	Single glazed		0.26–0.3	25
	Double glazed		0.52	
Pavement lights	Cast iron or concrete framed	100 mm	1.5	
Perspex	Corrugated sheets		0.05	12
Plaster	Lightweight	13 mm	0.11	9
	Wallboard and skim coat	13 mm	0.12	
	Lath and plaster	19 mm	0.25	
	Traditional lime plaster			20
	Traditional lath + plaster ceiling		0.5	
Plywood	Sheet			7
Polystyrene	Expanded sheet			0.2
Potatoes				7
Precast concrete planks	Beam and block plus 50 mm topping	150–225 mm	1.8–3.3	
	Hollowcore plank	150 mm	2.4	
	Hollowcore plank	200 mm	2.7	
	Solid plank and 50 mm topping	75–300 mm	3.7–7.4	
Quarry tiles	Including mortar bedding	12.5 mm	0.32	
Roofing tiles	Clay – plain		0.77	19
	Clay pantile		0.42	19
	Concrete		0.51	24
	Slate		0.30	28
Sand	Dry, loose			16
	Wet, compact			19
Screed	Sand/cement			22
Shingle	Coarse, graded, dry			19
Slate	Slab			28
Snow	Fresh		minimum 0.6	1
	Wet, compacted		minimum 0.6	3
Softwood				6
	Battens for slating and tiling		0.03	
	25 mm tongued and grooved boards on 100 × 50 timber joists at 400 c/c		0.23	
	25 mm tongued and grooved boards on 250 × 50 timber joists at 400 c/c		0.33	
Soils	Loose sand and gravels			16
	Dense sand and gravels			22
	Soft/firm clays and silts			18
	Stiff clays and silts			21

Typical weights of building materials – continued

Material	Description	Thickness/ quantity of unit	Unit load kN/m²	Bulk density kN/m³
Stainless steel roofing	Longstrip	0.4 mm	0.05	78
Steel	Mild			78
Stone				
Granite	Cornish (Cornwall)			26
	Rublislaw (Grampian)			25
Limestone	Bath (Wiltshire)			21
	Mansfield (Nottinghamshire)			22
	Portland (Dorset)			22
Marble	Italian			27
Sandstone	Bramley Fell (West Yorkshire)			22
	Forest of Dean (Gloucestershire)			24
	Darley Dale or Kerridge (Derbyshire)			23–25
Slate	Welsh			28
Terracotta				18
Terrazzo	Paving	20 mm	0.43	22
Thatch	Including battens	305 mm	0.45	
Timber	see Hardwood or Softwood			
Vehicles	London bus	73.6 kN		
	New Mini Cooper	11.4 kN		
	Rolls Royce	28.0 kN		
	Volvo estate	17.8 kN		
Water	Fresh			10
	Salt			10–12
Woodwool slabs				6
Zinc	Cast			72
	Longstrip roofing	0.8 mm	0.06	

Minimum imposed floor loads

The following table from BS 6399: Part 1 gives the normally accepted minimum floor loadings. Clients can consider sensible reductions in these loads if it will not compromise future flexibility. A survey by Arup found that office loadings very rarely even exceed the values quoted for domestic properties.

The gross live load on columns and/or foundations from sections A to D in the table, can be reduced in relation to the number of floors or floor area carried to BS 6399: Part 1. Live load reductions are not permitted for loads from storage and/or plant, or where exact live loadings have been calculated.

Type of activity/occupancy for part of the building or structure	Examples of specific use	UDL kN/m²	Point load kN
A Domestic and residential activities (also see category C)	All usages within self-contained dwelling units. Communal areas (including kitchens) in blocks of flats with limited use (see Note 1) (for communal areas in other blocks of flats, see C3 and below)	1.5	1.4
	Bedrooms and dormitories except those in hotels and motels	1.5	1.8
	Bedrooms in hotels and motels Hospital wards Toilet areas	2.0	1.8
	Billiard rooms	2.0	2.7
	Communal kitchens except in flats covered by Note 1	3.0	4.5
	Balconies — Single dwelling units and communal areas in blocks of flats with limited use (see Note 1)	1.5	1.4
	Guest houses, residential clubs and communal areas in blocks of flats except as covered by Note 1	Same as rooms to which they give access but with a minimum of 3.0	1.5/m run concentrated at the outer edge
	Hotels and motels	Same as rooms to which they give access but with a minimum of 4.0	1.5/m run concentrated at the outer edge
B Offices and work areas not covered elsewhere	Operating theatres, X-ray rooms, utility rooms	2.0	4.5
	Work rooms (light industrial) without storage	2.5	1.8
	Offices for general use	2.5	2.7
	Banking halls	3.0	2.7
	Kitchens, laundries, laboratories	3.0	4.5
	Rooms with mainframe computers or similar equipment	3.5	4.5
	Machinery halls, circulation spaces therein	4.0	4.5
	Projection rooms	5.0	Determine loads for specific use
	Factories, workshops and similar buildings (general industrial)	5.0	4.5
	Foundries	20.0	Determine loads for specific use
	Catwalks	–	1.0 at 1 mc/c
	Balconies	Same adjacent rooms but with a minimum of 4.0	1.5 kN/m run concentrated at the outer edge
	Fly galleries (load to be distributed uniformly over width)	4.5 kN/m run	–
	Ladders	–	1.5 rung load

Minimum imposed floor loads – continued

Type of activity/ occupancy for part of the building or structure	Examples of specific use		UDL kN/m²	Point load kN
C Areas where people may congregate	Public, institutional and communal dining rooms and lounges, cafes and restaurants (see Note 2)		2.0	2.7
C1 Areas with tables	Reading rooms with no book storage		2.5	4.5
	Classrooms		3.0	2.7
C2 Areas with fixed seats	Assembly areas with fixed seating (see Note 3)		4.0	3.6
	Places of worship		3.0	2.7
C3 Areas without obstacles for moving people	Corridors, hallways, aisles, stairs, landings, etc. in institutional type buildings (not subject to crowds or wheeled vehicles), hostels, guest houses, residential clubs, and communal areas in blocks of flats not covered by Note 1. (For communal areas in blocks of flats covered by Note 1, see A)	Corridors, hallways, aisles, etc. (foot traffic only)	3.0	4.5
		Stairs and landings (foot traffic only)	3.0	4.0
	Corridors, hallways, aisles, stairs, landings, etc. in all other buildings including hotels and motels and institutional buildings	Corridors, hallways, aisles, etc. (foot traffic only)	4.0	4.5
		Corridors, hallways, aisles, etc., subject to wheeled vehicles, trolleys, etc.	5.0	4.5
		Stairs and landings (foot traffic only)	4.0	4.0
	Industrial walkways (1 person access) Industrial walkways (2 way pedestrian access) Industrial walkways (dense pedestrian access)		3.0 5.0 7.5	2.0 3.6 4.5
	Museum floors and art galleries for exhibition purposes		4.0 (see Note 4)	4.5
	Balconies (except as specified in A)		Same as adjacent rooms but with a minimum of 4.0	1.5/m run concentrated at the outer edge
	Fly galleries		4.5 kN/m run distributed uniformly over width	–
C4 Areas with possible physical activities (see clause 9)	Dance halls and studios, gymnasia, stages		5.0	3.6
	Drill halls and drill rooms		5.0	9.0
C5 Areas susceptible to overcrowding (see clause 9)	Assembly areas without fixed seating, concert halls, bars, places of worship and grandstands (see note 4)		5.0	3.6
	Stages in public assembly areas		7.5	4.5
D Shopping areas	Shop floors for the sale and display of merchandise		4.0	3.6

E Warehousing and storage areas. Areas subject to accumulation of goods. Areas for equipment and plant	General areas for static equipment not specified elsewhere (institutional and public buildings)	2.0	1.8
	Reading rooms with book storage, e.g. libraries	4.0	4.5
	General storage other than those specified	2.4 per metre of storage height	7.0
	File rooms, filing and storage space (offices)	5.0	4.5
	Stack rooms (books)	2.4 per metre of storage height (6.5 kN/m² min)	7.0
	Paper storage for printing plants and stationery stores	4.0 per metre of storage height	9.0
	Dense mobile stacking (books) on mobile trolleys, in public and institutional buildings	4.8 per metre of storage height (9.6 kN/m² min)	7.0
	Dense mobile stacking (books) on mobile trucks, in warehouses	4.8 per metre of storage height (15 kN/m² min)	7.0
	Cold storage	5.0 per metre of storage height (15 kN/m² min)	9.0
	Plant rooms, boiler rooms, fan rooms, etc., including weight of machinery	7.5	4.5
	Ladders	–	1.5 rung load
F	Parking for cars, light vans, etc. not exceeding 2500 kg gross mass, including garages, driveways and ramps	2.5	9.0
G	Vehicles exceeding 2500 kg. Driveways, ramps, repair workshops, footpaths with vehicle access, and car parking	To be determined for specific use	

NOTES:
1. Communal areas in blocks of flats with limited use refers to blocks of flats not more than three storeys in height and with not more than four self-contained single family dwelling units per floor accessible from one staircase.
2. Where these same areas may be subjected to loads due to physical activities or overcrowding, e.g. a hotel dining room used as a dance floor, imposed loads should be based on occupancy C4 or C% as appropriate. Reference should also be made to Clause 9.
3. Fixed seating is seating where its removal and use of the space for other purposes is improbable.
4. For grandstands and stadia see the requirements of the appropriate certifying authority.
5. Museums, galleries and exhibition spaces often need more capacity than this, sometimes up to 10 kN/m².

Source: BS 6399: Part 1: 1996.

Typical unit floor and roof loadings

Permanent partitions shown on the floor plans should be considered as dead load. Flexible partitions which may be movable should be allowed for in imposed loads, with a minimum of 1 kN/m².

Timber floor	Live loading: domestic/office	1.5/2.5 kN/m²
	(Office partitions)	(1.0)
	Timber boards/plywood	0.15
	Timber joists	0.2
	Ceiling and services	0.15
	Domestic/office totals	**Total 2.0/4.0 kN/m²**
Timber flat roof	Snow and access	0.75 kN/m²
	Asphalt waterproofing	0.45
	Timber joists and insulation	0.2
	Ceiling and services	0.15
		Total 1.55 kN/m²
Timber pitched roof	Snow	0.6 kN/m²
	Slates, timber battens and felt	0.55
	Timber rafters and insulation	0.2
	Ceiling and services	0.15
		Total 1.5 kN/m²
Internal RC slab	Live loading: office/classroom/corridors, etc.	2.5/3.0/4.0 kN/m²
	Partitions	1.0 (minimum)
	50 screed/75 screed/raised floor	1.2/1.8/0.4
	Solid reinforced concrete slab	24t
	Ceiling and services	0.15
		Total – kN/m²
External RC slab	Live loading: snow and access/office/bar	0.75/2.5/5.0 kN/m²
	Slabs/paving	0.95
	Asphalt waterproofing and insulation	0.45
	50 screed	1.2
	Solid reinforced concrete slab	24t
	Ceiling and services	0.15
		Total – kN/m²
Metal deck roofing	Live loading: snow/wind uplift	0.6/–1.0 kN/m²
	Outer covering, insulation and metal deck liner	0.3
	Purlins – 150 deep at 1.5 m c/c	0.1
	Services	0.1
	Primary steelwork: light beams/trusses	0.5–0.8/0.7–2.4
		Total – kN/m²

Typical 'all up' loads

For very rough assessments of the loads on foundations, 'all up' loads can be useful. The best way is to 'weigh' the particular building, but very general values for small-scale buildings might be:

Steel clad steel frame	5–10 kN/m^2
Masonry clad timber frame	10–15 kN/m^2
Masonry walls and precast concrete floor slabs	15–20 kN/m^2
Masonry clad steel frame	15–20 kN/m^2
Masonry clad concrete frame	20–25 kN/m^2

Wind loading

BS 6399: Part 2 and BS EN 1991-1-5 give methods for determining the peak gust wind loads on buildings and their components. Structures susceptible to dynamic excitation fall outside the scope of the guidelines. While BS 6399 in theory allows for a very site-specific study of the many design parameters, it does mean that grossly conservative values can be calculated if the 'path of least resistance' is taken through the code. Unless the engineer is prepared to work hard and has a preferred 'end result' to aim for, the values from BS 6399 tend to be larger than those obtained from the now withdrawn wind code CP3: Chapter V: Part 2.

As wind loading relates to the size and shape of the building, the size and spacing of surrounding structures, altitude and proximity to the sea or open stretches of country, it is difficult to summarize the design methods. The following dynamic pressure values have been calculated (on a whole building basis) for an imaginary building $20\,m \times 20\,m$ in plan and $10\,m$ tall (with equal exposure conditions and no dominant openings) in different UK locations. The following values should not be taken as prescriptive, but as an idea of an 'end result' to aim for. Taller structures will tend to have slightly higher values and where buildings are close together, funnelling should be considered. Small buildings located near the bases of significantly taller buildings are unlikely to be sheltered as the wind speeds around the bases of tall buildings tends to increase.

Typical values of dynamic pressure, q in kN/m^2

Building location	Maximum q for prevailing south westerly wind kN/m^2	Minimum q for north easterly wind kN/m^2	Arithmetic mean q kN/m^2
Scottish mountain-top	3.40	1.81	2.60
Dover cliff-top	1.69	0.90	1.30
Rural Scotland	1.14	0.61	0.87
Coastal Scottish town	1.07	0.57	0.82
City of London high rise	1.03	0.55	0.80
Rural northern England	1.02	0.54	0.78
Suburban South-East England	0.53	0.28	0.45
Urban Northern Ireland	0.88	0.56	0.72
Rural Northern Ireland	0.83	0.54	0.74
Rural upland Wales	1.37	0.72	1.05
Coastal Welsh town	0.94	0.40	0.67
Conservative quick scheme value for most UK buildings	–	–	1.20

NOTE:
These are typical values which do not account for specific exposure or topographical conditions.

Barrier and handrail loadings

Minimum horizontal imposed loads for barriers, parapets, and balustrades, etc.

Type of occupancy for part of the building or structure	Examples of specific use	Line load kN/m	UDL on infill kN/m²	Point load on infill kN
A Domestic and residential activities	(a) All areas within or serving exclusively one single family dwelling including stairs, landings, etc. but excluding external balconies and edges of roofs (see C3 ix)	0.36	0.5	0.25
	(b) Other residential (but also see C)	0.74	1.0	0.5
B and E Offices and work areas not included elsewhere including storage areas	(c) Light access stairs and gangways not more than 600 mm wide	0.22	n/a	n/a
	(d) Light pedestrian traffic routes in industrial and storage buildings except designated escape routes	0.36	0.5	0.25
	(e) Areas not susceptible to overcrowding in office and institutional buildings. Also industrial and storage buildings except as given above	0.74	1.0	0.5
C Areas where people may congregate: C1/C2 areas with tables or fixed seating	(f) Areas having fixed seating within 530 mm of the barrier, balustrade or parapet	1.5	1.5	1.5
	(g) Restaurants and bars	1.5	1.5	1.5
C3 Areas without obstacles for moving people and not susceptible to overcrowding	(h) Stairs, landings, corridors, ramps	0.74	1.0	0.5
	(i) External balconies and edges of roofs. Footways and pavements within building curtilage adjacent to basement/sunken areas	0.74	1.0	0.5
C5 Areas susceptible to overcrowding	(j) Footways or pavements less than 3 m wide adjacent to sunken areas	1.5	1.5	1.5
	(k) Theatres, cinemas, discotheques, bars, auditoria, shopping malls, assembly areas, studios. Footways or pavements greater than 3 m wide adjacent to sunken areas	3.0	1.5	1.5
	(l) Grandstands and stadia	See requirements of the appropriate certifying authority		
D Retail areas	(m) All retail areas including public areas of banks/building societies or betting shops. For areas where overcrowding may occur, see C5	1.5	1.5	1.5
F/G Vehicular	(n) Pedestrian areas in car parks including stairs, landings, ramps, edges or internal floors, footways, edges of roofs	1.5	1.5	1.5
	(o) Horizontal loads imposed by vehicles	See clause 11. (Generally F ⩾ 150 kN)		

NOTE:
Line load applied at 1.1 m above datum (a finished level on which people may stand: on a floor, roof, wide parapet, balcony, ramp, pitch line of stairs (nosings), etc.).

Source: BS 6399: Part 1: 1996.

Minimum barrier heights

Use	Position	Height mm
Single family dwelling	(a) Barriers in front of a window	800
	(b) Stairs, landings, ramps, edges of internal floors	900
	(c) External balconies, edges of roofs	1100
All other uses	(d) Barrier in front of a window	800
	(e) Stairs	900
	(f) Balconies and stands, etc. having fixed seating within 530 mm of the barrier	800*
	(g) Other positions	1100

*Site lines should be considered as set out in clause 6.8 of BS 6180.

Source: BS 6180: 1999.

Selection of materials

Material	Advantage	Disadvantage
Aluminium	Good strength to dead weight ratio for long spans Good corrosion resistance Often from recycled sources	Cannot be used where stiffness is critical Stiffness is a third of that of steel About two to three times the price of steel
Concrete	Design is tolerant to small, late alterations Integral fire protection Integral corrosion protection Provides thermal mass if left exposed Client pays as the site work progresses: 'pay as you pour'	Dead load limits scope Greater foundation costs Greater drawing office and detailing costs Only precasting can accelerate site work Difficult to post-strengthen elements Fair faced finish needs very skilled contractors and carefully designed joints
Masonry	Provides thermal mass The structure is also the cladding Can be decorative by using a varied selection of bricks Economical for low rise buildings Inherent sound, fire and thermal properties Easy repair and maintenance	Skilled site labour required Long construction period Less economical for high rise Large openings can be difficult Regular movement joints Uniform appearance can be difficult to achieve
Steelwork	Light construction reduces foundation costs Intolerant to late design changes Fast site programme Members can be strengthened easily Ideal for long spans and transfer structures	Design needs to be fixed early Needs applied insulation, fire protection and corrosion protection Skilled workforce required Early financial commitment required from client to order construction materials Long lead-ins Vibrations can govern design
Timber	Traditional/low-tech option Sustainable material Cheap and quick with simple connections Skilled labour not an absolute requirement Easily handled	Limited to 4–5 storeys maximum construction height Requires fire protection Not good for sound insulation Must be protected against insects and moisture Connections can carry relatively small loads

NOTE: See sustainability chapter for additional considerations.

48

Selection of floor construction

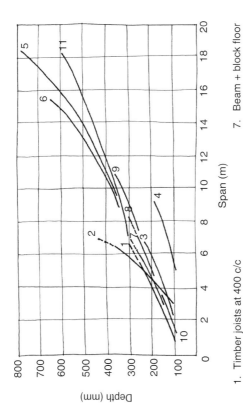

Depth (mm)

Span (m)

1. Timber joists at 400 c/c
2. Stressed skin ply panel
3. One way reinforced concrete slab
4. Precast prestressed concrete plank
5. Precast double tee beams
6. Coffered concrete slab
7. Beam + block floor
8. Reinforced concrete flat slab
9. Post tensioned flat slab
10. Concrete metal deck slab
11. Composite steel beams

Transportation

Although the transport of components is not usually the final responsibility of the design engineer, it is important to consider the limitations of the available modes of transport early in the design process using Department for Transport (DfT) information. Specific cargo handlers should be consulted for comment on sea and air transport, but a typical shipping container is 2.4 m wide, 2.4–2.9 m high and can be 6 m, 9 m, 12 m or 13.7 m in length. Transportation of items which are likely to exceed 20 m by 4 m should be very carefully investigated. Private estates may have additional and more onerous limitations on deliveries and transportation. Typical road and rail limitations are listed below as the most common form of UK transport, but the relevant authorities should be contacted to confirm the requirements for specific projects.

Rail transportation

Railtrack can carry freight in shipping containers or on flat bed wagons. The maximum load on a four axle flat wagon is 66 tonnes. The maximum height of a load is 3.9 m above the rails and wagons are generally between 1.4 and 1.8 m high. All special requirements should be discussed with Railtrack Freight or Network Rail.

Road transport

The four main elements of legislation which cover the statutory controls on length, width, marking, lighting and police notification for large loads are the Motor Vehicles (Construction & Use) Regulations 1986; the Motor Vehicles (Authorization of Special Types) General Order 1979, the Road Vehicles Lighting Regulations 1989 and the Road Traffic Act 1972. A summary of the requirements is set out below.

Height of load
There is no statutory limit governing the overall height of a load; however, where possible it should not exceed 4.95 m from the road surface to maximize use of the motorway and trunk road network (where the average truck flat bed is about 1.7 m). Local highway authorities should be contacted for guidance on proposed routes avoiding head height restrictions on minor roads for heights exceeding 3.0 m–3.6 m.

Weight of vehicle or load

Gross weight of vehicle, W kg	Notification requirements
$44\,000 < W \le 80\,000$ or has any axle weight greater than permitted by the Construction & Use Regulations	2 days' clear notice with indemnity to the Highway and Bridge Authorities
$80\,000 < W \le 150\,000$	2 days' clear notice to the police and 5 days' clear notice with indemnity to the Highway and Bridge Authorities
$W > 150\,000$	DfT Special Order BE16 (allow 10 weeks for application processing) plus 5 days' clear notice to the police and 5 days' clear notice with indemnity to the Highway and Bridge Authorities

Width of load

Total loaded width*, B m	Notification requirements
$B \leq 2.9$	No requirement to notify police
$2.9 < B \leq 5.0$	2 days' clear notice to police
$5.0 < B \leq 6.1$	DfT permission VR1 (allow 10 days for application processing) and 2 days' clear notice to police
$B > 6.1$	DfT Special Order BE16 (allow 8 weeks for application processing) and 5 days' clear notice to police and 5 days' clear notice with indemnity to Highway and Bridge Authorities

*A load may project over one or both sides by up to 0.305 m, but the overall width is still limited as above.

Loads with a width of over 2.9 m or with loads projecting more than 0.305 m on either side of the vehicle must be marked to comply with the requirements of the Road Vehicles Lighting Regulations 1989.

Length of load

Total loaded length, L m	Notification requirements
$L < 18.75$	No requirement to notify police
$18.75 \leq L < 27.4$	Rigid or articulated vehicles*. 2 days' clear notice to police
(rigid vehicle) $L > 27.4$	DfT Special Order BE16 (allow 8 weeks for application processing) and 5 days' clear notice to police and 5 days' clear notice with indemnity to Highway and Bridge Authorities
(all other trailers) $L > 25.9$	All other trailer combinations carrying the load. 2 days' clear notice to police

*The length of the front of an articulated motor vehicle is excluded if the load does not project over the front of the motor vehicle.

Projection of overhanging loads

Overhang position	Overhang length, L m	Notification requirements
Rear	$L < 1.0$	No special requirement
	$1.0 < L < 2.0$	Load must be made clearly visible
	$2.0 < L < 3.05$	Standard end marker boards are required
	$L > 3.05$	Standard end marker boards are required plus police notification and an attendant is required
Front	$L < 1.83$	No special requirement
	$2.0 < L < 3.05$	Standard end marker boards are required plus the driver is required to be accompanied by an attendant
	$L > 3.05$	Standard end marker boards are required plus police notification and the driver is required to be accompanied by an attendant

Typical vehicle sizes and weights

Vehicle type		Weight, W kg	Length, L m	Width, B m	Height, H m	Turning circle m
3.5 tonne van		3 500	5.5	2.1	2.6	13.0
7.5 tonne van		7 500	6.7	2.5	3.2	14.5
Single decker bus		16 260	11.6	2.5	3.0	20.0
Refuse truck		16 260	8.0	2.4	3.4	17.0

Typical vehicle sizes and weights – continued

Vehicle type		Weight, W kg	Length, L m	Width, B m	Height, H m	Turning circle m
2 axle tipper		16 260	6.4	2.5	2.6	15.0
Van (up to 16.3 tonnes)		16 260	8.1	2.5	3.6	17.5
Skiploader		16 260	6.5	2.5	3.7	14.0
Fire engine		16 260	7.0	2.4	3.4	15.0
Bendy bus		17 500	18.0	2.6	3.1	23.0

Temporary works toolkit

Steel trench prop load capacities

Better known as 'Acrow' props, these adjustable props should conform to BS 4704 or BS EN 1065. Verticality of the loads greatly affects the prop capacity and fork heads can be used to eliminate eccentricities. Props exhibiting any of the following defects should not be used:

- A tube with a bend, crease or noticeable lack of straightness.
- A tube with more than superficial corrosion.
- A bent head or base plate.
- An incorrect or damaged pin.
- A pin not properly attached to the prop by the correct chain or wire.

Steel trench 'acrow' prop sizes and reference numbers to BS 4074

Prop size/reference*	Height range Minimum m	Maximum m
0	1.07	1.82
1	1.75	3.12
2	1.98	3.35
3	2.59	3.96
4	3.20	4.87

*The props are normally identified by their length.

Steel trench prop load capacities

A prop will carry its maximum safe load when it is plumb and concentrically loaded as shown in the charts in BS 4074. A reduced safe working load should be used for concentric loading with an eccentricity, $e \leq 1.5°$ out of plumb as follows:

Capacity of props with e $\leq$ 1.5°
kN

Height m	≤2.75	3.00	3.25	3.50	3.75	4.00	4.25	4.50	4.75
Prop size 0, 1, 2 and 3	17	16	13	11	10	–	–	–	–
Prop size 4	–	–	17	14	11	10	9	8	7

Soldiers

Slim soldiers, also known as slimshors, can be used horizontally and vertically and have more load capacity than steel trench props. Lengths of 0.36m, 0.54m, 0.72m, 0.9m, 1.8m, 2.7m or 3.6m are available. Longer units can be made by joining smaller sections together. A connection between units with four M12 bolts will have a working moment capacity of about 12kNm, which can be increased to 20kNm if stiffeners are used.

Slimshor section properties

Area cm^2	I_{xx} cm^4	I_{yy} cm^4	Z_{xx} cm^3	Z_{yy} cm^3	r_x cm	r_y cm	$M_{max\,x}$ kNm	$M_{max\,y}$ kNm
19.64	1916	658	161	61	9.69	5.70	38	7.5

Slimshor compression capacity

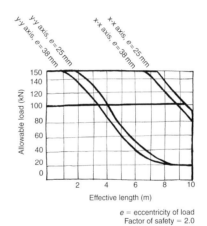

e = eccentricity of load
Factor of safety = 2.0

Slimshor moment capacity

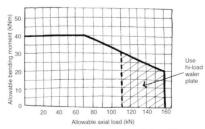

Factor of safety = 1.8

Source: RMD Kwikform (2002).

Ladder beams

Used to span horizontally in scaffolding or platforms, ladder beams are made in 48.3φ 3.2 CHS, 305 mm deep, with rungs at 305 mm centres. All junctions are saddle welded. Ladder beams can be fully integrated with scaffold fittings. Bracing of both the top and bottom chords is required to prevent buckling. Standard lengths are 3.353 m (11'), 4.877 m (16') and 6.400 m (21').

Manufacturers should be contacted for loading information. However, if the tension chord is tied at 1.5 m centres and the compression chord is braced at 1.8 m centres the moment capacity for working loads is about 8.5 kNm. If the compression chord bracing is reduced to 1.5 m centres, the moment capacity will be increased to about 12.5 kNm. The maximum allowable shear is about 12 kN.

Unit beams

Unit beams are normally about 615 mm deep, are about 2.5 times stronger than ladder beams and are arranged in a similar way to a warren girder. Loads should only be applied at the node points. May be used to span between scaffolding towers or as a framework for temporary buildings. As with ladder beams, bracing of both the top and bottom chords is required to prevent buckling, but diagonal plan bracing should be provided to the compression flange. Units can be joined together with M24 bolts to make longer length beams. Standard lengths are 1.8 m (6'), 2.7 m (9') and 3.6 m (12').

Manufacturers should be contacted for loading information. However, if the tension chord is tied at 3.6 m centres and the compression chord is braced at 2.4 m centres the moment capacity for working loads is about 13.5 kNm. If the compression bracing is reduced to 1.2 m centres, the moment capacity will be increased to about 27.5 kNm. The maximum allowable shear is about 14 kN.

4
Basic and Shortcut Tools for Structural Analysis

British Standard load factors and limit states

There are two design considerations: strength and stiffness. The structure must be strong enough to resist the worst loading conditions without collapse and be stiff enough to resist normal working conditions without excessive deflection or deformation. Typically the requirements for strength and stiffness are split between the following 'limit states':

Ultimate limit state (ULS) – Strength (including yielding, rupture, buckling and forming a mechanism), stability against overturning and swaying, fracture due to fatigue and brittle fracture.

Serviceability limit state (SLS) – Deflection, vibration, wind induced oscillation and durability.

A factor of safety against structural failure of 2.0 to 10.0 will be chosen depending on the materials and workmanship. There are three main methods of applying the factor of safety to structural design:

Allowable or permissible stress design – Where the ultimate strengths of the materials are divided by a factor of safety to provide design stresses for comparison with unfactored loads. Normally the design stresses stay within the elastic range. This method is not strictly applicable to plastic (e.g. steel) or semi-plastic (e.g. concrete or masonry) materials and there is one factor of safety to apply to all conditions of materials, loading and workmanship. This method has also been found to be unsafe in some conditions when considering the stability of structures in relation to overturning.

Load factor design – Where working loads are multiplied by a factor of safety for comparison with the ultimate strength of the materials. This method does not consider variability of the materials and as it deals with ultimate loads, it cannot be used to consider deflection and serviceability under working loads.

Ultimate loads or limit state design – The applied loads are multiplied by partial factors of safety and the ultimate strengths of the materials are divided by further partial factors of safety to cover variation in the materials and workmanship. This method allows a global factor of safety to be built up using the partial factors at the designer's discretion, by varying the amount of quality control which will be available for the materials and workmanship. The designer can therefore choose whether to analyse the structure with working loads in the elastic range, or in the plastic condition with ultimate loads. Serviceability checks are generally made with unfactored, working loads.

Eurocode introduction and load factors

When completed, Eurocodes will be the most technically advanced suite of design codes in the world and probably the most wide-ranging codification of structural design ever experienced. Eurocodes are intended to be a Kitemark, or CEN (Comité Européen de Normalisation), standard for buildings as a 'passport' for designers between EU countries. They will be used as an acceptable basis for meeting compliance with UK Building Regulations and the requirements of other public authorities. There are 10 codes in various states of publication, with each code being split into a number of subsections (not listed here):

BS EN 1990	**Basis of design**
1991	Actions
1992	Concrete
1993	Structural steelwork
1994	Composite steel and concrete
1995	Timber
1996	Masonry
1997	Foundations
1998	Seismic
1999	Aluminium

These codes cover the principles, rules and recommended values for ultimate limit state structural design. However, safety, economy and durability have been 'derogated' to the member states and for each code, there is a supplementary set of National Annexes. Allowing for the regional variations, the UK National Annexe is expected to use a decimal point (instead of the continental style ',' used in the CEN documents) and make some changes to partial safety factors. These are termed Nationally Determined Parameters (NDPs). The British Standards Institute cannot change any text in the core CEN document nor publish a version of the CEN document with the values of the NDPs pasted in, but the National Annex may include reference to Non-Contradictory Complementary Information (NCCI), such as national standards or guidance documents.

An additional complication is that British Standards included both design guidance alongside materials and workmanship requirements. The Eurocode intends to publish materials and workmanship as separate standards. Also several topics typically covered in significant detail in British Standards have been omitted, so that the provision of supplementary technical information is essential.

After completion of the Eurocode parts and National Annexes, there is no known programme for completion of Residual Standards, NCCI, other guidance material and updated British Standards to cover aspects omitted by the Eurocodes.

Progress has been at glacial pace, with the whole process integration expected to take 26 years to March 2010. The most recent programme proposed that the National Annexes would be published by 2005, allowing a three-year period for the Eurocodes to co-exist with British Standards, after which the British Standards would be withdrawn (withdrawn ≠ outlawed). However early in 2008, only a third of all the UK National Annexes are complete, so the end of the co-existence period may well be extended for some, or all, of the standards.

In addition to design codes, engineers rely on a large amount of support documentation (e.g. textbooks, software, proprietary products) and it will take some time for this information to become fully available.

The suite of Eurocodes will ultimately be an excellent resource for designers, after engineers get over the initial language barrier. Many of the vocabulary changes appear to stem from philosophical changes in the design approach, but some unnecessary garbling seems to have crept into the English versions – possibly as a result of drafting by international committees. A summary of terminology is listed below:

Eurocode	English translation
Action	Load
Effect of actions	Stress/strain/deflection/rotation
Permanent action	Dead load
Variable action	Live load
Execution	Construction process
Auxiliary construction works	Temporary works
"+"	To be combined with
,	Decimal point

Eurocode partial safety factors

One of the main differences between the Eurocodes and British Standards is the use of different partial safety factors and the option to refine/reduce load factors when different load cases are combined. Load combinations are expressed by equation 6.10 as follows:

$$\Sigma_{j \geq 1} \; \gamma_{G,j} \; G_{k,j} \; "+" \; \gamma_p \; P \; "+" \; \gamma_{Q,1} \; Q_{k,1} \; "+" \; \Sigma_{i>1} \; \gamma_{Q,i} \; \psi_{0,i} \; Q_{k,i}$$

In English this expression means:

Actions	Permanent "+" Prestress "+" Leading variable
	Actions Actions
	"+" Accompanying variable

Summary of Eurocode partial load factors

Limit state	Permanent actions		Variable actions					
			Imposed		Wind		Temperature induced	
	Unfavourable γ_{sup}	Favourable γ_{inf}	Leading γ	Accompanying* $\gamma\psi_0$	Leading γ	Accompanying $\gamma\psi_0$	Leading γ	Accompanying $\gamma\psi_0$
Static equilibrium	1.00	0.90	1.50	1.05	1.50	0.75	1.50	0.90
Structural strength	1.35	1.00	1.50	1.05	1.50	0.75	1.50	0.90
Geotechnical strength	1.00	1.00	1.30	0.91	1.30	0.65	1.30	0.78

NOTES:

1. Partial load factors for variable loads (either leading or accompanying) should be taken as $\gamma = 0$ for favourable (restoring).

2. Combination values (ψ_0) for accompanying loads are given for long-term structural situations. Alternative values for frequent and quasi-permanent combinations (suitable for temporary structures, installation or repair) should be used where appropriate.

3. *For storage loads, $\gamma\psi_0 = 1.5$.

Comparison of BS and Eurocode partial load factors

For one variable action (imposed or wind):

British Standards:	$1.4G_k + (1.4 \text{ or } 1.6)Q_k$
Eurocodes:	$1.35G_k + 1.5Q_k$

For one variable action (imposed or wind) with restoring permanent action:

British Standards:	$1.0G_k + (1.4 \text{ or } 1.6)Q_k$ for steel and concrete
	$0.9G_k + (1.4 \text{ or } 1.6)Q_k$ for masonry
Eurocodes:	$0.9G_k + 1.5Q_k$ for equilibrium
	$1.0G_k + 1.5Q_k$ for structural strength

For two or more variable actions (imposed and wind):

British Standards:	$1.2G_k + 1.2Q_{kl} + 1.2Q_{ka}$
Eurocodes:	$1.35G_k + 1.5Q_{kl} + 0.75Q_{ka}$

Geometric section properties

Section	A mm²	C_x mm	C_y mm	I_x cm⁴	I_y cm⁴	J (approx.) cm⁴
	b^2	$\dfrac{b}{2}$	$\dfrac{b}{2}$	$\dfrac{b^4}{12}$	$\dfrac{b^4}{12}$	$\dfrac{5b^4}{36}$
	bd	$\dfrac{d}{2}$	$\dfrac{b}{2}$	$\dfrac{bd^3}{12}$	$\dfrac{db^3}{12}$	$\dfrac{d^3}{3}\left[b - 0.63d\left(1 - \dfrac{d^4}{12b^4}\right)\right]$ for $d > b$
	$\dfrac{\pi d^2}{4}$	$\dfrac{d}{2}$	$\dfrac{d}{2}$	$\dfrac{\pi d^4}{64}$	$\dfrac{\pi d^4}{64}$	$\dfrac{\pi d^4}{32}$

Section	Area	c_x	c_y			
	$\dfrac{bd}{2}$	$\dfrac{d}{3}$	$\dfrac{b}{2}$	$\dfrac{bd^3}{36}$	$\dfrac{db^3}{48}$	$\dfrac{b^3 d^3}{(15b^2 + 20d^2)}$ for $\dfrac{2}{3} < \dfrac{b}{d} < \sqrt{3}$
	$b^2 - (b - 2t)^2$	$\dfrac{b}{2}$	$\dfrac{b}{2}$	$\dfrac{b^4 - (b - 2t)^4}{12}$	$\dfrac{b^4 - (b - 2t)^4}{12}$	$(b - t)^3 t$

Elastic modulus I/y, plastic modulus, S = sum of first moments of area about central axis, the shape factor = S/Z

Geometric section properties – continued

Section	A mm^2	C_x mm	C_y mm	I_x cm^4	I_y cm^4	J (approx.) cm^4
	$\dfrac{\pi(d^2 - (d - 2t)^2)}{4}$	$\dfrac{d}{2}$	$\dfrac{d}{2}$	$\dfrac{\pi(d^4 - (d - 2t)^4)}{64}$	$\dfrac{\pi(d^4 - (d - 2t)^4)}{64}$	$\dfrac{\pi(d - t)^3 t}{4}$
	$2bt_1 + t_2(d - 2t_1)$	$\dfrac{d}{2}$	$\dfrac{b}{2}$	$\dfrac{bd^3 - (b - t_2)(d - 2t_1)^3}{12}$	$\dfrac{2t_1 b^3 - (d - 2t_1)t_2^3}{12}$	$\dfrac{2t_1^3 b + t_2^3 d}{3}$
	$bd - 2bt_1 - (d - 2t_1)t_2$	$\dfrac{d}{2}$	$b^2 t_1 + \dfrac{1}{2}(d - 2t_1)t_2^2$	$\dfrac{bd^3 - (b - t_2)(d - 2t_1)^3}{12}$	$\dfrac{2t_1 b^3 - (d - 2t_1)t_2^3}{12} + 2bt_1\left(\dfrac{b}{2} - C_y\right)^2 + t_2(d - 2t_1)\left(C_y - \dfrac{t_2}{2}\right)^2$	$\dfrac{t^3(d + 2b)}{3}$

Section	A	c_x / c_y		I_x	I_y	J
T-section (dimensions d, b, t_1, t_2, c_x, c_y)	$bt_1 + (d-t_1)t_2$	$\dfrac{bt_1\left(d - \frac{t_1}{2}\right) + \frac{1}{2}(d-t_1)^2 t_2}{A}$	$\dfrac{b}{2}$	$\dfrac{bt_1^3 + t_2(d-t_1)^3}{12} + bt_1\left(d - C_x - \frac{t_1}{2}\right)^2 + t_2(d-t_1)\left(C_x - \frac{d-t_1}{2}\right)^2$	$\dfrac{t_1 b^3 - (d-t_1)t_2^3}{12}$	$\dfrac{t_1^3 b + t_2^3 d}{3}$
Angle section (dimensions d, b, t_1, t_2, c_x, c_y)	$dt_2 + (b-t_2)t_1$	$\dfrac{dt_2^2 + (b-t_2)t_1^2}{2A}$	$\dfrac{dt_2^2 + bt_1(b - t_2)}{2A}$	$\dfrac{t_2 d^3 + (b-t_2)t_1^3}{12} + dt_2\left(\frac{d}{2} - C_x\right)^2 + (b-t_2)\left(C_x - \frac{t_1}{2}\right)^2$	$\dfrac{t_1 b^3 + (d-t_1)t_2^3}{12} + bt_1\left(\frac{b}{2} - C_y\right)^2 + (d-t_1)\left(C_y - \frac{t_2}{2}\right)^2$	$\dfrac{t_1^3 b + t_2^3 d}{3}$

Elastic modulus I/y, plastic modulus, S = sum of first moments of area about central axis, the shape factor = S/Z

Parallel axis theorem

$$y = \frac{\sum A_i y_i}{\sum A} \qquad I_{xx} = \sum A_i (y - y_i)^2 + \sum I_c$$

Where:

y_i neutral axis depth of element from datum
y depth of the whole section neutral axis from the datum
A_i area of element
A area of whole section
I_{xx} moment of inertia of the whole section about the x-x axis
I_C moment of inertia of element

Composite sections

A composite section made of two materials will have a strength and stiffness related to the properties of these materials. An equivalent stiffness must be calculated for a composite section. This can be done by using the ratios of the Young's moduli to 'transform' the area of the weaker material into an equivalent area of the stronger material.

$$\alpha_E = \frac{E_1}{E_2} \qquad \text{For concrete to steel, } \alpha_E \cong 15$$
$$\text{For timber to steel, } \alpha_E \cong 35$$

Typically the depth of the material (about the axis of bending) should be kept constant and the breadth should be varied: $b_1 = \alpha_E b_2$. The section properties and stresses can then be calculated based on the transformed section in the stronger material.

Material properties

Homogeneous: same elastic properties throughout. Isotropic: same elastic properties in all directions. Anisotropic: varying elastic properties in two different directions. Orthotropic: varying elastic properties in three different directions. All properties are given for a temperature of 20°C.

Properties of selected metals

Material	Specific weight γ kN/m^3	Modulus of elasticity E kN/mm^2	Shear modulus of elasticity G kN/mm^2	Poisson's ratio ν	Proof or yield stress f_y N/mm^2	Ultimate strength* $f_{y\,ult}$ N/mm^2	Elongation at failure %
Aluminium pure	27	69	25.5	0.34	<25	<58	30–60
Aluminium alloy	27.1	70	26.6	0.32	130–250		
Aluminium bronze	77	120	46	0.30			
AB1					170–200	500–590	18–40
AB2					250–360	640–700	13–20
Copper	89	96	38	0.35	60–325	220–385	
Brass	84.5	102	37.3	0.35	290–300	460–480	
Naval brass (soft–hard)	84	100	39	0.34	170–140	410–590	30–15
Bronze	82–86	96–120	36–44	0.34	82–690	200–830	5–60
Phosphor bronze	88	116	43	0.33	–	410	15
Mild steel	78.5	205	82.2	0.3	275–355	430–620	20–22
Stainless steel 304L	78–80	180	76.9	0.3	210	520–720	45
Stainless steel duplex 2205	78–80	180	76.9	0.3	460	640–840	20
Grey cast iron	72	130	48	–	–	150/600c	–
Blackheart cast iron	73.5	170	68	0.26	180	260/780c	10–14
Wrought iron	74–78	190	75	0.3	210	340	35

*Ultimate tensile strength labelled c which denotes ultimate compressive stress.

Properties of selected stone, ceramics and composites

Material	Specific weight γ kN/m³	Modulus of elasticity E kN/mm²	Poisson's ratio ν	Characteristic crushing strength f_{cu} N/mm²	Ultimate tensile strength $f_{y\ ult}$ N/mm²
Carbon fibre (7.5 mmθ)	20	415			1750
Concrete	24	17–31	0.1–0.2	10–70	
Concrete blocks	5–20			3–20	
Clay brick	22.5–28	5–30		10–90	
Fibre glass	15	10*		150	100
Glass (soda)	24.8	74	0.22	1000	30–90
Glass (float)	25–25.6	70–74	0.2–0.27	1000	45 annealed 120–150 toughened
Granite	26	40–70	0.2–0.3	70–280	
Limestone	20–29	20–70	0.2–0.3	20–200	
Marble	26–29	50–100	0.2–0.3	50–180	

Properties of selected timber*

Material	Specific weight γ kN/m³	Modulus of elasticity E kN/mm²	Ultimate tensile strength $f_{y\ ult}$ N/mm²	Ultimate compressive strength $f_{cu\ ult}$ N/mm²	Ultimate shear strength $f_{v\ ult}$ N/mm²
Ash	6.5	10	60	48	10
Beech	7.4	10	60–110	27–54	8–14
Birch	7.1	15	85–90	67–74	13–18
English elm	5.6	11	40–54	17–32	8–11
Douglas fir	4.8–5.6	11–13	45–73	49–74	7.4–8.8
Mahogany	5.4	8	60	45	6
Oak	6.4–7.2	11–12	56–87	27–50	12–18
Scots pine	5.3	8–10	41.8	21–42	5.2–9.7
Poplar	4.5	7	40–43	20	4.8
Spruce	4.3	7–9	36–62	18–39	4.3–8
Sycamore	6.2	9–14	62–106	26–46	8.8–15

*These values are ultimate values. See the chapter on timber for softwood and hardwood design stresses.

Properties of selected polymers and plastics

Material	Specific weight γ kN/m^3	Modulus of elasticity E kN/mm^2	Ultimate tensile strength $f_{y\,ult}$ N/mm^2	Elongation at failure – %
Polythene HD	9–14	0.55–1	20–37	20–100
PVC	13–14	2.4–3.0	40–60	200
PVC plasticized	13–14	0.01	150	
Polystyrene	10–13	3–3.3	35–68	3
Perspex	12	3.3	80–90	6
Acrylic	11.7–12	2.7–3.2	50–80	2–8
PTFE	21–22	0.3–0.6	20–35	100
Polycarbonate	12	2.2–4	50–60	100–130
Nylon	11.5	2–3.5	60–110	50
Rubber	9.1	0.002–0.1	7–20	100–800
Epoxy resin	16–20	20	68–200	4
Neoprene		0.7–20	3.5–24	
Carbon fibre		240	3500	1.4
Kevlar 49		125	3000	2.8
Polyester fabric + PVC coat	14	14	900	14–20

Coefficients of linear thermal expansion

Amount of linear thermal expansion, $l_{thermal} = \alpha(t_{max} - t_{min})l_{overall}$. A typical internal temperature range for the UK might be: $-5°C$ to $35°C$. Externally this might be more like $-15°C$ to $60°C$ to allow for frost, wind chill and direct solar gain.

Material	α $10^{-6}/°C$
Aluminium	24
Aluminium bronze	17
Brass	18–19
Bronze	20
Copper	17
Float glass	8–9
Cast iron	10–11
Wrought iron	12
Mild steel	12
Stainless steel – austenitic	18
Stainless steel – ferritic	10
Lead	29
Wood – parallel to the grain	3
Wood – perpendicular to the grain	30
Zinc	26
Stone – granite	8–10
Stone – limestone	3–4
Stone – marble	4–6
Stone – sandstone	7–12
Concrete – dense gravel aggregate	10–14
Concrete – limestone aggregate	7–8
Plaster	18–21
Clay bricks	5–8
Concrete blocks	6–12
Polycarbonate	60–70
GRP (polyester/glass fibre)	18–25
Rigid PVC	42–72
Nylon	80–100
Asphalt	30–80

Coefficients of friction

The frictional force, $F = \mu N$, where N is the force normal to the frictional plane.

Materials	Coefficient of sliding friction μ
Metal on metal	0.15–0.60
Metal on hardwood	0.20–0.60
Wood on wood	0.25–0.50
Rubber on paving	0.70–0.90
Nylon on steel	0.30–0.50
PTFE on steel	0.05–0.20
Metal on ice	0.02
Masonry on masonry	0.60–0.70
Masonry on earth	0.50
Earth on earth	0.25–1.00

Sign conventions

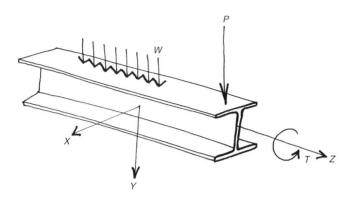

When members are cut into sections for the purpose of analysis, the cut section can be assumed to be held in equilibrium by the internal forces. A consistent sign convention like the following should be adopted:

A positive bending moment, M, results in tension in the bottom of the beam, causing the upper face of the beam to be concave. Therefore this is called a sagging moment. A negative bending moment is called a hogging moment. A tensile axial force, N, is normally taken as positive. Shear force, S, 'couples' are normally considered positive when they would result in a clockwise rotation of the cut element. A positive torque, T, is generally in an anti-clockwise direction.

Beam bending theory

Moment: $M = -EI\dfrac{d^2y}{dx^2}$

Shear: $Q = \dfrac{dM}{dx}$

Elastic constants

Hooke's law defines Young's modulus of elasticity, $E = \sigma/\varepsilon$. Young's modulus is an elastic constant to describe linear elastic behaviour, where σ is stress and ε is the resulting strain. Hooke's law in shear defines the shear modulus of elasticity, $G = \tau/\gamma$, where τ is the shear stress and γ is the shear strain. Poisson's ratio, $\nu = \varepsilon_{lateral}/\varepsilon_{axial}$, relates lateral strain over axial strain for homogeneous materials. The moduli of elasticity in bending and shear are related by: $G = E/(2(1 + \nu))$ for elastic isotropic materials. As ν is normally from 0 to 1.5, G is normally between 0.3 and 0.5 of E.

Elastic bending relationships

$$\frac{M}{I} = \frac{\sigma}{y} = \frac{E}{R}$$

Where M is the applied moment, I is the section moment of inertia, σ is the fibre bending stress, y is the distance from the neutral axis to the fibre and R is the radius of curvature. The section modulus, $Z = I/y$ and the general equation can be simplified so that the applied bending stress, $\sigma = M/Z$.

Horizontal shear stress distribution

$$\tau = \frac{QAy}{bI}$$

Where τ is the horizontal shear stress, Q is the applied shear, b is the breadth of the section at the cut line being considered and A is the area of the segment above the cut line; I is the second moment of area of whole section and y is the distance from centre of area above the cut line to centroid of whole section.

Horizontal shear stresses have a parabolic distribution in a rectangular section. The average shear stress is about 60% of the peak shear (which tends to occur near the neutral axis).

Deflection limits

Typical vertical deflection limits

Total deflection	span/250
Live load deflection	span/360
Domestic timber floor joists	span/330 or 14 mm
Deflection of brittle elements	span/500
Cantilevers	span/180
Vertical deflection of crane girders due to static vertical wheel loads from overhead travelling cranes	span/600
Purlins and sheeting rails (dead load only)	span/200
Purlins and sheeting rails (worst case dead, imposed, wind and snow)	span/100

Typical horizontal deflection limits

Sway of single storey columns	height/300
Sway of each storey of multi-storey column	height/300
Sway of columns with movement sensitive cladding	height/500
Sway of portal frame columns (no cranes)	to suit cladding
Sway of portal frame columns (supporting crane runways)	to suit crane runway
Horizontal deflection of crane girders (calculated on the top flange properties alone) due to horizontal crane loads	span/500
Curtain wall mullions and transoms (single glazed)	span/175
Curtain wall mullions and transoms (double glazed)	span/250

Beam bending and deflection formulae

P is a point load in kN, W is the total load in kN on a span of length L and w is a distributed load in kN/m.

Loading condition	Reactions	Maximum moments	Maximum deflection
	$R_A = R_B = \dfrac{P}{2}$	$M_{midspan} = \dfrac{PL}{4}$	$\delta_{midspan} = \dfrac{PL^3}{48EI}$
	$R_A = \dfrac{Pb}{L}$ $R_B = \dfrac{Pa}{L}$	$M_c = \dfrac{Pab}{L}$	When $a > b$, $\delta_x = \dfrac{Pab(L+b)}{27EIL}\sqrt{3a(L+b)}$ at $x = \sqrt{\dfrac{a(L+b)}{3}}$ from A

Beam bending and deflection formulae – continued

Loading condition	Reactions	Maximum moments	Maximum deflection
	$R_A = R_B = P$	$M_c = Pa$	$\delta_{midspan} = \dfrac{PL^3}{6EI}\left(\dfrac{3a}{4L} - \left(\dfrac{a}{L}\right)^3\right)$
	Third points: $a = \dfrac{L}{3}$	$M = \dfrac{PL}{3}$	$\delta = \dfrac{23PL^3}{648EI}$
	$R_A = R_B = \dfrac{W}{2}$	$M_{midspan} = \dfrac{WL}{8}$	$\delta_{midspan} = \dfrac{5WL^3}{384EI}$

$$R_A = \frac{W}{L}\left(\frac{b}{2} + c\right)$$

$$R_B = \frac{W}{L}\left(\frac{b}{2} + a\right)$$

$$M_{max} = \frac{W}{b}\left(\frac{x_1^2 - a^2}{2}\right)$$

When $x_1 = a + \frac{R_A b}{W}$

$$\delta_{max} = \frac{W}{384EI}(8L^3 - 4Lb^2 + b^3)$$

$$R_A = R_B = \frac{W}{2}$$

$$M_{midspan} = \frac{WL}{6}$$

$$\delta_{midspan} = \frac{WL^3}{60EI}$$

Beam bending and deflection formulae – continued

Loading condition	Reactions	Maximum moments	Maximum deflection
	$R_A = \dfrac{W}{3}$ $R_B = \dfrac{2W}{3}$	$M_x = \dfrac{Wx(L^2 - x^2)}{3L^2}$ maximum at $x = 0.5774L$	$\delta_{midspan} = \dfrac{5WL^3}{384EI}$ $\delta_{max} = \dfrac{0.01304WL^3}{EI}$ when $x = 0.5193L$
	$R_A = \dfrac{w}{2a}(a^2 - b^2)$ $R_B = \dfrac{w}{2a}(a + b)^2$	$M_B = \dfrac{wb^2}{2}$ $M_C = \dfrac{w(a + b)^2(a - b)^2}{8a^2}$ maximum at $x = \dfrac{a}{2}\left(1 - \dfrac{b^2}{a^2}\right)$	$\delta_c = \dfrac{w}{24EI}\left(x^4 - 2ax^3 + \dfrac{2b^2}{a}x^3 + a^3x - 2ab^2x\right)$ $\delta_{free\,tip} = \dfrac{wb}{24EI}(3b^3 + 4ab^2 - a^3)$

$$R_A = \frac{Pb^2(L+2a)}{L^3}$$

$$R_B = \frac{Pb^2(L+2b)}{L^3}$$

$$M_A = \frac{-Pab^2}{L^2}$$

$$M_B = \frac{-Pba^2}{L^2}$$

$$M_C = \frac{2Pa^2b^2}{L^3}$$

$$\delta_{max} = \frac{2Pa^3b^2}{3EI(L+2a)^2}$$

$$\text{when } x = \frac{L^2}{3L-2a}$$

$$\delta_c = \frac{Pa^3b^3}{3EIL^3}$$

$a > b$

$$R_A = R_B = \frac{W}{2}$$

$$M_A = M_B = \frac{-WL}{12}$$

$$M_C = \frac{WL}{24}$$

$$\delta_{midspan} = \frac{WL^3}{384EI}$$

$$R_A = P$$

$$M_A = -Pa$$

$$\delta_{tip} = \frac{Pa^3}{3EI}\left(1+\frac{3b}{2a}\right)$$

$$\delta_B = \frac{Pa^3}{3EI}$$

Beam bending and deflection formulae – continued

Loading condition	Reactions	Maximum moments	Maximum deflection
	$R_A = W$	$M_A = \dfrac{-WL}{2}$	$\delta_{midspan} = \dfrac{WL^3}{8EI}$
	$R_A = P - R_B$ $R_B = \dfrac{Pa^2(2L + b)}{2L^3}$	$M_A = \dfrac{-Pb(L^2 - b^2)}{2L^2}$ $M_C = \dfrac{Pb}{2}\left(2 - \dfrac{3b}{L} + \dfrac{b^3}{L^3}\right)$	$\delta_C = \dfrac{Pa^3b^2}{12EIL^3}(4L - a)$
	$R_A = \dfrac{5W}{8}$ $R_B = \dfrac{3W}{8}$	$M_A = \dfrac{-WL}{8}$ $M_D = \dfrac{9WL}{128}$ at 0.62L from A	$\delta_{max} = \dfrac{WL^3}{185EI}$ at 0.58L from A
	$R_A = \dfrac{-3Pb}{2a}$	$M_A = \dfrac{Pb}{2}$	$\delta_D = \dfrac{Pb^2}{4EI}\left(a + \dfrac{4b}{3}\right)$
	$R_B = \dfrac{P}{a}\left(a + \dfrac{3b}{2}\right)$	$M_B = -Pb = -2M_A$	$\delta_D = \dfrac{-Pa^2b}{27EI}$ at 0.66a

Clapeyron's equations of three moments

Clapeyron's equations can be applied to continuous beams with three supports, or to two-span sections of longer continuous beams.

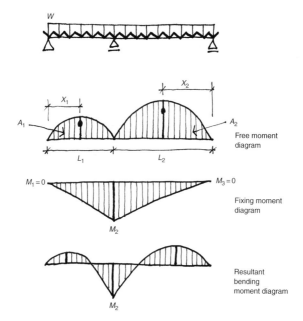

Free moment diagram

Fixing moment diagram

Resultant bending moment diagram

General equation

$$\frac{M_1 L_1}{I_1} + 2M_2\left(\frac{L_1}{I_1} + \frac{L_2}{I_2}\right) + \frac{M_3 L_2}{I_2} = 6\left(\frac{A_1 \bar{x}_1}{L_1 I_1} + \frac{A_2 \bar{x}_2}{L_2 I_2}\right) + 6E\left(\frac{y_2}{L_1} + \frac{(y_2 - y_3)}{L_2}\right)$$

Where:
M bending moment
A area of 'free' moment diagram if the span is treated as simply supported
L span length
$\bar{x}$ distance from support to centre of area of the moment diagram
I second moment of area
y deflections at supports due to loading

Usual case: level supports and uniform moment of area

$$M_1 L_1 + 2M_2(L_1 + L_2) + M_3 L_2 = 6\left(\frac{A_1 \bar{x}_1}{L_1} + \frac{A_2 \bar{x}_2}{L_2}\right)$$

where $y_1 = y_2 = y_3 = 0$ and $I_1 = I_2 = I_3$

M_1 and M_3 are either: unknown for fixed supports, zero for simple supports or known cantilever end moments, and can be substituted into the equation to provide a value for M_2.

Free ends: $M_1 = M_3 = 0$

$$2M_2(L_1 + L_2) = 3\left(\frac{A_1 \bar{x}_1}{L_1} + \frac{A_2 \bar{x}_2}{L_2}\right)$$

which can be further simplified to

$$M_2 = \frac{w(L_1^3 + L_2^3)}{8(L_1 + L_2)}$$

$$w = kN/m$$

Multiple spans

The general case can be applied to groups of three supports for longer continuous beams with n spans. This will produce $(n - 2)$ simultaneous equations which can be resolved to calculate the $(n - 2)$ unknown bending moments.

Continuous beam bending formulae

Moments of inertia are constant and all spans of *L* metres are equal. *W* is the total load on one span (in kN) from either distributed or point loads.

Reaction = coefficient × W

Moment = coefficient × W × L

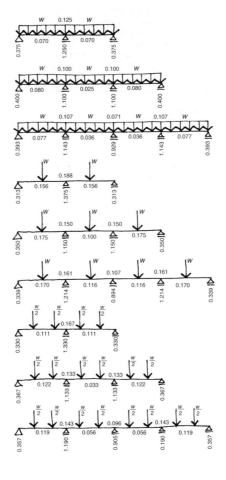

Struts

The critical buckling load of a strut is the applied axial load which will cause the strut to buckle elastically with a sideways movement. There are two main methods of determining this load: Euler's theory, which is simple to use or the Perry–Robertson theory, which forms the basis of the buckling tables in BS 449.

Effective length

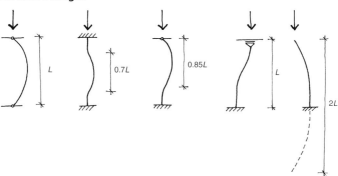

Euler

Euler critical buckling load: $P_E = \dfrac{\pi^2 EI}{L_e^2}$

Euler critical buckling stress: $\sigma_e = \dfrac{P_E}{A} = \dfrac{\pi^2 E r_y^2}{L_e^2} = E\left(\dfrac{\pi r_y}{L_e}\right)^2$

r_y and I are both for the weaker axis or for the direction of the effective length L_e under consideration.

Perry–Robertson

Perry–Robertson buckling load:

$$P_{PR} = A \left[\frac{(\sigma_c + \sigma_e(K + 1))}{2} - \sqrt{\left(\frac{\sigma_c + \sigma_e(K + 1)}{2} \right)^2 - \sigma_c \sigma_e} \right]$$

where $K = 0.3 \left(\dfrac{L_e}{100 r_y} \right)^2$

σ_e is the Euler critical stress as calculated above and σ_c is the yield stress in compression.

Pinned strut with uniformly distributed lateral load

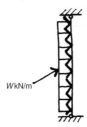

W kN/m

Maximum bending moment, where $\alpha = \sqrt{\dfrac{P}{EI}}$

$$M_{max} = \frac{wEI}{P} \left[\sec\left(\frac{\alpha L}{2} \right) - 1 \right]$$

Maximum compressive stress $\sigma_{c\ max} = \dfrac{My}{I} + \dfrac{P}{A}$

Maximum deflection $\delta_{max} = \dfrac{-M}{P} + \dfrac{wL^2}{8P}$

Rigid frames under lateral loads

Rigid or plane frames are generally statically indeterminate. A simplified method of analysis can be used to estimate the effects of lateral load on a rigid frame based on its deflected shape, and assumptions about the load, shared between the columns. The method assumes notional pinned joints at expected points of contraflexure, so that the equilibrium system of forces can be established by statics. The vertical frame reactions as a result of the lateral loads are calculated by taking moments about the centre of the frame.

The following methods deal with lateral loads on frames, but similar assumptions can be made for vertical analysis (such as treating beams as simply supported) so that horizontal and vertical moments and forces can be superimposed for use in the sizing and design of members.

Rigid frame with infinitely stiff beam

It is assumed that the stiffness of the top beam will spread the lateral load evenly between the columns. From the expected deflected shape, it can be reasonably assumed that each column will carry the same load. Once the column reactions have been assumed, the moments at the head of the columns can be calculated by multiplying the column height by its horizontal base reaction. As the beam is assumed to be infinitely stiff, it is assumed that the columns do not transfer any moment into the beam.

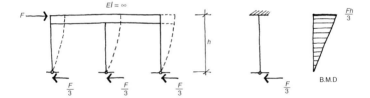

Rigid frame constant stiffness (*EI*)

As the top beam is not considerably stiffer than the columns, it will tend to flex and cause a point of contraflexure at mid span, putting extra load on the internal columns. It can be assumed that the internal columns will take twice the load (and therefore moment) of the external columns. As before, the moments at the head of columns can be calculated by multiplying the column height by its horizontal base reaction. The maximum moment in the beam due to horizontal loading of the frame is assumed to equal the moment at the head of the external columns.

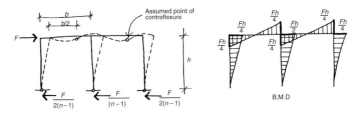

n = number of columns

Multi-storey frame with beams and columns of constant stiffness

For a multi-storey frame, points of contraflexure can be assumed at mid points on beams and columns. Each storey is considered in turn as a separate subframe between the column points of contraflexure. The lateral shears are applied to the subframe columns in the same distribution as the single storey frames, so that internal columns carry twice the load of the external columns. As analysis progresses down the building, the total lateral shear applied to the top of each subframe should be the sum of the lateral loads applied above the notional point of contraflexure. The shears are combined with the lateral load applied to the subframe, to calculate lateral shear reactions at the bottom of each subframe. The frame moments in the columns due to the applied lateral loads increase towards the bottom of the frame. The maximum moments in the beam due to lateral loading of the frame are assumed to equal the difference between the moments at the external columns.

86

Multi-storey frame – continued

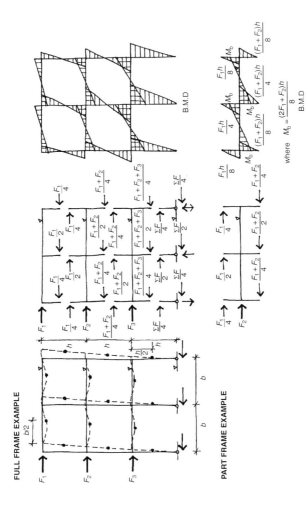

FULL FRAME EXAMPLE

PART FRAME EXAMPLE

where $M_b = \dfrac{(2F_1 + F_2)h}{8}$

B.M.D

Plates

Johansen's yield line theory studies the ultimate capacity of plates. Deflection needs to be considered in a separate elastic analysis. Yield line analysis is a powerful tool which should not be applied without background reading and a sound understanding of the theory.

The designer must try to predict a series of failure crack patterns for yield line analysis by numerical or virtual work methods. Crack patterns relate to the expected deflected shape of the slab at collapse. For any one slab problem there may be many potential modes of collapse which are geometrically and statically possible. All of these patterns should be investigated separately. It is possible for the designer to inadvertently omit the worst case pattern for analysis which could mean that the resulting slab might be designed with insufficient strength. Crack patterns can cover whole slabs, wide areas of slabs or local areas, such as failure at column positions or concentrated loads. Yield line moments are typically calculated as kNm/m width of slab.

The theory is most easily applied to isotropic plates which have the same material properties in both directions. An isotropic concrete slab is of constant thickness and has the same reinforcement in both directions. The reinforcement should be detailed to suit the assumptions of yield line analysis. Anisotropic slabs can be analysed if the 'degree of anisotropy' is selected before a standard analysis. As in the analysis of laterally loaded masonry panels, the results of the analysis can be transformed on completion to allow for the anisotropy.

The simplest case to consider is the isotropic rectangular slab:

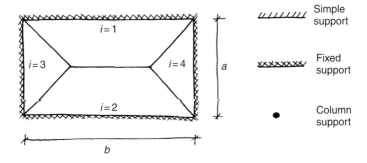

The designer must decide on the amount of fixity, i, at each support position. Generally $i = 0$ for simple support and $i = 1$ for fixed or encastre supports. The amount of fixity determines how much moment is distributed to the top of the slab m', where $m' = im$ and m is the moment in the bottom of the slab.

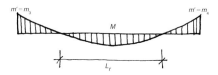

Fixed supports reduce the sagging moments, m, in the bottom of the slab. The distance between the points of zero moment can be considered as a 'reduced effective length', L_r.

$$L_r = \frac{2L}{\left[\sqrt{(1 + i_1)} + \sqrt{(1 + i_2)}\right]} \quad \text{and where } L_r < L : M = \frac{wL_r^2}{8}$$

For the rectangular slab L_r should be calculated for both directions:

$$a_r = \frac{2a}{\left[\sqrt{(1 + i_1)} + \sqrt{(1 + i_2)}\right]} \quad b_r = \frac{2b}{\left[\sqrt{(1 + i_3)} + \sqrt{(1 + i_4)}\right]}$$

So that the design moment is: $M = \dfrac{wa_r b_r}{8\left(1 + \dfrac{a_r}{b_r} + \dfrac{b_r}{a_r}\right)}$

For fixity on all sides of a square slab (where $a = b = L$) the design moment, $M = wL^2/24\,\text{kNm/m}$.

For a point load or column support, $M = P/2\pi\,\text{kNm/m}$.

Selected yield line solutions

These patterns are some examples of those which need to be considered for a given slab. Yield line analysis must be done on many different crack patterns to try to establish the worst case failure moment. Both top and bottom steel should be considered by examining different failure patterns with sagging and hogging crack patterns.

$$a_r = \frac{2a}{\sqrt{1+i_2} + \sqrt{1+i_4}} \quad b_r = \frac{2b}{\sqrt{1+i_1} + \sqrt{1+i_3}}$$

$$m = \frac{wa_r b_r}{8\left(1 + \frac{a_r}{b_r} + \frac{b_r}{a_r}\right)}$$

$$b_r = \frac{2b}{\sqrt{1+i_1} + \sqrt{1+i_3}} \quad m = \frac{wa_r b_r}{3 + 12\frac{a}{b_r} + 2i_2\left(1 + \frac{b_r}{a}\right)}$$

$$a \le b_r$$

Top steel also required.

$$b_r = \frac{b}{\sqrt{1+i_1}} \quad m = \frac{wa_r b_r}{\frac{3}{2} + 3\frac{a}{b_r} + i_2\left(1 + 2\frac{b_r}{a}\right)}$$

For opposite case, exchange a and b, i_1 and i_2.

$$F = 0.6\frac{(a+c)i_1 + (b+d)i_2}{a+b+c+d} \quad m_0 = \frac{3wab}{8\left(2 + \frac{a}{b} + \frac{b}{a}\right)}$$

$$m = \frac{m_0 - 0.15wcd}{1+F} \quad\quad a \le b \le 2a$$

$$m' = \frac{w}{6}(c^2 + d^2)$$

Bottom steel required for main span.

$$\text{if } c = 0.35a, \, m = m' = \frac{wa^2}{16}$$

$$m = \frac{wh^2}{3}\left(0.39 - \frac{c}{h}\right), \ m' = \frac{wc^2}{6}$$

If $c = 0.33h$, $m = m' = \dfrac{wa^2}{55}$

$$m = \frac{wh^2}{2}\left(0.33 - \left(\frac{c}{2h}\right)^{\frac{2}{3}}\right)$$

Point or
concentrated
load

$$m = \frac{P}{2\pi}$$

All moments are in kNm/m.

Torsion

Elastic torsion of circular sections: $\dfrac{T}{J} = \dfrac{\tau}{r} = \dfrac{G\phi}{L}$

Where, T is the applied torque, J is the polar moment of inertia, τ is the torsional shear stress, r is the radius, ϕ is the angle of twist, G is the shear modulus of elasticity and L is the length of member.

The shear strain, γ, is constant over the length of the member and $r\phi$ gives the displacement of any point along the member. Materials yield under torsion in a similar way to bending. The material has a stress/strain curve with gradient G up to a limiting shear stress, beyond which the gradient is zero.

The torsional stiffness of a member relies on the ability of the shear stresses to flow in a loop within the section shape which will greatly affect the polar moment of area, which is calculated from the relationship $J = \int r^2 dA$. This can be simplified in some closed loop cases to $J = I_{zz} = I_{xx} + I_{yy}$.

Therefore for a solid circular section, $J = \pi d^4/32$ for a solid square bar, $J = 5d^4/36$ and for thin walled circular tubes, $J = \pi(d_{outer}^4 - r_{inner}^4)/32$ or $J = 2\pi r^3 t$ and the shear stress, $\tau = \dfrac{T}{2At}$ where t is the wall thickness and A is the area contained within the tube.

Thin walled sections of arbitrary and open cross sections have less torsional stiffness than solid sections or tubular thin walled sections which allow shear to flow around the section. In thin walled sections the shear flow is only able to develop within the thickness of the walls and so the torsional stiffness comes from the sum of the stiffness of its parts: $J = \frac{1}{3} \int_{section} t^3 ds$. This can be simplified to $J \cong \sum (bt^3/3)$, where $\tau = Tt/J$.

J for thick open sections are beyond the scope of this book, and must be calculated empirically for the particular dimensions of a section. For non-square and circular shapes, the effect of the warping of cross sections must be considered in addition to the elastic effects set out above.

Taut wires, cables and chains

The cables are assumed to have significant self-weight. Without any externally applied loads, the horizontal component of the tension in the cable is constant and the maximum tension will occur where the vertical component of the tension reaches a maximum. The following equations are relevant where there are small deflections relative to the cable length.

L	Span length
h	Cable sag
A	Area of cable
ΔLs	Cable elongation due to axial stress
C	Length of cable curve
E	Modulus of elasticity of the cable
$s = h/L$	Sag ratio
w	Applied load per unit length
V	Vertical reaction
y	Equation for the deflected shape
D	Height of elevation
H	Horizontal reaction
T_{max}	Maximum tension in cable
x	Distance along cable

Uniformly loaded cables with horizontal chords

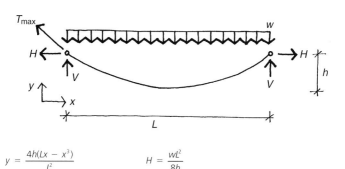

$$y = \frac{4h(Lx - x^3)}{L^2}$$

$$H = \frac{wL^2}{8h}$$

$$V = \frac{wL}{2}$$

$$T_{max} = H\sqrt{1 + 16s^2}$$

$$C \cong L\left(1 + \frac{8}{3}s^2 - \frac{32}{5}s^4 + \cdots\right)$$

$$\Delta L_s \cong \frac{HL(1 + \frac{16s}{3})}{AE}$$

Uniformly loaded cables with inclined chords

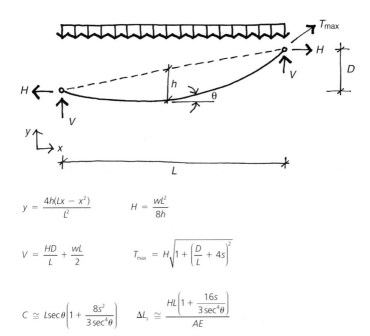

$$y = \frac{4h(Lx - x^2)}{L^2} \qquad H = \frac{wL^2}{8h}$$

$$V = \frac{HD}{L} + \frac{wL}{2} \qquad T_{max} = H\sqrt{1 + \left(\frac{D}{L} + 4s\right)^2}$$

$$C \cong Lsec\theta\left(1 + \frac{8s^2}{3\sec^4\theta}\right) \qquad \Delta L_s \cong \frac{HL\left(1 + \frac{16s}{3\sec^4\theta}\right)}{AE}$$

Vibration

When using long spans and lightweight construction, vibration can become an important issue. Human sensitivity to vibration has been shown to depend on frequency, amplitude and damping. Vibrations can detract from the use of the structure or can compromise the structural strength and stability.

Vibrations can be caused by wind, plant, people, adjacent building works, traffic, earthquakes or wave action. Structures will respond differently depending on their mass and stiffness. Damping is the name given to the ability of the structure to dissipate the energy of the vibrations – usually by friction in structural and non-structural components. While there are many sources of advice on vibrations in structures, assessment is not straightforward. In simple cases, structures should be designed so that their natural frequency is greater than 4.5 Hz to help prevent the structure from being dynamically excitable. Special cases may require tighter limits.

A simplified method of calculating the natural frequency of a structure (f in Hz) is related to the static dead load deflection of the structure, where g is the acceleration due to gravity, δ is the static dead load deflection estimated by normal elastic theory, k is the stiffness ($k = EI/L$), m is a UDL and M is a concentrated load. E is the modulus of elasticity, I is the moment of inertia and L is the length of the member. This method can be used to check the results of more complex analysis.

Member		Estimate of natural frequency, α_f
General rule for structures with concentrated mass		$f = \dfrac{1}{2\pi}\sqrt{\dfrac{g}{\delta}}$
General rule for most structures with distributed loads		$f = \dfrac{1}{2\pi}\sqrt{\dfrac{k}{m}}$
Simplified rule for most structures		$f = \dfrac{18}{\sqrt{\delta}}$
Simply supported, mass concentrated in the centre		$f = \dfrac{1}{2\pi}\sqrt{\dfrac{48EI}{ML^3}}$
Simply supported, sagging, mass and stiffness distributed		$f = \dfrac{\pi}{2}\sqrt{\dfrac{EI}{mL^4}}$
Simply supported, contraflexure, mass and stiffness distributed		$f = 2\pi\sqrt{\dfrac{EI}{mL^4}}$
Cantilever, mass concentrated at the end		$f = \dfrac{1}{2\pi}\sqrt{\dfrac{3EI}{ML^3}}$
Cantilever, mass and stiffness distributed		$f = 0.56\sqrt{\dfrac{EI}{mL^4}}$
Fixed ends, mass and stiffness distributed		$f = 3.56\sqrt{\dfrac{EI}{mL^4}}$

For normal floors with span/depth ratios of 25 or less, there are unlikely to be any vibration problems. Typically problems are encountered with steel and lightweight floors with spans over about 8 m.

SOURCE: BOLTON, A. (1978).

5
Geotechnics

Geotechnics is the engineering theory of soils, foundations and retaining walls. This chapter is intended as a guide which can be used alongside information obtained from local building control officers, for feasibility purposes and for the assessment of site investigation results. Scheme design should be carried out on the basis of a full site investigation designed specifically for the site and structure under consideration.

The relevant codes of practice are:

- BS 5930 for Site Investigation.
- BS 8004 for Foundation Design and BS 8002 for Retaining Wall Design.
- Eurocode 7 for Geotechnical Design.

The following issues should be considered for all geotechnical problems:

- UK (and most international codes) use unfactored loads, while Eurocodes use factored loads.
- All values in this chapter are based on unfactored loads.
- Engineers not familiar with site investigation tests and their implications, soil theory and bearing capacity equations should not use the information in this chapter without using the sources listed in 'Further Reading' for information on theory and definitions.
- The foundation information included in this chapter allows for simplified or idealized soil conditions. In practice, soil layers and variability should be allowed for in the foundation design.
- All foundations must have an adequate factor of safety (normally $\gamma_f = 2$ to 3) applied to the ultimate bearing capacity to provide the allowable bearing pressure for design purposes.
- Settlement normally controls the design and allowable bearing pressures typically limit settlement to 25 mm. Differential settlements should be considered. Cyclic or dynamic loading can cause higher settlements to occur and therefore require higher factors of safety.
- Foundations in fine grained soils (such as clay, silt and chalk) need to be taken down to a depth below which they will not be affected by seasonal changes in the moisture content of the soil, frost action and the action of tree roots. Frost action is normally assumed to be negligible from 450 mm below ground level. Guidelines on trees and shallow foundations in fine grained soils are covered later in the chapter.
- Ground water control is key to the success of ground and foundation works and its effects must be considered, both during and after construction. Dealing with water within a site may reduce the water table of surrounding areas and affect adjoining structures.
- It is nearly always cheaper to design wide shallow foundations to a uniform and predetermined depth, than to excavate narrow foundations to a depth which might be variable on site.

Selection of foundations and retaining walls

The likely foundation arrangement for a structure needs to be considered so that an appropriate site investigation can be specified, but the final foundation arrangement will normally only be decided after the site investigation results have been returned.

Foundations for idealized structure and soil conditions

Foundations must always follow the building type – i.e. a large-scale building needs large-scale/deep foundations. Pad and strip foundations cannot practically be taken beyond 3 m depth and these are grouped with rafts in the classification 'shallow foundations', while piles are called deep foundations. They can have diameters from 75 mm to 2000 mm and be 5 m to 100 m in length. The smaller diameters and lengths tend to be bored cast in-situ piles, while larger diameters and lengths are driven steel piles.

Idealized extremes of structure type	Idealized soil conditions				
	Firm, uniform soil in an infinitely thick stratum	Firm stratum of soil overlying an infinitely thick stratum of soft soil	Soft, uniform soil in an infinitely thick stratum	High water table and/or made ground	Soft stratum of soil overlying an infinitely thick stratum of firm soil or rock
Light flexible structure	Pad or strip footings	Pad or strip footings	Friction piles or surface raft	Piles or surface raft	Bearing piles or piers
Heavy rigid structure	Pad or strip footings	Buoyant raft or friction piles	Buoyant raft or friction piles	Buoyant raft or piles	Bearing piles or piers

Retaining walls for idealized site and soil conditions

Idealized site conditions	Idealized soil types		
	Dry sand and gravel	Saturated sand and gravel	Clay and silt
Working space* available	• Gravity or cantilever retaining wall • Reinforced soil, gabion or crib wall	• Dewatering during construction of gravity or cantilever retaining wall	• Gravity or cantilever retaining wall
Limited working space	• King post or sheet pile as temporary support • Contiguous piled wall • Diaphragm wall • Soil nailing	• Sheet pile and dewatering • Secant bored piled wall • Diaphragm wall	• King post or sheet pile as temporary support • Contiguous piled wall • Soil nailing • Diaphragm wall
Limited working space and special controls on ground movements	• Contiguous piled wall • Diaphragm wall	• Secant bored piled wall • Diaphragm wall	• Contiguous piled wall • Diaphragm wall

*Working space available to allow the ground to be battered back during wall construction.

Site investigation

In order to decide on the appropriate form of site investigation, the engineer must have established the position of the structure on the site, the size and form of the structure, and the likely foundation loads.

BS 5390: Part 2 suggests that the investigation is taken to a depth of 1.5 times the width of the loaded area for shallow foundations. A loaded area can be defined as the width of an individual footing area, the width of a raft foundation, or the width of the building (if the foundation spacing is less than three times the foundation breadth). An investigation must be conducted to prove bedrock must be taken down 3 m beyond the top of the bedrock to ensure that rock layer is sufficiently thick.

Summary of typical site investigation requirements for idealized soil types

Soil type	Type of geotechnical work		
	Excavations	**Shallow footings and rafts**	**Deep foundations and piles**
Sand	• Permeability for dewatering and stability of excavation bottom • Shear strength for loads on retaining structures and stability of excavation bottom	• Shear strength for bearing capacity calculations • Site loading tests for assessment of settlements	• Test pile for assessment of allowable bearing capacity and settlements • Deep boreholes to probe zone of influence of piles
Clay	• Shear strength for loads on retaining structure and stability of excavation bottom • Sensitivity testing to assess strength and stability and the possibility of reusing material as backfill	• Shear strength for bearing capacity calculations • Consolidation tests for assessment of settlements • Moisture content and plasticity tests to predict heave potential and effects of trees	• Long-term test pile for assessment of allowable bearing capacity and settlements • Shear strength and sensitivity testing to assess bearing capacity and settlements • Deep boreholes to probe zone of influence of piles

Soil classification

Soil classification is based on the sizes of particles in the soil as divided by the British Standard sieves.

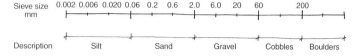

Soil description by particle size

As soils are not normally uniform, standard descriptions for mixed soils have been defined by BS 5930. The basic components are boulders, cobbles, gravel, sand, silt and clay and these are written in capital letters where they are the main component of the soil. Typically soil descriptions are as follows:

Slightly sandy GRAVEL	up to 5% sand	Sandy GRAVEL	5%–20% sand
Very sandy GRAVEL	20%–50% sand	GRAVEL/SAND	equal proportions
Very gravelly SAND	20%–50% gravel	Slightly gravelly SAND	up to 5% gravel
Slightly silty SAND (or GRAVEL)	up to 5% silt	Silty SAND (or GRAVEL)	5%–15% silt
Very silty SAND (or GRAVEL)	15%–35% silt	Slightly clayey SAND (or GRAVEL)	up to 5% clay
Clayey SAND (or GRAVEL)	5%–15% clay	Very clayey SAND (or GRAVEL)	15%–35% clay
Sandy SILT (or CLAY)	35%–65% sand	Gravelly SILT (or CLAY)	35%–65% gravel
Very coarse	over 50% cobbles and boulders		

Soil description by consistency

Homogeneous	A deposit consisting of one soil type.
Heterogeneous	A deposit containing a mixture of soil types.
Interstratified	A deposit containing alternating layers, bands or lenses of different soil types.
Weathered	Coarse soils may contain weakened particles and/or particles sorted according to their size. Fine soils may crumble or crack into a 'column' type structure.
Fissured clay	Breaks into multifaceted fragments along fissures.
Intact clay	Uniform texture with no fissures.
Fibrous peat	Recognizable plant remains present, which retains some strength.
Amorphous peat	Uniform texture, with no recognizable plant remains.

Typical soil properties

The presence of water is critical to the behaviour of soil and the choice of shear strength parameters (internal angle of shearing resistance, ϕ and cohesion, c) are required for geotechnical design.

If water is present in soil, applied loads are carried in the short term by pore water pressures. For granular soils above the water table, pore water pressures dissipate almost immediately as the water drains away and the loads are effectively carried by the soil structure. However, for fine grained soils, which are not as free draining, pore water pressures take much longer to dissipate. Water and pore water pressures affect the strength and settlement characteristics of soil.

The engineer must distinguish between undrained conditions (short-term loading, where pore water pressures are present and design is carried out for total stresses on the basis of ϕ_u and C_u) and drained conditions (long-term loading, where pore water pressures have dissipated and design is carried out for effective stresses on the basis of ϕ' and c').

Drained conditions, $\phi' > 0$

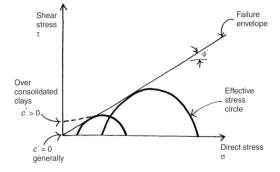

Approximate correlation of properties for drained granular soils

Description	SPT[*] N blows	Effective internal angle of shearing resistance ϕ'	Bulk unit weight γ_{bulk} kN/m^3	Dry unit weight γ_{dry} kN/m^3
Very loose	0–4	26–28	<16	<14
Loose	4–10	28–30	16–18	14–16
Medium dense	10–30	30–36	18–19	16–17
Dense	30–50	36–42	19–21	17–19
Very dense	>50	42–46	21	19

*An approximate conversion from the standard penetration test to the Dutch cone penetration test: $C_r \approx 400\ N$ kN/m^2.

For saturated, dense, fine or silty sands, measured N values should be reduced by: $N = 15 + 0.5(N - 15)$.

Approximate correlation of properties for drained cohesive soils

The cohesive strength of fine grained soils normally increases with depth. Drained shear strength parameters are generally obtained from very slow triaxial tests in the laboratory. The effective internal angle of shearing resistance, ϕ', is influenced by the range and distribution of fine particles, with lower values being associated with higher plasticity. For a normally consolidated clay the effective (or apparent) cohesion, c', is zero but for an overconsolidated clay it can be up to 30 kN/m^2.

Soil description	Typical shrinkability	Plasticity index PI %	Bulk unit weight γ_{bulk} kN/m^3	Effective internal angle of shearing resistance ϕ'
Clay	High	>35	16–22	18–24
Silty clay	Medium	25–35	16–20	22–26
Sandy clay	Low	10–25	16–20	26–34

Undrained conditions, $\phi_u = 0$

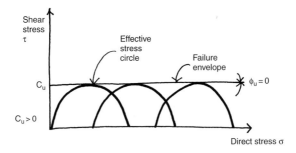

Approximate correlation of properties for undrained cohesive soils

Description	Undrained shear strength C_u kN/m²	Bulk unit weight γ_{bulk} kN/m³
Very stiff and hard clays	>150	19–22
Stiff clays	100–150	
Firm to stiff clays	75–100	17–20
Firm clays	50–75	
Soft to firm clays	40–50	
Soft clays and silts	20–40	16–19
Very soft clays and silts	20	

It can be assumed that $C_u \approx 4.5\,N$ if the clay plasticity index is greater than 30, where N is the number of Standard Penetration Test (SPT) blows.

Typical values of Californian Bearing Ratio (CBR)

Type of soil	Plasticity index	Predicted CBR %
Heavy clay	70	1.5–2.5
	60	1.5–2.5
	50	1.5–2.5
	40	2.0–3.0
Silty clay	30	2.5–6.0
Sandy clay	20	2.5–8.0
	10	1.5–8.0
Silt	–	1.0–2.0
Sand (poorly graded)	–	20
Sand (well graded)	–	40
Sandy gravel (well graded)	–	60

Source: Highways Agency.

Typical angle of repose for selected soils

The angle of repose is very similar to, and often confused with, the internal angle of shearing resistance. The internal angle of shearing resistance is calculated from laboratory tests and indicates the theoretical internal shear strength of the soil for use in calculations while the angle of repose relates to the expected field behaviour of the soil. The angle of repose indicates the slope which the sides of an excavation in the soil might be expected to stand at. The values given below are for short-term, unweathered conditions.

Soil type	Description	Typical angle of repose	Description	Typical angle of repose
Top soil	Loose and dry	35–40	Loose and saturated	45
Loam	Loose and dry	40–45	Loose and saturated	20–25
Peat	Loose and dry	15	Loose and saturated	45
Clay/Silt	Firm to moderately firm	17–19	Puddle clay	15–19
	Sandy clay	15	Silt	19
	Loose and wet	20–25	Solid naturally moist	40–50
Sand	Compact	35–40	Loose and dry	30–35
	Sandy gravel	35–45	Saturated	25
Gravel	Uniform	35–45	Loose shingle	40
	Sandy compact	40–45	Stiff boulder/ hard shale	19–22
	Med coarse and dry	30–45	Med coarse and wet	25–30
Broken rock	Dry	35	Wet	45

Preliminary sizing

Typical allowable bearing pressures under static loads

Description	Safe bearing capacity[1] kN/m²	Field description/notes
Strong igneous rocks and gneisses	10000	Footings on unweathered rock
Strong limestones and hard sandstones	4000	
Schists and slates	3000	
Strong shales and mudstones	2000	
Hard block chalk	80–600	Beware of sink holes and hollowing as a result of water flow
Compact gravel and sandy gravel[2]	>600	Requires pneumatic tools for excavation
Medium dense gravel and sandy gravel[2]	200–600	Hand pick – resistance to shovelling
Loose gravel and sandy gravel[2]	<200	Small resistance to shovelling
Compact sand[2]	>300	Hand pick – resistance to shovelling
Medium dense sand[2]	100–300	Hand pick – resistance to shovelling
Loose sand[2]	<100	Small resistance to shovelling
Very stiff and hard clays	300–600	Requires pneumatic spade for excavation but can be indented by the thumbnail
Stiff clays	150–300	Hand pick – cannot be moulded in hand but can be indented by the thumb
Firm clays	75–150	Can be moulded with firm finger pressure
Soft clays and silts	<75	Easily moulded with firm finger pressure
Very soft clays and silts	Nil	Extrudes between fingers when squeezed
Firm organic material/medieval fill	20–40	Can be indented by thumbnail. Only suitable for small-scale buildings where settlements may not be critical
Unidentifiable made ground	25–50	Bearing values depend on the likelihood of voids and the compressibility of the made ground
Springy organic material/peats	Nil	Very compressible and open structure
Plastic organic material/peats	Nil	Can be moulded in the hand and smears the fingers

NOTES:
1. This table should be read in accordance with the limitations of BS 8004.
2. Values for granular soil assume that the footing width, B, is not less than 1 m and that the water table is more than B below the base of the foundation.

Source: BS 8004: 1986.

Quick estimate design methods for shallow foundations

General equation for allowable bearing capacity after Brinch Hansen

Factor of safety against bearing capacity failure, $\gamma_f = 2.0$ to 3.0, q_o' is the effective overburden pressure, γ is the unit weight of the soil, B is the width of the foundation, c is the cohesion (for the drained or undrained case under consideration) and N_c, N_q and N_γ are shallow bearing capacity factors.

Strip footings: $q_{allowable} = \dfrac{cN_c + q_o'N_q + 0.5\gamma BN_\gamma}{\gamma_f}$

Pad footings: $q_{allowable} = \dfrac{1.3cN_c + q_o'N_q + 0.4\gamma BN_\gamma}{\gamma_f}$

Approximate values for the bearing capacity factors N_c, N_q and $N\gamma$ are set out below in relation to ϕ.

Internal angle of shear ϕ	Bearing capacity factors[*]		
	N_c	N_q	N_γ
0	5.0	1.0	0.0
5	6.5	1.5	0.0
10	8.5	2.5	0.0
15	11.0	4.0	1.4
20	15.5	6.5	3.5
25	21.0	10.5	8.0
30	30.0	18.5	17.0
35	45.0	34.0	40.0
40	75.0	65.0	98.0

*Values from charts by Brinch Hansen (1961).

Simplified equations for allowable bearing capacity after Brinch Hansen

For very preliminary design, Terzaghi's equation can be simplified for uniform soil in thick layers.

Spread footing on clay

$q_{allowable} = 2C_u$ Spread footing on undrained cohesive soil ($\gamma_f = 2.5$)

Spread footing on gravel

$q_{allowable} = 10N$ Pad footing on dry soil ($\gamma_f = 3$)

$q_{allowable} = 7N$ Strip footing on dry soil ($\gamma_f = 3$)

$q_{wet\ allowable} = q_{allowable}/2$ Spread foundation at or below the water table

Where N is the SPT value.

Quick estimate design methods for deep foundations

Concrete and steel pile capacities

Concrete piles can be cast in situ or precast, prestressed or reinforced. Steel piles are used where long or lightweight piles are required. Sections can be butt welded together and excess can be cut away. Steel piles have good resistance to lateral forces, bending and impact, but they can be expensive and need corrosion protection.

Typical maximum allowable pile capacities can be 300 to 1800 kN for bored piles (diameter 300 to 600 mm), 500 to 2000 kN for driven piles (275 to 400 mm square precast or 275 to 2000 mm diameter steel), 300 to 1500 kN for continuous flight auger (CFA) piles (diameter 300 to 600 mm) and 50 to 500 kN for mini piles (diameter 75 to 280 mm and length up to 20 m). The minimum pile spacing achievable is normally about three diameters between the pile centres.

Working pile loads for CFA piles in granular soil ($N = 15$), $\phi = 30°$

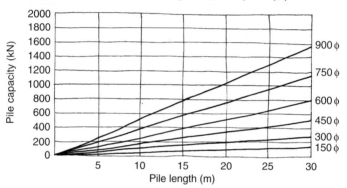

Working pile loads for CFA piles in granular soil ($N = 25$), $\phi = 35°$

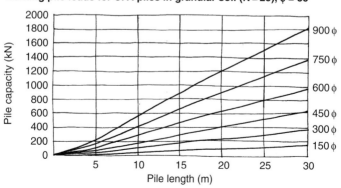

Working pile loads for CFA piles in cohesive soil ($C_u = 50$)

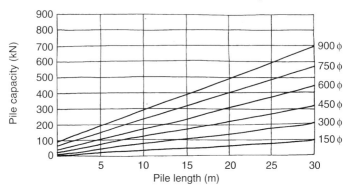

Working pile loads for CFA piles in cohesive soil ($C_u = 100$)

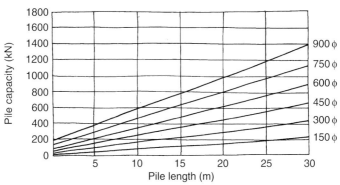

Single bored piles in clay

$$Q_{allow} = \frac{N_c A_b c_{base}}{\gamma_{f\,base}} + \frac{\alpha \bar{c} A_s}{\gamma_{f\,shaft}}$$

Where A_b is the area of the pile base, A_s is the surface area of the pile shaft in the clay, $\bar{c}$ is the average value of shear strength over the pile length and is derived from undrained triaxial tests, where $\alpha = 0.3$ to 0.6 depending on the time that the pile boring is left open. Typically $\alpha = 0.3$ for heavily fissured clay and $\alpha = 0.45–0.5$ for firm to stiff clays (e.g. London clay). $N_c = 9$ where the embedment of the tip of the pile into the clay is more than five diameters. The factors of safety are generally taken as 2.5 for the base and 3.0 for the shaft.

Group action of bored piles in clay
The capacity of groups of piles can be as little as 25 per cent of the collective capacity of the individual piles.

A quick estimate of group efficiency:

$$E = 1 - \left(\tan^{-1} \frac{D}{S} \right) \frac{[m(n-1) + n(m-1)]}{90mn}$$

Where D is the pile diameter, S is the pile spacing and m and n represent the number of rows in two directions of the pile group.

Negative skin friction
Negative skin friction occurs when piles have been installed through a compressible material to reach firm strata. Cohesion in the soft soil will tend to drag down on the piles as the soft layer consolidates and compresses causing an additional load on the pile. This additional load is due to the weight of the soil surrounding the pile. For a group of piles a simplified method of assessing the additional load per pile can be based on the volume of soil which would need to be supported on the pile group. $Q_{skin\ friction} = AH\gamma/N_p$ where A is the area of the pile group, H is the thickness of the layer of consolidating soil or fill which has a bulk density of γ, and N_p is the number of piles in the group. The chosen area of the pile group will depend on the arrangement of the piles and could be the area of the building or part of the building. This calculation can be applied to individual piles, although it can be difficult to assess how much soil could be considered to contribute to the negative skin friction forces.

Piles in granular soil

Although most methods of determining driven pile capacities require information on the resistance of the pile during driving, capacities for both driven and bored piles can be estimated by the same equation. The skin friction and end bearing capacity of bored piles will be considerably less than driven piles in the same soil as a result of loosening caused by the boring and design values of γ, N and $k_s \tan\delta$ should be selected for loose conditions.

$$Q_{allow} = \frac{N_q^* A_b q_o' + A_s q_{o\,mean}' \, k_s \tan\delta}{\gamma_f}$$

Where N_q^* is the pile bearing capacity factor based on the work of Berezantsev, A_b is the area of the pile base, A_s is the surface area of the pile shaft in the soil, q_o' is the effective overburden pressure, k_s is the horizontal coefficient of earth pressure, K_o is the coefficient of earth pressure at rest, δ is the angle of friction between the soil and the pile face, ϕ' is the effective internal angle of shearing resistance and the factor of safety, $\gamma_f = 2.5$ to 3.

Typical values of N_q^*	Pile length Pile diameter		
ϕ	5	20	70
25	16	11	7
30	29	24	20
35	69	53	45
40	175	148	130

*Berezantsev (1961) values from charts for N_q based on ϕ calculated from uncorrected N values.

Typical values of δ and k_s for sandy soils can therefore be determined based on work by Kulhawy (1984) as follows:

Pile face/soil type	Angle of pile/soil friction δ/ϕ'
Smooth (coated) steel/sand	0.5–0.9
Rough (corrugated) steel/sand	0.7–0.9
Cast in place concrete/sand	1.0
Precast concrete/sand	0.8–1.0
Timber/sand	0.8–0.9

Installation and pile type	Coefficients of horizontal soil stress/earth pressure at rest k_s/k_o
Driven piles large displacement	1.00–2.00
Driven piles small displacement	0.75–1.25
Bored cast in place piles	0.70–1.00
Jetted piles	0.50–0.70

Although pile capacities improve with depth, it has been found that at about 20 pile diameters, the skin friction and base resistances stop increasing and 'peak' for granular soils. Generally the peak value for base bearing capacity is 110 000 kN/m² for a pile length of 10 to 20 pile diameters and the peak values for skin friction are 10 kN/m² for loose granular soil, 10 to 25 kN/m² for medium dense granular soil, 25 to 70 kN/m² for dense granular soil and 70 to 110 kN/m² for very dense granular soil.

Source: Kulhawy, F.H. (1984). Reproduced by permission of the ASCE.

Pile caps

Pile caps transfer the load from the superstructure into the piles and take up tolerances on the pile position (typically ±75 mm). The pile cap normally projects 150 mm in plan beyond the pile face and if possible, only one depth of pile cap should be used on a project to minimize cost and labour. The Federation of Piling Specialists suggest the following pile cap thicknesses which generally will mean that the critical design case will be for the sum of all the pile forces to one side of the cap centre line, rather than punching shear:

Pile diameter (mm)	300	350	400	450	500	550	600	750
Pile cap depth (mm)	700	800	900	1000	1100	1200	1400	1600

Retaining walls

Rankine's theory on lateral earth pressure is most commonly used for retaining wall design, but Coulomb's theory is easier to apply for complex loading conditions. The most difficult part of Rankine's theory is the appropriate selection of the coefficient of lateral earth pressure, which depends on whether the wall is able to move. Typically where sufficient movement of a retaining wall is likely and acceptable, 'active' and 'passive' pressures can be assumed, but where movement is unlikely or unacceptable, the earth pressures should be considered 'at rest'. Active pressure will be mobilized if the wall moves 0.25–1 per cent of the wall height, while passive pressure will require movements of 2–4 per cent in dense sand or 10–15 per cent in loose sand. As it is normally difficult to assume that passive pressure will be mobilized, unless it is absolutely necessary for stability (e.g. embedded walls), the restraining effects of passive pressures are often ignored in analysis. The main implications of Rankine's theory are that the engineer must predict the deflected shape, to be able to predict the forces which will be applied to the wall.

Rankine's theory assumes that movement occurs, that the wall has a smooth back, that the retained ground surface is horizontal and that the soil is cohesionless, so that: $\sigma_h = k\sigma_v$

For soil at rest, $k = k_o$, for active pressure, $k = k_a$ and for passive pressure, $k = k_p$.

$$k_o \approx 1 - \sin\phi \qquad k_a = \frac{(1 - \sin\phi)}{(1 + \sin\phi)} \qquad k_p = \frac{1}{k_a} = \frac{(1 + \sin\phi)}{(1 - \sin\phi)}$$

For cohesive soil, k_o should be factored by the overconsolidation ratio,

$$OCR = \sqrt{\frac{\text{pre-consolidation pressure}}{\text{effective overburden pressure}}}.$$

Typical k_o values are 0.35 for dense sand, 0.6 for loose sand, 0.5 to 0.6 for normally consolidated clay and 1.0 to 2.8 for overconsolidated clays such as London clay. The value of k_o depends on the geological history of the soil and should be obtained from a geotechnical engineer.

Rankine's theory can be adapted for cohesive soils, which can shrink away from the wall and reduce active pressures at the top of the wall as a tension 'crack' forms. Theoretically the soil pressures over the height of the tension crack can be omitted from the design, but in practice the crack is likely to fill with water, rehydrate the clay and remobilize the lateral pressure of the soil. The height of crack is $h_c = 2c'/(\gamma\sqrt{k_a})$ for drained conditions and $h_c = 2C_u/\gamma$ for undrained conditions.

Preliminary sizing of retaining walls

Gravity retaining walls – Typically have a base width of about 60−80 per cent of the retained height.

Propped embedded retaining walls – There are 16 methods for the design of these walls depending on whether they are considered flexible (sheet piling) or rigid (concrete diaphragm). A reasonable approach is to use BS 8002 Free Earth Support Method which takes moments about the prop position, followed by the Burland & Potts Method as a check. Any tension crack height is limited to the position of the prop.

Embedded retaining walls – Must be designed for fixed earth support where passive pressures are generated on the rear of the wall, at the toe. An approximate design method is to design the wall with free earth support by the same method as the propped wall but with moments taken at the foot of the embedded wall, before adding 20 per cent extra depth as an estimate of the extra depth required for the fixed earth condition.

Trees and shallow foundations

Trees absorb water from the soil which can cause consolidation and settlements in fine grained soils. Shallow foundations in these conditions may be affected by these settlements and the National House Building Council (NHBC) publish guidelines on the depth of shallow foundations on silt and clay soils to take the foundation to a depth beyond the zone of influence of tree roots. The information reproduced here is current in 2002, but the information may change over time and amendments should be checked with NHBC.

The effect depends on the plasticity index of the soil, the proximity of the tree to the foundation, the mature height of the tree and its water demand. The following suggested minimum foundation depths are based on the assumption that low water demand trees are located 0.2 times the mature height from the building, moderate water demand trees at 0.5 times the mature height and high water demand trees at 1.25 times the mature height of the tree. Where the plasticity index of the soil is not known, assume high plasticity.

Plasticity index, PI = Liquid limit − Plastic limit		Minimum foundation depth with no trees m
Low	10–20%	0.75
Medium	10–40%	0.9
High	>40%	1.0

Source: NHBC (2007). The information may change at any time and revisions should be checked with NHBC.

Water demand and mature height of selected UK trees

The following common British trees are classified as having high, moderate or low water demand. Where the tree cannot be identified, assume high water demand.

Water demand	Broad leaved trees				Conifers	
	Species	Mature height* m	Species	Mature height* m	Species	Mature height* m
High	Elm	18–24	Poplar	25–28	Cypress	18–20
	Eucalyptus	18	Willow	16–24		
	Oak	16–24	Hawthorn	10		
Moderate	Acacia false	18	Lime	22	Cedar	20
	Alder	18	Maple	8–18	Douglas fir	20
	Apple	10	Mountain ash	11	Pine	20
	Ash	23	Pear	12	Spruce	18
	Bay laurel	10	Plane	26	Wellingtonia	30
	Beech	20	Plum	10	Yew	12
	Blackthorn	8	Sycamore	22		
	Cherry	9–17	Tree of heaven	20		
	Chestnut	20–24	Walnut	18		
			Whitebeam	12		
Low	Birch	14	Hornbeam	17		
	Elder	10	Laburnum	12		
	Fig	8	Magnolia	9		
	Hazel	8	Mulberry	9		
	Holly	12	Tulip tree	20		
	Honey locust	14				

*For range of heights within species, see the full NHBC source table for full details.

NOTES:

1. Where species is known, but the subspecies is not, the greatest height should be assumed.
2. Further information regarding trees and water demand is available from the Arboricultural Association or the Arboricultural Advisory and Help Service.

Source: NHBC (2007). The information may change at any time and revisions should be checked with NHBC.

Suggested depths for foundations on cohesive soil

If D is the distance between the tree and the foundation, and H is the mature height of the tree, the following three charts (based on soil shrinkability) will estimate the required foundation depth for different water demand classifications. The full NHBC document allows for a reduction in the foundation depth for climatic reasons, for every 50 miles from the South-East of England.

Suggested depths for foundations on highly shrinkable soil

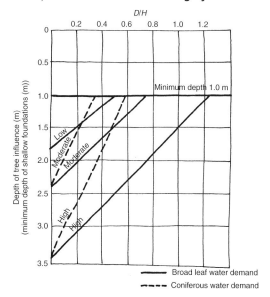

Source: NHBC (2007). The information may change at any time and revisions should be checked with NHBC.

Suggested depths for foundations on medium shrinkable soil

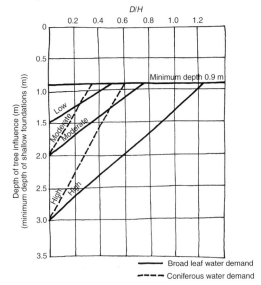

Suggested depths for foundations on low shrinkability soil

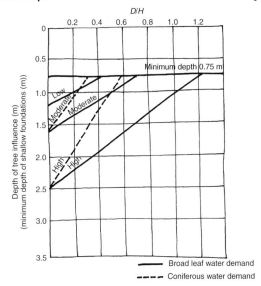

Source: NHBC (2007). The information may change at any time and revisions should be checked with NHBC.

Contaminated land

Contamination can be present as a result of pollution from previous land usage or movement of pollutants from neighbouring sites by air or ground water. The main categories of contamination are chemical, biological (pathological bacteria) and physical (radioactive, flammable materials, etc.).

The Environmental Protection Act 1990 (in particular Part IIA) is the primary legislation covering the identification and remediation of contaminated land. The Act defines contamination as solid, liquid, gas or vapour which might cause harm to 'targets'. This can mean harm to the health of living organisms or property, or other interference with ecological systems. The contamination can be on, in or under the land. The Act applies if the contamination is causing, or will cause, significant harm or results in the pollution of controlled waters including coastal, river and ground water. In order to cause harm the pollution must have some way (called a 'pathway') of reaching the 'target'. The amount of harm which can be caused by contamination will depend on the proposed use for the land. Remediation of contaminated land can remove the contamination, reduce its concentrations below acceptable levels, or remove the 'pathway'.

The 1990 Act set up a scientific framework for assessing the risks to human health from land contamination. This has resulted in Contaminated Land Exposure Assessment (CLEA) and development of Soil Guideline Values for residential, allotment or industrial/commercial land use. Where contaminant concentration levels exceed the Soil Guideline Values, further investigation and/or remediation is required. Reports are planned for a total of 55 contaminants and some are available on the Environment Agency website. Without the full set, assessment is frequently made using Guideline Values from the Netherlands. Other frequently mentioned publications are Kelly and the now superseded ICRCL list. Zero Environment has details of the ICRCL, Kelly and Dutch lists on its website.

Before developing a 'brownfield site' (i.e. a site which has previously been used) a desk study on the history of the site should be carried out to establish its previous uses and therefore likely contaminants. Sampling should then be used to establish the nature and concentration of any contaminants. Remedial action may be dictated by law, but should be feasible and economical on the basis of the end use of the land.

Common sources of contamination

Specific industries can be associated with particular contaminants and the site history is invaluable in considering which soil tests to specify. The following list is a summary of some of the most common sources of contamination.

Common contaminants	Possible sources of contaminants
'Toxic or heavy metals' (cadmium, lead, arsenic, mercury, etc.)	Metal mines; iron and steelworks; foundries and electroplating
'Safe' metals (copper, nickel, zinc, etc.)	Anodizing and galvanizing; engineering/ship/scrap yards
Combustible materials such as coal and coke dust	Gas works; railways; power stations; landfill sites
Sulphides, chlorides, acids and alkalis	Made-up ground
Oily or tarry deposits and phenols	Chemical refineries; chemical plants; tar works
Asbestos	Twentieth century buildings

Effects of contaminants

Effect of contaminant	Typical contaminants
Toxic/narcotic gases and vapours	Carbon monoxide or dioxide, hydrogen sulphide, hydrogen cyanide, toluene, benzene
Flammable and explosive gases	Acetylene, butane, hydrogen sulphide, hydrogen, methane, petroleum hydrocarbons
Flammable liquids and solids	Fuel oils, solvents; process feedstocks, intermediates and products
Combustible materials	Coal residues, ash timber, variety of domestic commercial and industrial wastes
Possible self-igniting materials	Paper, grain, sawdust – microbial degradation of large volumes if sufficiently damp
Corrosive substances	Acids and alkalis; reactive feedstocks, intermediates and products
Zootoxic metals and their salts	Cadmium, lead, mercury, arsenic, beryllium and copper
Other zootoxic metals	Pesticides, herbicides
Carcinogenic substances	Asbestos, arsenic, benzene, benzo(a)pyrene
Substances resulting in skin damage	Acids, alkalis, phenols, solvents
Phytotoxic metals	Copper, zinc, nickel, boron
Reactive inorganic salts	Sulphate, cyanide, ammonium, sulphide
Pathogenic agents	Anthrax, polio, tetanus, Weils
Radioactive substances	Waste materials from hospitals, mine workings, power stations, etc.
Physically hazardous materials	Glass, blades, hospital wastes – needles, etc.
Vermin and associated pests	Rats, mice and cockroaches (contribute to pathogenic agents)

(Where zootoxic means toxic to animals and phytotoxic means toxic to plants.)

Site investigation and sampling

Once a desk study has been carried out and the most likely contaminants are known, an assessment must be carried out to establish the risks associated with the contaminants and the proposed land use. These two factors will determine the maximum concentrations of contaminants which will be acceptable. These maximum concentrations are the Soil Guideline Values published by the Department for the Environment, Food and Rural Affairs (DEFRA) as part of the CLEA range of documents.

Once the soil guideline or trigger values have been selected, laboratory tests can be commissioned to discover if the selected soil contaminants exist, as well as their concentration and their distribution over the site. Reasonably accurate information can be gathered about the site using a first stage of sampling and testing to get a broad picture and a second stage to define the extents of localized areas of contamination.

Sampling on a rectangular grid with cores of 100 mm diameter, it is difficult to assess how many samples might be required to get a representative picture of the site. British Standards propose 25 samples per 10 000 m² which is only 0.002 per cent of the site area. This would only give a 30 per cent confidence of finding a 100 m² area of contamination on the site, while 110 samples would give 99 per cent confidence. It is not easy to balance the cost and complexity of the site investigations and the cost of any potential remedial work, without an appreciation of the extent of the contamination on the site!

Remediation techniques

There are a variety of techniques available depending on the contaminant and target user. The chosen method of treatment will not necessarily remove all of a particular contaminant from a site as in most circumstances it may be sufficient to reduce the risk to below the predetermined trigger level. In some instances it may be possible to change the proposed layout of a building to reduce the risk involved. However, if the site report indicates that the levels of contaminant present in the soil are too high, four main remediation methods are available:

Excavation

Excavation of contaminated soil for specialist disposal or treatment (possibly in a specialist landfill site) and reconstruction of the site with clean fill material. This is expensive and the amount of excavated material can sometimes be reduced, by excavating down to a limited 'cut-off' level, before covering the remaining soil with a barrier and thick granular layer to avoid seepage/upward migration. Removal of soil on restricted sites might affect existing, adjacent structures.

Blending

Clean material is mixed into the bulk of the contaminated land to reduce the overall concentrations taking the test samples below trigger values. This method can be cost effective if some contaminated soil is removed and replaced by clean imported fill, but it is difficult to implement and the effects on adjacent surfaces and structures must be taken into account.

Isolation

Isolation of the proposed development from the contaminants can be attempted by displacement sheet piling, capping, horizontal/vertical barriers, clay barriers, slurry trenches or jet grouting. Techniques should prevent contaminated soil from being brought out of the ground to contaminate other areas.

Physical treatment

Chemical or biological treatment of the soil so that the additives bond with and reduce the toxicity of, or consume, the contaminants.

6
Timber and Plywood

Commercial timbers are defined as hardwoods and softwoods according to their botanical classification rather than their physical strength. Hardwoods are from broad leaved trees which are deciduous in temperate climates. Softwoods are from conifers, which are typically evergreen with needle shaped leaves.

Structural timber is specified by a strength class which combines the timber species and strength grade. Strength grading is the measurement or estimation of the strength of individual timbers, to allow each piece to be used to its maximum efficiency. This can be done visually or by machine. The strength classes referred to in Eurocode 5 and BS 5268 are C14 to C40 for softwoods (C is for coniferous) and D30 to D70 for hardwoods (D is for deciduous). The number refers to the ultimate bending strength in N/mm^2 before application of safety factors for use in design. The Eurocodes use Limit State Design with factored design loads. The British Standards use permissible stresses and grade stresses are modified by load factors according to the design conditions. C16 is the most commonly available softwood, followed by the slightly stronger C24. Specification of C24 should generally be accompanied by checks to confirm that it has actually been used on site in preference to the more readily available C16.

Timber products

Wood-based sheet materials are the main structural timber products, containing substantial amounts of wood in the form of strips, veneers, chips, flakes or fibres. These products are normally classified as:

Laminated panel products – Plywood, laminated veneered lumber (LVL) and glue laminated timber (glulam) for structural use. Made out of laminations 2 to 43 mm thick depending on the product.

Particleboard – Chipboard, orientated strandboard (OSB) and wood-wool. Developed to use forest thinnings and sawmill waste to create cheap panelling for building applications. Limited structural uses.

Fibreboard – Such as hardboard, medium density fibreboard (MDF). Fine particles bonded together with adhesive to form general, non-structural, utility boards.

Summary of material properties

Density   1.2 to 10.7 kN/m³. Softwood is normally assumed to be between 4 and 6 kN/m³.

Moisture content   After felling, timber will lose moisture to align itself with atmospheric conditions and becomes harder and stronger as it loses water. In the UK the atmospheric humidity is normally about 14%. Seasoning is the name of the controlled process where moisture content is reduced to a level appropriate for the timber's proposed use. Air seasoning within the UK can achieve a moisture content of 17–23% in several months for softwood, and over a period of years for hardwoods. Kiln drying can be used to achieve the similar moisture contents over several days for softwoods or two to three weeks for hardwoods.

Moisture content should be lower than 20% to stop fungal attack.

Shrinkage   Shrinkage occurs as a result of moisture loss. Typical new structural softwood will reduce in depth across the grain by as much as 3–4% once it is installed in a heated environment. Shrinkage should be allowed for in structural details.

BS 5268: Part 2 sets out Service Classes 1, 2 and 3 which define timber as having moisture contents of <12%, <20% and >20% respectively.

Sizes and processing of timber

Sawn   Most basic cut of timber (rough or fine – although rough is most common) for use where tolerances of ±3 mm are not significant.

Regularized   Two parallel faces planed where there is a dimensional requirement regarding depth or width. Also known as 'Surfaced 2 sides' (S2S).

Planed All Round   Used for exposed or dimensional accuracy on all four sides. Also known as 'Surfaced 4 sides' (S4S).

Planing and processing reduces the sawn or 'work size' (normally quoted at 20% moisture content) to a 'target size' (for use in calculations) ignoring permitted tolerances and deviations.

Timber section sizes

Selected timber section sizes and section properties

Timber over the standard maximum length, of about 5.5 m, is more expensive and must be pre-ordered.

Basic size*		Area	Z_{XX}	Z_{YY}	I_{XX}	I_{YY}	r_{XX}	r_{YY}
D mm	B mm	10^2 mm^2	10^3 mm^3	10^3 mm^3	10^6 mm^4	10^6 mm^4	mm	mm
100	38	38	63.3	24.1	3.17	0.46	28.9	11.0
100	50	50	83.3	41.7	4.17	1.04	28.9	14.4
100	63	63	105.0	66.2	5.25	2.08	28.9	18.2
100	75	75	125.0	93.8	6.25	3.52	28.9	21.7
100	100	100	166.7	166.7	8.33	8.33	28.9	28.9
150	38	57	142.5	36.1	10.69	0.69	43.3	11.0
150	50	75	187.5	62.5	14.06	1.56	43.3	14.4
150	63	94	236.3	99.2	17.72	3.13	43.3	18.2
150	75	112	281.3	140.6	21.09	5.27	43.3	21.7
150	100	150	375.0	250.0	28.13	12.50	43.3	28.9
150	150	225	562.5	562.5	42.19	42.19	43.3	43.3
175	38	66	194.0	42.1	16.97	0.80	50.5	11.0
175	50	87	255.2	72.9	22.33	1.82	50.5	14.4
175	63	110	321.6	115.8	28.14	3.65	50.5	18.2
175	75	131	382.8	164.1	33.50	6.15	50.5	21.7
200	38	76	253.3	48.1	25.33	0.91	57.7	11.0
200	50	100	333.3	83.3	33.33	2.08	57.7	14.4
200	63	126	420.0	132.3	42.00	4.17	57.7	18.2
200	75	150	500.0	187.5	50.00	7.03	57.7	21.7
200	100	200	666.7	333.3	66.67	16.67	57.7	28.9
200	150	300	1000.0	750.0	100.00	56.25	57.7	43.3
200	200	400	1333.3	1333.3	133.33	133.33	57.7	57.7
225	38	85	320.6	54.2	36.07	1.03	65.0	11.0
225	50	112	421.9	93.8	47.46	2.34	65.0	14.4
225	63	141	531.6	148.8	59.80	4.69	65.0	18.2
225	75	168	632.8	210.9	71.19	7.91	65.0	21.7
250	50	125	520.8	104.2	65.10	2.60	72.2	14.4
250	75	187	781.3	234.4	97.66	8.79	72.2	21.7
250	100	250	1041.7	416.7	130.21	20.83	72.2	28.9
250	250	625	2604.2	2604.2	325.52	325.52	72.2	72.2
300	50	150	750.0	125.0	112.50	3.13	86.6	14.4
300	75	225	1125.0	281.3	168.75	10.55	86.6	21.7
300	100	300	1500.0	500.0	225.00	25.00	86.6	28.9
300	150	450	2250.0	1125.0	337.50	84.38	86.6	43.3
300	300	900	4500.0	4500.0	675.00	675.00	86.6	86.6

*Under dry exposure conditions.

Source: BS 5268: Part 2: 1991.

Tolerances on timber cross sections

BS EN 336 sets out the customary sizes of structural timber. Class 1 timbers are 'sawn' and Class 2 timbers are 'planed'. The permitted deviations for tolerance Class 1 are −1 mm to +3 mm for dimensions up to 100 mm and −2 mm to +4 mm for dimensions greater than 100 mm. For Class 2, the tolerance for dimensions up to 100 mm is ±1 mm and ±1.5 mm for dimensions over 100 mm. Structural design to BS 5268 allows for these tolerances and therefore analysis should be carried out for a 'target' section. It is the dimensions of the target section which should be included in specifications and on drawings.

Laminated timber products

Plywood

Plywood consists of veneers bonded together so that adjacent plies have the grain running in orthogonal directions. Plywoods in the UK generally come from America, Canada, Russia, Finland or the Far East, although the Russian and Far East plywood is not listed in BS 5268 and therefore is not proven for structural applications. The type of plywood available is dependent on the import market. It is worthwhile calling around importers and stockists if a large or special supply is required. UK sizes are based on the imperial standard size of 8′ × 4′ (2.440 × 1.220 m). The main sources of imported plywood in the UK are:

Canada and America   The face veneer generally runs parallel to the longer side. Mainly imported as Douglas fir 18 mm ply used for concrete shuttering, although 9 and 12 mm are also available. Considered a specialist structural product by importers.

Finland   The face veneer can be parallel to the short or long side. Frequently spruce, birch or birch faced ply. Birch plys are generally for fair faced applications, while spruce 9, 12, 18 and 24 mm thick is for general building use, such as flooring and roofing.

Glue laminated timber

Timber layers, normally 43 mm thick, are glued together to build up deep beam sections. Long sections can be produced by staggering finger joints in the layers. Standard beam widths vary from 90 mm to 240 mm although widths up to 265 mm and 290 mm are available. Beam heights and lengths are generally limited to 2050 mm and 31 m respectively. Column sections are available with widths of 90–200 mm and depths of 90–420 mm. Tapered and curved sections can also be manufactured. Loads are generally applied at 90° to the thickness of the layers.

Laminated veneered lumber (LVL)

LVL is similar to plywood but is manufactured with 3 mm veneers in a continuous pro-duction line to create panels 1.8 m wide, up to 26 m in length. It is quite a new product, with relatively few UK suppliers. Beam sections for long spans normally have all their laminations running longitudinally, while smaller, panel products tend to have about a fifth of the laminations cross bonded to improve lateral bending strength. Finnforest produce Kerto-S LVL for beams and Kerto-Q LVL for panels. Standard sections are as follows:

Depth/width (mm)	Thickness of panel (mm)								
	27	33	39	45	51	57	63	69	75
200	●	●	●	●	●	●	●	●	●
225	●	●	●	●	●	●	●	●	●
260		●	●	●	●	●	●	●	●
300			●	●	●	●	●	●	●
360				●	●	●	●	●	●
400					●	●	●	●	●
450						●	●	●	●
500							●	●	●
600								●	●
Kerto type	S/Q	S/Q	S/Q	S/Q	S/Q	S/Q	S/Q	S/Q	S

Source: Finnforest (2002).

Durability and fire resistance

Durability

Durability of timber depends on its resistance to fungal decay. Softwood is more prone to weathering and fungal attack than hardwood. Some timbers (such as oak, sweet chestnut, western red cedar and Douglas fir) are thought to be acidic and may need to be isolated from materials such as structural steelwork. The durability of timber products (such as plywood, LVL and glulam) normally depends on the stability and water resistance of the glue.

Weathering

On prolonged exposure to sunlight, wind and rain, external timbers gradually lose their natural colours and turn grey. Repeated wetting and drying cycles raise the surface grain, open up surface cracks and increase the risk of fungal attack, but weathering on its own generally causes few structural problems.

Fungal attack

For growth in timber fungi need oxygen, a minimum moisture content of 20% and temperatures between 20°C and 30°C. Kiln drying at temperatures over 40°C will generally kill fungi, but fungal growth can normally be stopped by reducing the moisture content. Where structural damage has occurred, the affected timber should be cut away and replaced by treated timber. The remaining timber can be chemically treated to limit future problems. Two of the most common destructive fungi are:

'Dry rot' – *Serpula lacrimans* Under damp conditions, white cotton wool strands form over the surface of the timber. Under drier conditions, a grey-white layer forms over the timber with occasional patches of yellow or lilac. Fruiting bodies are plate-like forms which disperse red spores. As a result of an attack, the timber becomes dry and friable (hence the name dry rot) and breaks up into cube-like pieces both along and across the grain.

'Wet rot' – *Coniophora puteana* Known as cellar fungus, this fungus is the most common cause of timber decay in the UK. It requires high moisture contents of 40–50% which normally result from leaks or condensation. The decayed timber is dark and cracked along the grain. The thin strands of fungus are brown or black, but the green fruiting bodies are rarely seen in buildings. The decay can be hidden below the timber surface.

Insect attack

Insect attack on timber in the UK is limited to a small number of species and tends to be less serious than fungal attack. The reverse is generally true in hotter climates. Insects do not depend on damp conditions although some species prefer timber which has already suffered from fungal attack. Treatment normally involves removal of timber and treatment with pesticides. Some common insect pests in the UK are:

'Common furniture beetle' – *Nobium punctatum* This beetle is the most widespread. It attacks hardwoods and softwoods, and can be responsible for structural damage in severe cases. The brown beetle is 3–5 mm long; leaves flight holes of approximately 2 mm in diameter between May and September, and is thought to be present in up to 20 per cent of all buildings.

'Wood boring weevils' – *Pentarthum huttonii and Euophrym confine* Wood boring beetles attack timber previously softened by fungal decay. *Pentarthum huttonii* is the most common of the weevils and produces damage similar in appearance to the common furniture beetle. The beetles are 3−5 mm long and leave 1 mm diameter flight holes.

'Powder post beetle' – *Lyctus brunneus* The powder post beetle attacks hardwoods, particularly oak and ash, until the sapwood is consumed. The extended soaking of vulnerable timbers in water can reduce the risk of attack but this is not normally commercially viable. The 4 mm reddish-brown beetle leaves flight holes of about 1.5 mm diameter.

'Death watch beetle' – *Xestobium rufovillosum* The death watch beetle characteristically attacks partly decayed hardwoods, particularly oak, and is therefore responsible for considerable damage to old or historic buildings. The beetles typically make tapping noises during their mating season between March and June. Damp conditions encourage infestation. The brown beetle is approximately 8 mm long and leaves a flight hole of 3 mm diameter.

'Longhorn beetle' – *Hylotrupes bajulus* The house long horn beetle is a serious pest, mainly present in parts of southern England. The beetle can infest and cause significant structural damage to the sapwood of seasoned softwood. Affected timbers bulge where tunnelling has occurred just below the surface caused by larvae that can be up to 35 mm long. The flight holes of the black beetle are oval and up to 10 mm across.

Source: BRE Digests 299, 307 and 345. Reproduced with permission by Building Research Establishment.

Fire resistance

Timber is an organic material and is therefore combustible. As timber is heated, water is driven off as vapour. By the time it reaches 230–250°C, the timber has started to break down into charcoal, producing carbon monoxide and methane (which cause flaming). The charcoal will continue to smoulder to carbon dioxide and ash. However, despite its combustibility, large sections of timber can perform better in fire than the equivalent sections of exposed steel or aluminium. Timber has a low thermal conductivity which is further protected by the charred surface, preventing the interior of the section from burning.

BS 5268: Part 4 details the predicted rates of charring for different woods which allows them to be 'fire engineered'. Most timbers in BS 5268 have accepted charring rates of 20 mm in 30 minutes and 40 mm in 60 minutes. The exceptions are western red cedar which chars more quickly at 25 mm in 30 minutes and 50 mm in 60 minutes, and oak, utile, teak, jarrah and greenheart which all char slower at 15 mm in 30 minutes and 30 mm in 60 minutes. Linear extrapolation is permitted for periods between 15 and 90 minutes.

Preliminary sizing of timber elements

Typical span/depth ratios for softwoods

Description	Typical depth (mm)
Domestic floor (50 mm wide joists at 400 mm c/c)	$L/24 + 25$ to 50
Office floors (50 mm wide joists at 400 mm c/c)	$L/15$
Rafters (50 mm wide joists at 400 mm c/c)	$L/24$
Beams/purlins	$L/10$ to 15
Independent posts	Min. 100 mm square
Triangular trusses	$L/5$ to 8
Rectangular trusses	$L/10$ to 15
Plywood stressed skin panels	$L/30$ to 40

Connections which rely on fixings (rather than dead bearing) to transfer the load in and out of the timber can often control member size and for preliminary sizing. Highly stressed individual members should be kept at about 50 per cent capacity until the connections can be designed in detail.

Plywood stress skin panels

Stress skin panels can be factory made using glue and screws or on site just using screws. The screws tend to be at close centres to accommodate the high longitudinal shear stresses. Plywood can be applied to the top, or top and bottom, of the internal softwood joists (webs). The webs are spaced according to the width of the panel and the point loads that the panel will need to carry. The spacing is normally about 600 mm for a UDL of 0.75 kN/m^2, about 400 mm for a UDL of 1.5 kN/m^2 or about 300 mm for a UDL of more than 1.5 kN/m^2. The direction of the face grain of the plywood skin will depend on the type of plywood chosen for the panel. The top ply skin will need to be about 9–12 mm thick for a UDL of 0.75 kN/m^2 or about 12–18 mm thick for a UDL of 1.5 kN/m^2. The bottom skin, if required, is usually 8–9 mm. The panel design is normally controlled by deflection and for economy the EI of the trial section should be about $4.4WL^2$.

Domestic floor joist capacity chart

See the graph below for an indication of the load carrying capacity of various joist sizes in grade C16 timber spaced at 400 mm centres.

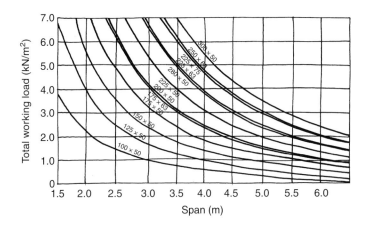

Span tables for solid timber tongued and grooved decking

UDL* kN/m²	Single span (m) for decking thickness				Double span (m) for decking thickness			
	38 mm	50 mm	63 mm	75 mm	38 mm	50 mm	63 mm	75 mm
1.0	2.2	3.0	3.8	4.7	3.0	3.9	5.2	6.3
1.5	1.9	2.6	3.3	4.0	2.6	3.4	4.5	5.4
2.0	1.7	2.3	3.0	3.6	2.4	3.1	4.0	4.9
2.5	1.6	2.2	2.8	3.4	2.2	2.9	3.7	4.6
3.0	1.5	2.1	2.6	3.2	2.1	2.7	3.5	4.3

*These loads limit the deflection to span/240 as the decking is not normally used with a ceiling.

Timber design to BS 5268

BS 5268: Part 2: 2002 gives guidance on the basis of permissible stresses in the timber. All applied loads for analysis should be unfactored as it is the timber grade stresses
which are factored to represent the design conditions being considered.

Notation for BS 5268: Part 2

Symbols	Subscripts						
	Type of force/stress etc.		Significance		Geometry		
σ Stress	c	Compression	a	Applied	$\parallel$	Parallel to the grain
τ Shear stress	m	Bending	grade	Grade	$\perp$	Perpendicular to the grain
E Modulus of elasticity	t	Tension	adm	Permissible	α	Angle to the grain
i Radius of gyration			mean	Arithmetic mean		
			min	Minimum		

Source: BS 5268: Part 2: 2002.

Selected timber grade stresses and E for timber in service classes 1 and 2

| Strength class | Bending parallel to grain σ_b N/mm² | Tension parallel to grain $\sigma_{t\parallel}$ N/mm² | Compression parallel to grain $\sigma_{c\parallel}$ N/mm² | Compression perpendicular to the grain[1] | | Shear parallel to the grain $\tau_\parallel$ N/mm² | Modulus of elasticity | | Average density ρ kg/m³ |
				Mean $\sigma_{c\perp}$ N/mm²	Minimum $\sigma_{c\perp}$ N/mm²		Mean E_{mean} N/mm²	Minimum E_{min} N/mm²	
C16	5.3	3.2	6.8	2.2	1.7	0.67	8800	5800	370
C24	7.5	4.5	7.9	2.4	1.9	0.71	10800	7200	420
D40	12.5	7.5	12.6	3.9	3.0	2.00	10800	7500	700
D50	16.0	9.6	15.2	4.5	3.5	2.20	15000	12600	780

NOTES:

1. Wane is where the timber has rounded edges when it has been cut from the edge of the tree, causing the timber section size to be slightly reduced. Where the specification specifically prohibits wane at bearing areas the higher values of compression perpendicular to the grain should be used, otherwise the lower values apply.

2. The moisture contents for service classes are: Class 1 <12%, Class 2 <20% and Class 3 >20%.

3. For Class 3, timber grade stresses need to be reduced by the factors for K_2 given in Table 13 in BS 5268, but normally in wet conditions the type and preservation of the timber must be carefully selected.

4. In the absence of specific data, properties perpendicular to the grain can be assumed as:
   Tension perpendicular to the grain, torsional shear and rolling shear = $\tau_\parallel/3$
   Modulus of elasticity perpendicular to the grain = $E/20$
   Shear modulus = $E/16$

Source: BS 5268: Part 2: 2002.

Horizontally glue laminated grade stresses

The following modification factors should be applied to the grade stresses for C16 or C24 timber to obtain the equivalent glue laminated (glulam) grade stresses for members built up in 43 mm thick horizontal laminations.

Number of laminations	Bending parallel to grain K_{15}	Tension parallel to grain K_{16}	Compression parallel to grain K_{17}	Compression perpendicular to the grain K_{18}	Shear parallel to the grain K_{19}	Modulus of elasticity K_{20}
4	1.26	1.26	1.04	1.55	2.34	1.07
5	1.34	1.34				
7	1.39	1.39				
≥10	1.43	1.43				

Source: BS 5268: Part 2: 2002.

133

Selected plywood grade stresses and E for service classes 1 and 2

Plywood type	Thickness mm	Bending face grain parallel to span $\sigma_{b\parallel}$ N/mm^2	Tension parallel to face grain $\sigma_{t\parallel}$ N/mm^2	Compression parallel to face grain $\sigma_{c\parallel}$ N/mm^2	Bearing on face N/mm^2	Rolling shear N/mm^2	Modulus of elasticity in bending with face grain parallel to span N/mm^2	Panel shear N/mm^2	Shear modulus (for panel shear) N/mm^2
American unsanded	9.5 12.5 18.0	6.80 7.09 7.24	3.64 3.69 4.53	5.71 5.66 4.73	2.67	0.44	5300 4650 4150	1.58	250
Canadian Douglas Fir unsanded	9.5 12.5 18.5	12.20 10.90 15.30	6.55 4.97 6.89	9.90 7.49 10.40	2.16	0.39	6145 5490 5920	1.72	260
Finnish birch faced sanded	9.0 12.0 18.0 24.0	19.60 18.32 17.58 17.14	19.75 19.16 18.62 18.32	10.00 9.80 9.60 9.50	3.93	1.23	5200 4900 4600 4450	4.83	320
Finnish conifer unsanded	9.0 12.0 18.0 24.0	11.08 10.44 9.80 9.50	9.01 8.77 8.52 8.37	8.13 7.93 7.58 7.73	1.88	0.79	4100 3850 3650 3500	3.74	270

Source: BS 5268: Part 2: 2002.

Selected plywood properties to BS 5268 for all service classes

BS 5628 lists the properties of many different types of plywood. The properties listed here are extracts from BS 5268 for the plywoods most commonly available from UK timber importers in 2002.

Nominal thickness mm	Number of plies mm	Minimum thickness mm	Section properties for 1 m plywood width			Approximate weight kg/m²
			Area cm²	Z cm³	I cm⁴	
American construction and industrial plywood: unsanded						
9.5	3	8.7	87	12.6	5.49	540
12.5	4 & 5	11.9	119	23.6	14.04	730
18.0	4 & 5	17.5	175	51.0	44.66	1080
Canadian Douglas fir and softwood plywood: unsanded						
9.5	3	9.0	90	13.5	6.08	440
12.5	4 & 5	12.0	120	24.0	14.40	580
18.5	5, 6 & 7	18.0	180	54.0	48.60	850
Finnish birch-faced plywood: sanded						
9.0	7	8.8	88	12.9	5.68	620
12.0	7 & 9	11.5	115	22.0	12.70	810
18.0	11 & 13	17.1	171	48.7	41.70	1160
24.0	13, 15 & 17	22.9	229	87.4	100.10	1520
Finnish conifer plywood: sanded						
9.0	3, 5 & 7	8.6	86	12.3	5.30	530
12.0	4, 5, 7 & 9	11.5	115	22.0	12.50	660
18.0	6, 7, 9, 11 & 13	17.1	171	48.7	41.70	980
24.0	8, 9, 11, 13 & 17	22.9	229	87.4	100.10	1320

Selected LVL grade stresses for service classes 1 and 2

Strength class	Bending parallel to grain	Tension parallel to grain	Compression parallel to grain	Compression perpendicular to grain		Shear parallel to grain		Modulus of elasticity
	$\sigma_{b\parallel}$ N/mm²	$\sigma_{t\parallel}$ N/mm²	$\sigma_{c\parallel}$ N/mm²	Edge	Flat	Edge	Flat	Minimum
				$\sigma_{c\perp}$ N/mm²	$\sigma_{c\perp}$ N/mm²	$\tau_{\parallel}$ N/mm²	$\tau_{\parallel}$ N/mm²	E_{min} N/mm²
Kerto S	17.5	12.1	14.8	2.9	1.4	2.0	1.5	11500
Kerto Q	13.2	8.9	11.1	1.5	1.4	2.3	0.6	8360

NOTE: The average LVL density is about 510 kg/m³.

Sources: BS 5268: Part 2: 2002 Finnforest (2002).

Slenderness – maximum depth to breadth ratios

Degree of lateral support	Maximum d/b
No lateral support	2
Ends held in position	3
Ends held in position and members held in line as by purlins or tie rods at centres <$30b$	4
Ends held in position and compression edge held in line as by direct connection of sheathing, deck or joists	5
Ends held in position and compression edge held in line as by direct connection of sheathing, deck or joists together with adequate blocking spaced at centres <$6d$	6
Ends held in position and both edges held firmly in line	7

Modification factors

Duration of load K_3 factor

Duration of loading	K_3
Long term (dead + permanent imposed[1]) – Normally includes all live loads except for corridors, hallways and stairs, where the live load can sometimes be considered short term	1.00
Medium term (dead + snow, dead + temporary imposed)	1.25
Short term (dead + imposed + wind[2], dead + imposed + snow + wind[2])	1.50
Very short term (dead + imposed + snow + wind[3])	1.75

NOTES:
1. For uniformly distributed imposed load to BS 6399 $K_3 = 1.0$, except for type C3 occupancy (areas without obstacles for moving people) where $K_3 = 1.5$.
2. Short-term wind applies to either a 15 second gust, or where the largest diagonal dimension of the loaded area >50m.
3. Very short-term wind applies to either a 5 second gust, or where the largest diagonal dimension of the loaded area <50m.

Source: BS 5268: Part 2: 2002.

Depth factor for flexural members K_7 or width factor for tension members K_{14}

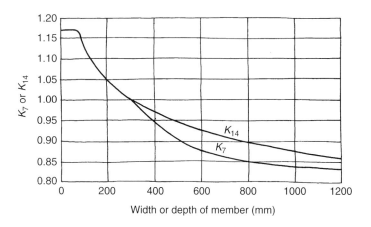

Width or depth of member (mm)

Load-sharing system factor K_8

Where a structural arrangement consists of four or more members such as rafters, joists, trusses or wall studs spaced at a maximum of 610 mm, $K_8 = 1.1$ may be used to increase the grade stresses. K_8 can also be applied to trimmer joists and lintels consisting of two or more timber elements connected in parallel.

Effective length of compression members

End conditions	L_e/L
Restrained at both ends in position and in direction	0.7
Restrained at both ends in position and one end in direction	0.85
Restrained at both ends in position but not in direction	1.0
Restrained at one end in position and in direction and at the other end in direction but not in position	1.5
Restrained at one end in position and direction and free at the other end	2.0

Generally the slenderness should be less than 180 for members carrying compression, or less than 250 where compression would only occur as a result of load reversal due to wind loading.

Compression buckling factor K_{12}

The stress in compression members should be less than the grade stress for compression parallel to the grain modified for service class, load sharing, duration of load and K_{12} for slenderness. The following graph of K_{12} has been calculated on the basis of E_{min} and $\sigma_{c\|}$ based on long-terms loads.

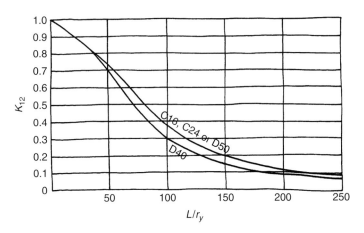

Source: BS 5268: Part 2: 2002.

Deflection and stiffness factor K_9

Generally the limit on deflection of timber structure is 0.003 × span or height. If this requirement is met, both the elastic and shear deflections are considered to be controlled. In domestic situations the total deflection must also be less than 14 mm. E_{mean} can be used in load-sharing situations. Elsewhere E_{min} should be used, modified by K_9 for trimmer joists and lintels. Glulam can be pre-cambered to compensate for deflections.

Number of pieces of timber making up the element	K_9	
	Softwoods	Hardwoods
1	1.00	1.00
2	1.14	1.06
3	1.21	1.08
≥4	1.24	1.10

Source: BS 5268: Part 2: 2002.

Timber joints

The code deals with nailed, screwed, bolted, dowelled and glued joints. Joint positions and fixing edge distances can control member sizes, as a result of the reduced timber cross section at the joint positions. Joint slip (caused by fixings moving in pre-drilled holes) can cause rotations which will have a considerable effect on overall deflections.

Nailed joints

The values given for nailed joints are for nails made from steel wire driven at right angles to the grain. In hardwood, holes normally need to be pre-drilled not bigger than $0.8d$ (where d is the fixing diameter).

Minimum nail spacings for timber to timber joints

The following nail spacings can be reduced for all softwoods (except Douglas fir) by multiplying by 0.8. However, the minimum allowable edge distance should never be less than $5d$. Reduction in spacing can be achieved for pre-drilled holes.

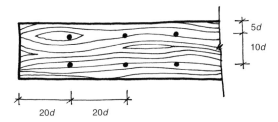

Permissible load for a nailed joint in service classes 1 and 2

$$F_{adm} = F \times K_{50} \times n \times \text{the number of shear planes}$$

n = the total number of nails in the joint

K_{50} = 0.9 for more than 10 nails in a line parallel to the action of the load. For nails driven into the end grain of the timber a further factor of 0.7 should be used. For pre-drilled holes a factor of 1.15 applies.

Basic single shear loads for nails in timber to timber joints

Nail diameter mm	SWG	Basic shear load kN Softwoods (not pre-drilled)			Hardwoods (pre-drilled)		
		Standard penetration mm	Strength class		Minimum penetration mm	Strength class	
			C16	C24		D40	D50
3	11	36	0.306	0.326	24	0.465	0.515
3.4	10	44	0.377	0.400	27	0.582	0.644
4.5	7	55	0.620	0.659	37	0.996	1.103
5.5	5	66	0.833	0.885	44	1.368	1.515

Basic single shear loads for nails in timber to plywood joints

Nominal plywood thickness mm	Nail diameter mm	Nail length mm	Basic shear load* kN		
			Softwood (not pre-drilled)		Hardwoods (pre-drilled)
			C16	C24	D40 and D50
6	3	50	0.256	0.267	0.295
	4.5	75	0.373	0.373	0.373
12	3	50	0.286	0.296	0.352
	4.5	75	0.461	0.478	0.589
18	3	50	0.344	0.355	0.417
	4.5	75	0.515	0.533	0.642
21	3	50	0.359	0.374	0.456
	4.5	75	0.550	0.569	0.682

*Additional capacity can be achieved for joints into Finnish birch and birch-faced plywoods, see BS 5268: Part 2: 2002: Table 63.

Basic withdrawal loads for nails in timber

Nail diameter mm	Basic load per pointside penetration N/mm			
	Strength class			
	C16	C24	D40	D50
3	1.71	2.31	6.52	10.86
3.4	1.93	2.62	7.39	12.31
4.5	2.62	3.54	10.00	16.65
5.5	3.13	4.21	11.95	19.91

Source: BS 5268: Part 2: 2002: Appendix G.

Screwed joints

The values given for screwed joints are for screws which conform to BS 1210 in pre-drilled holes. The holes should be drilled with a diameter equal to that of the screw shank (ϕ) for the part of the hole to contain the shank, reducing to a pilot hole (with a diameter of $\phi/2$) for the threaded portion of the screw. Where the standard head-side thickness is less than the values in the table, the basic load must be reduced by the ratio: actual/standard thickness. The headside thickness must be greater than 2ϕ twice the shank diameter. The following tables give values for UK screws rather than the European screws quoted in the latest British Standard.

Minimum screw spacings

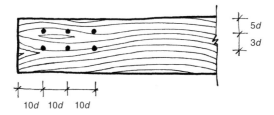

Permissible load for a screwed joint for service classes 1 and 2

$$F_{adm} = F \times K_{54} \times n$$

n = the total number of screws in the joint

$K_{54} = 0.9$ for more than 10 of the same diameter screws in a line parallel to the action of the load. For screws inserted into the end grain of the timber a further factor of 0.7 should be used.

Basic single shear loads for screws in timber to timber joints

Screw		Standard penetration		Basic single shear kN			
				Softwoods		Hardwoods	
Screw reference	Shank diameter mm	Headside mm	Pointside mm	C16	C24	D40	D50
No. 6	3.45	12	25	0.280	0.303	0.397	0.457
No. 8	4.17	15	30	0.360	0.390	0.513	0.592
No. 10	4.88	18	35	0.451	0.486	0.641	0.742
No. 12	5.59	19	39	0.645	0.698	0.921	1.065
No. 14	6.30	23	46	0.760	0.821	1.086	1.259
No. 16	7.01	25	50	1.010	1.092	1.454	1.690

Basic single shear loads for screws in timber to plywood joints

Nominal plywood thickness mm	Screw			Basic single shear* kN			
				Softwoods		Hardwoods	
	Screw reference	Shank diameter mm	Minimum screw length mm	C16	C24	D40	D50
12	No. 6	3.45	26	0.224	0.238	0.298	0.337
	No. 8	4.17	28	0.267	0.285	0.367	0.419
	No. 10	4.88	30	0.315	0.339	0.445	0.513
	No. 12	5.59	34	0.429	0.466	0.631	0.736
	No. 14	6.30	40	0.644	0.707	0.982	1.026
18	No. 6	3.45	32	0.300	0.312	0.365	0.398
	No. 8	4.17	34	0.343	0.359	0.431	0.476
	No. 10	4.88	36	0.390	0.412	0.504	0.563
	No. 12	5.59	40	0.499	0.533	0.679	0.772
	No. 14	6.30	46	0.703	0.760	1.009	1.168

*Extra capacity can achieved for joints into Finnish birch and birch-faced plywoods, see BS 5268: Part 2: 2002: Table 68.

Basic withdrawal loads for screws in timber

Screw reference	Shank diameter mm	Basic load per pointside penetration N/mm			
		Strength class			
		C16	C24	D40	D50
No. 6	3.45	12.2	14.7	27.7	37.3
No. 8	4.17	13.5	16.3	30.6	41.2
No. 10	4.88	14.7	17.8	33.5	45.0
No. 12	5.59	17.1	20.7	38.9	52.3
No. 14	6.30	18.2	22.1	41.5	55.9
No. 16	7.01	20.5	24.8	46.6	62.7

Source: BS 5268: Part 2: 2002: Appendix G.

Bolted joints

The values given for bolted joints are for black bolts which conform to BS EN 20898-1 with washers which conform to BS 4320. Bolt holes should not be drilled more than 2 mm larger than the nominal bolt diameter. Washers should have a diameter or width of three times the bolt diameter with a thickness of 0.25 times the bolt diameter and be fitted under the head and nut of each bolt. At least one complete thread should protrude from a tightened nut.

Minimum bolt spacings

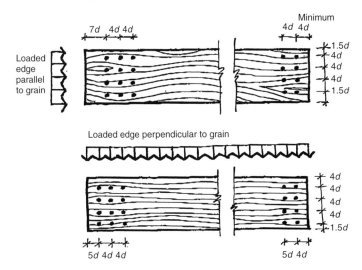

Permissible load for a bolted joint for service classes 1 and 2

$F_{adm} = F \times K_{57} \times n$

n = the total number of bolts in the joint

$K_{57} = 1 - (3(n - 1))/100$ for less than 10 of the same diameter bolts in a line parallel to the action of the load

$K_{57} = 0.7$ for more than 10 of the same diameter bolts in a line parallel to the action of the load

$K_{57} = 1.0$ for all other loading cases where more than one bolt is used in a joint

If a steel plate of minimum thickness 0.3 times the bolt diameter (or 2.5 mm) is bolted to the timber, the basic load can be multiplied by a factor of 1.25. Further improvements on the loads in bolts can be made by using toothed connectors, but these require larger spacings (hence fewer fixings) and correct installation can be difficult.

Basic single shear loads for one grade 4.6 bolt in a two member timber connection

Timber grade	Minimum member thickness (mm)	Basic single shear load for selected grade 4.6 bolt diameters in a two member* timber connection (kN)							
		Direction of loading							
		Parallel to the grain				Perpendicular to the grain			
		M8	M12	M16	M20	M8	M12	M16	M20
C16	47	1.22	1.80	2.30	2.73	1.13	1.56	1.91	2.19
	72	1.46	2.68	3.52	4.19	1.39	2.39	2.93	3.36
	97	1.46	3.13	4.63	5.64	1.39	2.79	3.94	4.52
C24	47	1.33	2.04	2.59	3.09	1.23	1.76	2.16	2.47
	72	1.55	2.93	3.97	4.73	1.47	2.64	3.30	3.79
	97	1.55	3.42	5.05	6.37	1.47	3.07	4.43	5.11
D40	47	1.83	3.08	3.92	4.67	1.83	3.08	3.92	4.67
	72	1.91	4.02	5.98	7.16	1.91	4.02	5.98	7.16
	97	1.91	4.21	6.93	9.32	1.91	4.21	6.93	9.32
D50	47	2.12	3.78	4.81	5.73	2.12	3.78	4.81	5.73
	72	2.12	4.66	6.92	8.78	2.12	4.66	6.92	8.78
	97	2.12	4.66	8.09	10.82	2.12	4.66	8.09	10.82

*Extra capacity for three member connections can be achieved, see BS 5268: Part 2: 2002: Tables 76, 77, 79 and 80.

Source: BS 5268: Part 2: 2002: Appendix G.

7
Masonry

Masonry, brought to the UK by the Romans, became a popular method of construction as the units could originally be lifted and placed with one hand. Masonry has orthotropic material properties relating to the bed or perpend joints of the masonry units. The compressive strength of the masonry depends on the strength of the masonry units and on the mortar type. Masonry is good in compression and has limited flexural strength. Where the flexural strength of masonry 'parallel to the bed joints' can be developed, the section is described as 'uncracked'. A cracked section (e.g. due to a damp proof course or a movement joint) relies on the dead weight of the masonry to resist tensile stresses. The structure should be arranged to limit tension or buckling in slender members, or crushing of stocky structures.

Summary of material properties

Clay bricks   The wide range of clays in the UK result in a wide variety of available brick strengths, colours and appearance. Bricks can be hand or factory made. Densities range between 22.5 and $28\,kN/m^3$. Clay bricks tend to expand due to water absorption. Engineering bricks have low water absorption, high strength and good durability properties (Class A strength $>70\,N/mm^2$; water absorption $\leq4.5\%$ by mass. Class B: strength $>50\,N/mm^2$; water absorption $\leq7.0\%$ by mass).

Calcium silicate bricks   Calcium silicates are low cost bricks made from sand and slaked lime rarely used due to their tendency to shrink and crack. Densities range between 17 and $21\,kN/m^3$.

Concrete blocks   Cement bound blocks are available in densities ranging between 5 and $20\,kN/m^3$. The lightest blocks are aerated; medium dense blocks contain slag, ash or pumice aggregate; dense blocks contain dense natural aggregates. Blocks can shrink by 0.01–0.09%, but blocks with shrinkage rates of no more than 0.03% are preferable to avoid the cracking of plaster and brittle finishes on the finished walls.

Stone masonry   Stone as rubble construction, bedded blocks or as facing to brick or blockwork is covered in BS 5390. Thin stone used as cladding or facing is covered by BS 8298.

Cement mortar   Sand, dry hydrate of lime and cement are mixed with water to form mortar. The cement cures on contact with water. It provides a bond strong enough that the masonry can resist flexural tension, but structural movement will cause cracking.

Lime mortar   Sand and non-hydraulic lime putty form a mortar, to which some cement (or other pozzolanic material) can be added to speed up setting. The mortar needs air, and warm, dry weather to set. Lime mortar is more flexible than cement mortar and therefore can resist considerably more movement without visible cracking.

Typical unit strengths of masonry

Material (relevant BS)	Class	Water absorption	Typical unit compressive strength N/mm^2
Fired clay bricks (BS 3921)	Engineering A	<4.5%	>70
	Engineering B	7.0%	>50
	Facings (bricks selected for appearance)	10–30%	10–50
	Commons (Class 1 to 15)	20–24%	
	Class 1		7
	Class 2		14
	Class 3		20
	Class 15		105
	Flettons	15–25%	15–25
	Stocks (bricks without frogs)	20–40%	3–20
Calcium silicate bricks (BS 187)	Classes 2 to 7		14–48.5
Concrete bricks (BS 6073: Pt 1)			7–20
Concrete blocks (BS 6073: Pt 1)	Dense solid		7, 10–30
	Dense hollow		3.5, 7, 10
	Lightweight		2.8, 3.5, 4, 7
Reconstituted stone (BS 6457)	Dense solid		Typically as dense concrete blocks
Natural stone (BS 5390 and BS 8298)	Structural quality. Strength is dependent on the type of stone, the quality, the direction of the bed, the quarry location		15–100

Geometry and arrangement

Brick and block sizes

The standard UK brick is 215 × 102 × 65 mm which gives a co-ordinating size of 225 × 112 × 75 mm. The standard UK block face size is 440 × 215 mm in thicknesses from 75 to 215 mm, giving a co-ordinating size of 450 × 225 mm. This equates to two brick stretchers by three brick courses.

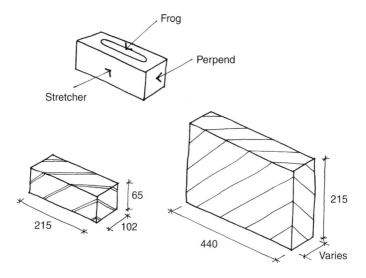

The Health and Safety Executive (HSE) require designers to specify blocks which weigh less than 20 kg to try to reduce repetitive strain injuries in bricklayers. Medium dense and dense blocks of 140 mm thick, or more, often exceed 20 kg. The HSE prefers designers to specify half blocks (such as Tarmac Topcrete) rather than rely on special manual handling (such as hoists) on site. In addition to this, the convenience and speed of block laying is reduced as block weight increases.

Non-hydraulic lime mortar mixes for masonry

Mix constituents	Approximate proportions by volume	Notes on general application
Lime putty:coarse sand	1:2.5 to 3	Used where dry weather and no frost are expected for several months
Pozzolanic*:lime putty:coarse sand	1:2 to 3:12	Used where an initial mortar set is required within a couple of days

*Pozzolanic material can be cement, fired china dust or ground granulated blast furnace slag (ggbfs).

The actual amount of lime putty used depends on the grading of the sand and the volume of voids. Compressive strength values for non-hydraulic lime mortar masonry can be approximated using the values for Class IV cement mortar. Due to the flexibility of non-hydraulic lime mortar, thermal and moisture movements can generally be accommodated by the masonry without cracking of the masonry elements or the use of movement joints. This flexibility also means that resistance to lateral load relies on mass and dead load rather than flexural strength. The accepted minimum thickness of walls with non-hydraulic lime mortar is 215 mm and therefore the use of lime mortar in standard single leaf cavity walls is not appropriate.

Cement mortar mixes for masonry

Mortar class	Compressive strength class	Type of mortar (proportions by volume)			Compressive strengths at 28 days N/mm^2
		Cement: lime:sand	Masonry cement: sand	Cement:sand with or without air entrainment	
Dry pack	–	1:0:3	–	–	
I	M12	1:0 to ¼:3	–	–	12
II	M6	1:½:4 to 4½	1:2½ to 3½	1:3 to 4	6
III	M4	1:1:5 to 6	1:4 to 5	1:5 to 6	4
IV	M2	1:2:8 to 9	1:5½ to 6½	1:7 to 8	2

NOTES:
1. Mix proportions are given by volume. Where sand volumes are given as variable amounts, use the larger volume for well-graded sand and the smaller volume for uniformly graded sand.
2. As the mortar strength increases, the flexibility reduces and likelihood of cracking increases.
3. Cement:lime:sand mortar provides the best bond and rain resistance, while cement:sand and plasticizer is more frost resistant.

Source: BS 5628: Part 1: 2005.

Selected bond patterns

For strength, perpends should not be less than one quarter of a brick from those in an adjacent course.

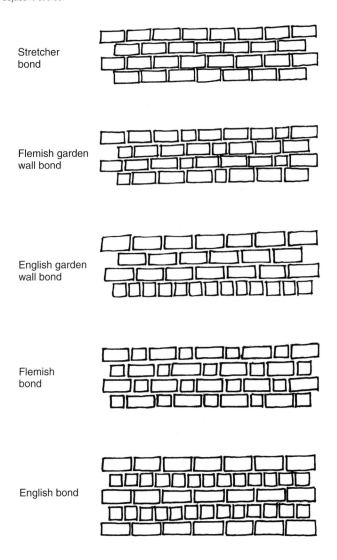

Stretcher bond

Flemish garden wall bond

English garden wall bond

Flemish bond

English bond

Movement joints in masonry with cement-based mortar

Movement joints to limit the lengths of walls built in cement mortar are required to minimize cracking due to deflection, differential settlement, temperature change and shrinkage or expansion. In addition to long wall panels, movement joints are also required at points of weakness, where stress concentrations might be expected to cause cracks (such as at steps in height or thickness or at the positions of large chases). Typical movement joint spacings are as follows:

Material	Approximate horizontal joint spacing[2] and reason for provision	Typical joint thickness mm	Maximum suggested panel length:height ratio[1]
Clay bricks	12 m for expansion	16	3:1
	15–18 m with bed joint reinforcement at 450 mm c/c	22	
	18–20 m with bed joint reinforcement at 225 mm c/c	25	
Calcium silicate bricks	7.5–9 m for shrinkage	10	3:1
Concrete bricks	6 m for shrinkage	10	2:1
Concrete blocks	6–7 m for shrinkage	10	2:1
	15–18 m with bed joint reinforcement at 450 mm c/c	22	
	18–20 m with bed joint reinforcement at 225 mm c/c	25	
Natural stone cladding	6 m for thermal movements	10	3:1

NOTES:
1. Consider bed joint reinforcement for ratios beyond the suggested maximum.
2. The horizontal joint spacing should be halved for joints which are spaced off corners.

Vertical joints are required in cavity walls every 9 m or three storeys for buildings over 12 m or four storeys. This vertical spacing can be increased if special precautions are taken to limit the differential movements caused by the shrinkage of the internal block and the expansion of the external brick. The joint is typically created by supporting the external skin on a proprietary stainless steel shelf angle fixed back to the internal structure. Normally 1 mm of joint width is allowed for each metre of masonry (with a minimum of 10 mm) between the top of the masonry and underside of the shelf angle support.

Durability and fire resistance

Durability

Durability of masonry relies on the selection of appropriate components detailed to prevent water and weather penetration. Wet bricks can suffer from spalling as a result of frost attack. Bricks of low porosity are required in positions where exposure to moisture and freezing is likely. BS 5268: Part 3 gives guidance on recommended combinations of masonry units and mortar for different exposure conditions as summarized:

Durability issues for selection of bricks and mortar

Application		Minimum strength of masonry unit	Mortar class[1]	Brick frost resistance[2] and soluble salt content[3]
Internal walls generally/external walls above DPC		Any block/15 N/mm² brick	III	FL, FN, ML, MN, OL or ON
External below DPC/freestanding walls/parapets		7 N/mm² dense block/ 20 N/mm² brick	III	FL or FN (ML or MN if protected from saturation)
Brick damp proof courses in buildings (BS 743)		Engineering brick A	I	FL or FN
Earth retaining walls		7 N/mm² dense block/ 30 N/mm² brick	I or II	FL or FN
Planter boxes		Engineering brick/ 20 N/mm² commons	I or II	FL or FN
Sills and copings		Selected block/ 30 N/mm² brick	I	FL or FN
Manholes and inspection chambers	Surface water	Engineering brick/ 20 N/mm² commons	I or II	FL or FN (ML or MN if more than 150 mm below ground level)
	Foul drainage	Engineering brick A	I or II	

NOTES:
1. Sulphate resisting mortar is advised where soluble sulphates are expected from the ground, saturated bricks or elsewhere.
2. F indicates that the bricks are frost resistant, M indicates moderate frost resistance and O indicates no frost resistance.
3. N indicates that the bricks have normal soluble salt content and L indicates low soluble salt content.
4. Retaining walls and planter boxes should be waterproofed on their retaining faces to improve durability and prevent staining.

Fire resistance

As masonry units have generally been fired during manufacture, their performance in fire conditions is generally good. Perforated and cellular bricks have a lesser fire resistance than solid units of the same thickness. The fire resistance of blocks is dependent on the grading of the aggregate and cement content of the mix, but generally 100 mm solid blocks will provide a fire resistance of up to 2 hours if load bearing and 4 hours if non-load bearing. Longer periods of fire resistance may require a thicker wall than is required for strength. Specific product information should be obtained from masonry manufacturers.

Preliminary sizing of masonry elements

Typical span/thickness ratios

Description	Typical thickness		
	Freestanding/ cantilever	Element supported on two sides	Panel supported on four sides
Lateral loading			
Solid wall with no piers – uncracked section	$H/6–8.5$[1]	$H/20$ or $L/20$	$H/22$ or $L/25$
Solid wall with no piers – cracked section	$H/4.5–6.4$[1]	$H/10$ or $L/20$	$H/12$ or $L/25$
External cavity wall[2] panel	–	$H/20$ or $L/30$	$H/22$ or $L/35$
External cavity wall[2] panel with bed joint reinforcement	–	$H/20$ or $L/35$	$H/22$ or $L/40$
External diaphragm wall panel	$H/10$	$H/14$	–
Reinforced masonry retaining wall (bars in pockets in the walls)	$H/10–15$	–	–
Solid masonry retaining wall (thickness at base)	$H/2.5–4$	–	–
Vertical loading			
Solid wall	$H/8$	$H/18–22$	–
Cavity wall	$H/11$	$H/5.5$	–
Masonry arch/vault	–	$L/20–30$	$L/30–60$
Reinforced brick beam depth	–	$L/10–16$	–

NOTES:
1. Depends on the wind exposure of the wall.
2. The spans or distances between lateral restraints are L in the horizontal direction and H in the vertical direction.
3. In cavity walls, the thickness is the sum of both leaves excluding the cavity width.

Vertical load capacity wall charts

Vertical load capacity of selected walls less than 150 mm thick

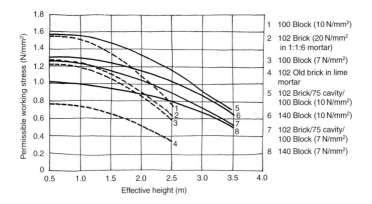

1 100 Block (10 N/mm²)

2 102 Brick (20 N/mm² in 1:1:6 mortar)

3 100 Block (7 N/mm²)

4 102 Old brick in lime mortar

5 102 Brick/75 cavity/ 100 Block (10 N/mm²)

6 140 Block (10 N/mm²)

7 102 Brick/75 cavity/ 100 Block (7 N/mm²)

8 140 Block (7 N/mm²)

Vertical load capacity of selected solid walls greater than 150 mm thick

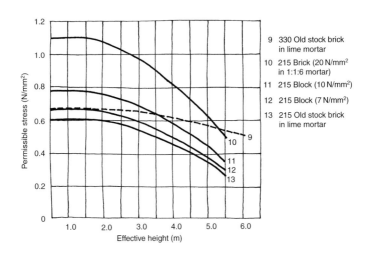

9 330 Old stock brick in lime mortar

10 215 Brick (20 N/mm² in 1:1:6 mortar)

11 215 Block (10 N/mm²)

12 215 Block (7 N/mm²)

13 215 Old stock brick in lime mortar

Preliminary sizing of external cavity wall panels

The following approach can be considered for cavity wall panels in non-load bearing construction up to about 3.5 m tall in buildings of up to four storeys high, in areas which have many windbreaks. Without major openings, cavity wall panels can easily span up to 3.5 m if spanning horizontally, while panels supported on four sides can span up to about 4.5 m. Load bearing wall panels can be larger as the vertical loads pre-compress the masonry and give it much more capacity to span vertically. Gable walls can be treated as rectangular panels and their height taken as the average height of the wall.

Detailed calculations for masonry around openings can sometimes be avoided if:

1. The openings are completely framed by lateral restraints.
2. The total area of openings is less than the lesser of: 10% of the maximum panel area (see the section on the design of external wall panels to BS 5628 under 'Lateral load' later in this chapter) or 25% of the actual wall area.
3. The opening is more than half its maximum dimension from any edge of the wall panel (other than its base) and from any adjacent opening.

Internal non-loadbearing masonry partition chart

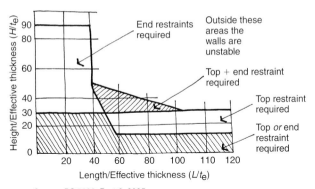

Source: BS 5628: Past 3: 2005

Source: BS 5628: Part 3: 2005.

Typical ultimate strength values for stone masonry

	Crushing N/mm^2	Tension N/mm^2	Shear N/mm^2	Bending N/mm^2
Basalt	8.5	8.6	4.3	–
Chalk	1.1	–	–	–
Granite	96.6	3.2	5.4	10.7
Limestone	53.7	2.7	4.3	6.4
Limestone soft	10.7	1.0	3.8	5.4
Marble	64.4	3.2	5.4	–
Sandstone	53.7	1.1	3.2	5.4
Sandstone soft	21.5	0.5	1.1	2.1
Slate	85.8	1.1	3.2	5.4

The strength values listed above assume that the stone is of good average quality and that the factor of safety commonly used will be 10. While this seems sensible for tension, shear and bending it does seem conservative for crushing strength. Better values can be achieved on the basis of strength testing. These values can be used in preliminary design, but where unknown stones or unusual uses are proposed, strength testing is advised. The strength of stone varies between sources and samples, and also depends on the mortar and the manner of construction. The British Stone website has listings of stone tests carried out by the Building Research Establishment (BRE).

As compressive load can be accompanied by a shear stress of up to half the compressive stress, shear stresses normally control the design of slender items such as walls and piers. Safe wall and pier loads are generally obtained by assuming a safe working compressive stress equal to twice the characteristic shear stress.

Source: Howe, J.A. (1910).

Masonry design to BS 5628

Partial load safety factors

Load combination	Load type			
	Dead	Imposed	Wind	Earth and water
Dead and imposed	1.4 or 0.9	1.6	–	1.2
Dead and wind	1.4 or 0.9	–	1.4*	1.2
Dead and wind (freestanding walls)	1.4 or 0.9	–	1.2*	–
Dead, imposed or wind	1.2	1.2	1.2*	1.2
Accidental damage	0.95 or 1.05	0.35 or 1.05	0.35	–

*Buildings should be capable of resisting a horizontal load equal to 1.5% of the total characteristic dead load (i.e. $0.015G_k$ above any level. In some instances $0.015G_k$ can be greater than the applied wind loadings.

Partial material safety factors

The factor of safety for the compressive strength of materials is generally taken as $\gamma_{mc} = 3.5$ while the factor of safety for flexural strength of materials has recently been reduced to $\gamma_{mf} = 3.0$ (assuming normal control of manufacture and construction). Tables 4a and 4b in BS 5628 allow these material safety factors to be reduced if special controls on manufacture and construction are in place.

Notation for BS 5628: Part 1

Symbols	Subscripts					
	Type of stress		Significance		Geometry	
f Stress	k	Compression	a	Applied	‖	Parallel to the bed joints
	kx	Bending	adm	Permissible	⊥	Perpendicular to the bed joints
	v	Shear				

In addition:

μ   The orthogonal ratio is the ratio of the flexural strengths in different directions, $\mu = fkx_{\parallel}/fkx\perp$.

α   Panel factor (determined by μ and panel size) which attempts to model how a panel with orthogonal properties distributes lateral load between the stronger (perpendicular to the bed joints) and the weaker (parallel to the bed joints) directions.

Source: BS 5628: Part 1: 2005.

Vertical load

Selected characteristic compressive masonry strengths for standard format brick masonry (N/mm^2)

Mortar class	Compressive strength of unit N/mm^2								
	5	10 Stock	15	20 Fletton	30	40	50 Class B	75 Class A	100 Class A
I m12	2.5	4.0	6.3	6.4	8.3	10.0	11.6	15.2	18.3
II m6	2.5	3.8	4.8	5.6	7.1	8.4	9.5	12.0	14.2
III m4	2.5	3.4	4.3	5.0	6.3	7.4	8.4	10.5	12.3
IV m2	2.5	2.8	3.6	4.1	5.1	6.1	7.1	9.0	10.5

Selected characteristic compressive masonry strengths for concrete block masonry (N/mm^2)

Mortar class and type of unit	Compressive strength of unit N/mm^2				
	2.9	3.6	5.2	7.3	10.4
100 mm solid or concrete filled hollow blocks					
II / M6	2.8	3.5	5.0	6.8	8.8
III / M4	2.8	3.5	5.0	6.4	8.4
140 mm solid or concrete filled hollow blocks					
II / M6	2.8	3.5	5.0	6.4	8.4
III / M4	2.8	3.5	5.0	6.4	8.2
215 mm solid concrete block wall					
(made up with 100 mm solid blocks laid on side)					
II / M6	1.4	1.7	2.5	3.2	4.2
III / M4	1.4	1.7	2.5	3.2	4.1

NOTES to both tables:
1. For columns or piers with cross sectional area, $A < 0.2\,m^2$ f_k should be multiplied by $\gamma_{col} = (0.7 + 1.5A)$.
2. Where a brick wall is 102 mm thick f_k can be multiplied by $\gamma = 1.15$, but wide format bricks need a reduction factor in accordance with BS 5628 cl. 23.1.4.
3. Natural stone masonry can be taken on the same values as concrete blocks of the same strength, but random rubble masonry in cement mortar should be considered to have 75% of this strength. Random rubble masonry in lime mortar can be considered to have a characteristic strength of 50% of the equivalent concrete blocks in Class IV mortar.

Source: BS 5628: Part 1: 2005.

Effective thickness

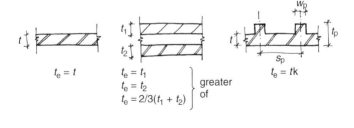

$$t_e = t$$

$$t_e = t_1$$
$$t_e = t_2$$
$$t_e = 2/3(t_1 + t_2)$$
greater of

$$t_e = tk$$

Stiffness coefficient *k* for walls stiffened by piers

Ratio of pier spacing (centre to centre) to pier width S_p/W_p	Ratio of pier thickness to wall thickness, t_p/t		
	1	2	3
≤6	1.0	1.4	2.0
10	1.0	1.2	1.4
≥20	1.0	1.0	1.0

Linear interpolation (but not extrapolation) is permitted.

Source: BS 5628: Part 1: 2005.

Effective height or length

Lateral supports should be able to carry any applied horizontal loads or reactions plus 2.5 per cent of the total vertical design load in the element to be laterally restrained. The effective height of a masonry wall depends on the horizontal lateral restraint provided by the floors or roofs supported on the wall. The effective length of a masonry wall depends on the vertical lateral restraint provided by cross walls, piers or returns. BS 5268: Part 1: clause 28.2 and Appendix C define the types of arrangement which provide simple or enhanced lateral restraint. The slenderness ratio of a masonry element should generally not exceed 27. For walls of less than 90 mm thick in buildings of two storeys or more, the slenderness should not exceed 20.

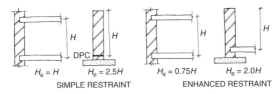

Horizontal resistance to lateral movement

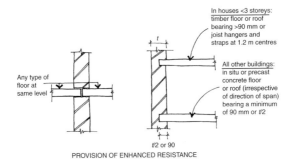

PROVISION OF ENHANCED RESISTANCE

Vertical resistance to lateral movement

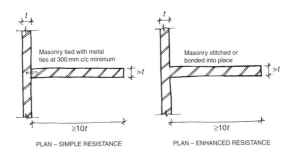

PLAN – SIMPLE RESISTANCE   PLAN – ENHANCED RESISTANCE

Sources: BS 5628: Part 1: 2005;
BS 5268: Part 3: 2005.

Slenderness reduction factor β

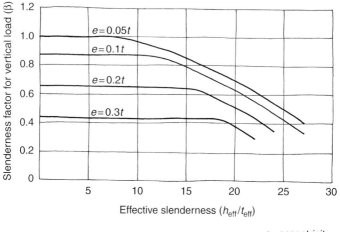

$e =$ eccentricity
$t =$ thickness

Vertical load resistance of walls and columns

Wall: $F_{k_{adm}} = \dfrac{(\beta t f_k)}{\gamma_m}$

Column: $F_{k_{adm}} = \dfrac{(\beta b t f_k)\gamma_{col}}{\gamma_m}$

t is the actual (not the effective thickness), b is the column partial width, β is the slenderness reduction factor, γ_{col} is the column reduction factor, γ_m is the partial factor of safety for materials and workmanship and f_k is the characteristic compressive strength of the masonry.

Additional factors (as listed in the notes below the tabulated values of f_k) may also be applied for the calculation of load resistance.

Shear strength

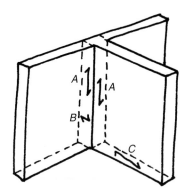

$$f_v = \frac{f_{vko}}{\gamma_m}$$

A Complementary shear acting in the vertical direction of the vertical plane:

$f_{vko} = 0.7\,\text{N/mm}^2$ for bricks in I and II (m12 and m6) mortar or $0.5\,\text{N/mm}^2$ for bricks in III and IV (m4 and m2) mortar

$f_{vko} = 0.35\,\text{N/mm}^2$ for $7\,\text{N/mm}^2$ blocks in I, II and III (m12, m6 and m4) mortar.

B Complementary shear acting in the horizontal direction of the vertical plane:

C Shear acting in the horizontal direction of the horizontal plane:

$f_{vko} = (0.35 + 0.6G_a)\text{N/mm}^2$

to a maximum of $1.75\,\text{N/mm}^2$ for I and II (m12 and m6) mortar, or $f_{vko} = (0.15 + 0.6G_a)$ N/mm^2 to a maximum of $1.4\,\text{N/mm}^2$ for III and IV (m4 and m2) mortar.

Lateral load

Wind zone map

The sizes of external wall panels are limited depending on the elevation of the panel above the ground and the typical wind speeds expected for the site.

Maximum permitted panel areas for external cavity wall panels * **(m^2)**

Wind zone	Height above ground m	Panel type A	B	C	D	E	F	G	H	I
1	**5.4**	11.0	17.5	26.5	20.5	32.0	32.0	8.5	14.0	19.5
	10.8	9.0	13.0	17.5	15.5	24.0	32.0	7.0	10.0	15.5
2	**5.4**	9.5	14.0	21.0	17.5	27.0	32.0	7.5	10.5	17.0
	10.8	8.0	11.5	13.5	13.0	19.0	28.0	6.0	9.0	13.0
3	**5.4**	8.5	12.5	15.5	14.5	22.0	30.5	6.5	9.5	14.5
	10.8	7.0	10.0	11.5	11.0	14.5	24.5	5.0	7.5	11.5
4	**5.4**	8.0	11.0	13.0	12.5	18.0	27.0	6.0	8.5	12.5
	10.8	6.5	9.0	10.5	9.5	12.5	21.5	4.0	6.5	10.0

*The values in the table are given for cavity walls with leaves of 100 mm block inner skin. If either leaf is increased to 140 mm the maximum permitted areas in the table can be increased by 20%.

No dimension of unreinforced masonry panels should generally exceed 50 × effective thickness. Reinforced masonry panels areas can be about 20% bigger than unreinforced panels and generally no dimension should exceed 60 × effective thickness.

Source: BS 5268: Part 3: 2005.

Characteristic flexural strength of masonry

The characteristic flexural strength of masonry works on the assumption that the masonry section is uncracked and therefore can resist some tensile stresses. If the wall is carrying some compressive load, then the wall will have an enhanced resistance to tensile stresses. Where tension develops which exceeds the tensile resistance of the masonry (e.g. at a DPC or crack location) the forces must be resisted by the dead loads alone and therefore the wall will have less capacity than an uncracked section.

Type of masonry unit	Plane of failure				Basic orthogonal ratio, μ^*	
	Parallel to the bed joints $f_{kx\parallel}$		Perpendicular to the bed joints $f_{kx\perp}$			
	Mortar I M12 N/mm^2	II and III M6 and M4 N/mm^2	I M12 N/mm^2	II and III M6 and M4 N/mm^2	I M12	II and III M6 and M4
Clay bricks having a water absorption of:						
less than 7%	0.7	0.5	2.0	1.5	0.35	0.33
between 7% and 12%	0.5	0.4	1.5	1.1	0.33	0.36
over 12%	0.4	0.3	1.1	0.9	0.36	0.33
Calcium silicate or concrete bricks		0.3		0.9		0.33
7.3 N/mm^2 concrete blocks:						
100 mm wide wall		0.25		0.60		0.41
140 mm wide wall		0.22		0.55		0.40
215 mm wide wall		0.17		0.45		0.38
Any thickness concrete blocks walls:						
10.5 N/mm^2		0.25		0.75		0.33
$\geq$14.0 N/mm^2		0.25		0.90		0.27

*The orthogonal ratio, $\mu = f_{kx\parallel}/f_{kx\perp}$, can be improved if $f_{kx\parallel}$ is enhanced by the characteristic dead load: $f_{kx\parallel|enhanced} = f_{kx\parallel}/\gamma_m + 0.9\,G_k$.

Source: BS 5628: Part 1: 2005.

Ultimate moments applied to wall panels

The flexural strength of masonry parallel to the bed joints is about a third of the flexural strength perpendicular to the bed joints. The overall flexural capacity of a panel depends on the dimensions, orthogonal strength ratio and support conditions of that panel. BS 5628: Part 1 uses α (which is based on experimental data and yield line analysis) to estimate how a panel will combine these different orthogonal properties to distribute the applied lateral loads and express this as a moment in one direction:

$$M_{a\parallel} = \mu\alpha\gamma_f W_k L^2 \text{ per unit length, or}$$

$$M_{a\perp} = \alpha\gamma_f W_k L^2 \text{ per unit height}$$

W_k is the applied distributed lateral load panel, L is the horizontal panel length and $M_{a\perp} = M_{a\parallel}/\mu$. Therefore the applied moment need only be checked against the flexural strength in one direction.

Ultimate flexural strength of an uncracked wall spanning horizontally

$$M_{adm\perp} = \left(\frac{f_{kx\perp}}{\gamma_m}\right) Z$$

Ultimate flexural strength of an uncracked wall spanning vertically

$$M_{adm\parallel} = \left(\frac{f_{kx\parallel}}{\gamma_m} + 0.9G_k\right) Z$$

The flexural strength parallel to the bed joints varies with the applied dead load and the height of the wall. Therefore the top half of non-load bearing walls are normally the critical case. There are published tables for non-load bearing and freestanding walls which greatly simplify calculations.

Ultimate flexural strength of a cracked wall spanning vertically

$$M_{adm\parallel} = \frac{\gamma_w t^2 h}{2\gamma_f}$$

The dead weight of the wall is used to resist lateral loads. Tension can be avoided if the resultant force is kept within the middle third of the wall. Where γ_f is an appropriate factor of safety against overturning.

Bending moment coefficients in laterally loaded wall panels

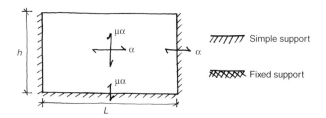

Panel support conditions	Orthogonal ration μ	Values of panel factor, α						
		h/L						
		0.30	0.50	0.75	1.00	1.25	1.50	1.75
A	1.00	0.031	0.045	0.059	0.071	0.079	0.085	0.090
	0.90	0.032	0.047	0.061	0.073	0.081	0.087	0.092
	0.80	0.034	0.049	0.064	0.075	0.083	0.089	0.093
	0.70	0.035	0.051	0.066	0.077	0.085	0.091	0.095
	0.60	0.038	0.053	0.069	0.080	0.088	0.093	0.097
	0.50	0.040	0.056	0.073	0.083	0.090	0.095	0.099
	0.40	0.043	0.061	0.077	0.087	0.093	0.098	0.101
	0.35	0.045	0.064	0.080	0.089	0.095	0.100	0.103
	0.30	0.048	0.067	0.082	0.091	0.097	0.101	0.104
B	1.00	0.024	0.035	0.046	0.053	0.059	0.062	0.065
	0.90	0.025	0.036	0.047	0.055	0.060	0.063	0.066
	0.80	0.027	0.037	0.049	0.056	0.061	0.065	0.067
	0.70	0.028	0.039	0.051	0.058	0.062	0.066	0.068
	0.60	0.030	0.042	0.053	0.059	0.064	0.067	0.069
	0.50	0.031	0.044	0.055	0.061	0.066	0.069	0.071
	0.40	0.034	0.047	0.057	0.063	0.067	0.070	0.072
	0.35	0.035	0.049	0.059	0.065	0.068	0.071	0.073
	0.30	0.037	0.051	0.061	0.066	0.070	0.072	0.074
C	1.00	0.020	0.028	0.037	0.042	0.045	0.048	0.050
	0.90	0.021	0.029	0.038	0.043	0.046	0.048	0.050
	0.80	0.022	0.031	0.039	0.043	0.047	0.049	0.051
	0.70	0.023	0.032	0.040	0.044	0.048	0.050	0.051
	0.60	0.024	0.034	0.041	0.046	0.049	0.051	0.052
	0.50	0.025	0.035	0.043	0.047	0.050	0.052	0.053
	0.40	0.027	0.038	0.044	0.048	0.051	0.053	0.054
	0.35	0.029	0.039	0.045	0.049	0.052	0.053	0.054
	0.30	0.030	0.040	0.046	0.050	0.052	0.054	0.055
D	1.00	0.013	0.021	0.029	0.035	0.040	0.043	0.045
	0.90	0.014	0.022	0.031	0.036	0.040	0.043	0.046
	0.80	0.015	0.023	0.032	0.038	0.041	0.044	0.047
	0.70	0.016	0.025	0.033	0.039	0.043	0.045	0.047
	0.60	0.017	0.026	0.035	0.040	0.044	0.046	0.048
	0.50	0.018	0.028	0.037	0.042	0.045	0.048	0.050
	0.40	0.020	0.031	0.039	0.043	0.047	0.049	0.051
	0.35	0.022	0.032	0.040	0.044	0.048	0.050	0.051
	0.30	0.023	0.034	0.041	0.046	0.049	0.051	0.052

E								
1.00	0.008	0.018	0.030	0.042	0.051	0.059	0.066
0.90	0.009	0.019	0.032	0.044	0.054	0.062	0.068
0.80	0.010	0.021	0.035	0.046	0.056	0.064	0.071
0.70	0.011	0.023	0.037	0.049	0.059	0.067	0.073
0.60	0.012	0.025	0.040	0.053	0.062	0.070	0.076
0.50	0.014	0.028	0.044	0.057	0.066	0.074	0.080
0.40	0.017	0.032	0.049	0.052	0.071	0.078	0.084
0.35	0.018	0.035	0.052	0.064	0.074	0.081	0.086
0.30	0.020	0.038	0.055	0.068	0.077	0.083	0.089

F								
1.00	0.008	0.016	0.026	0.034	0.041	0.046	0.051
0.90	0.008	0.017	0.027	0.036	0.042	0.048	0.052
0.80	0.009	0.018	0.029	0.037	0.044	0.049	0.054
0.70	0.010	0.020	0.031	0.039	0.046	0.051	0.055
0.60	0.011	0.022	0.033	0.042	0.048	0.053	0.057
0.50	0.013	0.024	0.036	0.044	0.051	0.056	0.059
0.40	0.015	0.027	0.039	0.048	0.054	0.058	0.062
0.35	0.016	0.029	0.041	0.050	0.055	0.060	0.063
0.30	0.018	0.031	0.044	0.052	0.057	0.062	0.065

G								
1.00	0.007	0.014	0.022	0.028	0.033	0.037	0.040
0.90	0.008	0.015	0.023	0.029	0.034	0.038	0.041
0.80	0.008	0.016	0.024	0.031	0.035	0.039	0.042
0.70	0.009	0.017	0.026	0.032	0.037	0.040	0.043
0.60	0.010	0.019	0.028	0.034	0.038	0.042	0.044
0.50	0.011	0.021	0.030	0.036	0.040	0.043	0.046
0.40	0.013	0.023	0.032	0.038	0.042	0.045	0.047
0.35	0.014	0.025	0.033	0.039	0.043	0.046	0.048
0.30	0.016	0.026	0.035	0.041	0.044	0.047	0.049

H								
1.00	0.005	0.011	0.018	0.024	0.029	0.033	0.036
0.90	0.006	0.012	0.019	0.025	0.030	0.034	0.037
0.80	0.006	0.013	0.020	0.027	0.032	0.035	0.038
0.70	0.007	0.014	0.022	0.028	0.033	0.037	0.040
0.60	0.008	0.015	0.024	0.030	0.035	0.038	0.041
0.50	0.009	0.017	0.025	0.032	0.036	0.040	0.043
0.40	0.010	0.019	0.028	0.034	0.039	0.042	0.045
0.35	0.011	0.021	0.029	0.036	0.040	0.043	0.046
0.30	0.013	0.022	0.031	0.037	0.041	0.044	0.047

I								
1.00	0.004	0.009	0.015	0.021	0.026	0.030	0.033
0.90	0.004	0.010	0.016	0.022	0.027	0.031	0.034
0.80	0.005	0.010	0.017	0.023	0.028	0.032	0.035
0.70	0.005	0.011	0.019	0.025	0.030	0.033	0.037
0.60	0.006	0.013	0.020	0.026	0.031	0.035	0.038
0.50	0.007	0.014	0.022	0.028	0.033	0.037	0.040
0.40	0.008	0.016	0.024	0.031	0.035	0.039	0.042
0.35	0.009	0.017	0.026	0.032	0.037	0.040	0.043
0.30	0.010	0.019	0.028	0.034	0.038	0.042	0.044

Bending moment coefficients in laterally loaded wall panels – continued

Panel support conditions	Orthogonal ratio μ	Values of panel factor, α						
		h/L						
		0.30	0.50	0.75	1.00	1.25	1.50	1.75
J	1.00	0.009	0.023	0.046	0.071	0.096	0.122	0.151
	0.90	0.010	0.026	0.050	0.076	0.103	0.131	0.162
	0.80	0.012	0.028	0.054	0.083	0.111	0.142	0.175
	0.70	0.013	0.032	0.060	0.091	0.121	0.156	0.191
	0.60	0.015	0.036	0.067	0.100	0.135	0.173	0.211
	0.50	0.018	0.042	0.077	0.133	0.153	0.195	0.237
	0.40	0.021	0.050	0.090	0.131	0.177	0.225	0.272
	0.35	0.024	0.055	0.098	0.144	0.194	0.244	0.296
	0.30	0.027	0.062	0.108	0.160	0.214	0.269	0.325
K	1.00	0.009	0.021	0.038	0.056	0.074	0.091	0.108
	0.90	0.010	0.023	0.041	0.060	0.079	0.097	0.113
	0.80	0.011	0.025	0.045	0.065	0.084	0.103	0.120
	0.70	0.012	0.028	0.049	0.070	0.091	0.110	0.128
	0.60	0.014	0.031	0.054	0.077	0.099	0.119	0.138
	0.50	0.016	0.035	0.061	0.085	0.109	0.130	0.149
	0.40	0.019	0.041	0.069	0.097	0.121	0.144	0.164
	0.35	0.021	0.045	0.075	0.104	0.129	0.152	0.173
	0.30	0.024	0.050	0.082	0.112	0.139	0.162	0.183
L	1.00	0.006	0.015	0.029	0.044	0.059	0.073	0.088
	0.90	0.007	0.017	0.032	0.047	0.063	0.078	0.093
	0.80	0.008	0.018	0.034	0.051	0.067	0.084	0.099
	0.70	0.009	0.021	0.038	0.056	0.073	0.090	0.106
	0.60	0.010	0.023	0.042	0.061	0.080	0.098	0.115
	0.50	0.012	0.027	0.048	0.068	0.089	0.108	0.126
	0.40	0.014	0.032	0.055	0.078	0.100	0.121	0.139
	0.35	0.016	0.035	0.060	0.084	0.108	0.129	0.148
	0.30	0.018	0.039	0.066	0.092	0.116	0.138	0.158

NOTES:

1. Linear interpolation of μ and h/L is permitted.
2. When the dimensions of a wall are outside the range of h/L given in this table, it will usually be sufficient to calculate the moments on the basis of a simple span. For example, a panel of type A having h/L less than 0.3 will tend to act as a freestanding wall, while the same panel having h/L greater than 1.75 will tend to span horizontally.

Source: BS 5628: Part 1: 2005.

Concentrated loads

Increased stresses are permitted under and close to the bearings of concentrated loads. The load is assumed to be spread uniformly beneath the bearing. The effect of this bearing pressure in combination with the stresses in the wall due to other loads should be less than the design bearing strength.

Location	Design bearing strength N/mm^2	Notes
Directly below bearing	$\dfrac{1.25f_k}{\gamma_m}$	Higher bearing strengths can be achieved depending on the configuration of the concentrated load as clause 34, BS 5628
At 0.4h below bearing	$\dfrac{\beta tf_k}{\gamma_m}$	The concentrated load should be distributed using a 45° load spread to 0.4h below the bearing, where h is the clear height of the wall

Source: BS 5628: Part 1: 2005.

Masonry design to CP111

CP111 is the 'old brick code' which uses permissible stresses and has been withdrawn. Although it is now not appropriate for new construction, it can be helpful when refurbishing old buildings as the ultimate design methods used in BS 5628 are not appropriate for use with old masonry materials.

Basic compressive masonry strengths for standard format bricks (N/mm²)

Description of mortar proportions by volume cement:lime:sand (BS 5628 mortar class)	Hardening time	Basic compressive stress of unit[1] N/mm²								
		2.8	7.0	10.5	20.5	27.5	34.5	52.0	69.0	$\geq$96.5
				Stock	Fletton				Class B	Class A
Dry pack – 1:0:3 (I)	7	0.28	0.70	1.05	1.65	2.05	2.50	3.50	4.55	5.85
Cement lime – 1:1:6 (III)	14	0.28	0.70	0.95	1.30	1.60	1.85	2.50	3.10	3.80
Cement lime – 1:2:9 (IV)	14	0.28	0.55	0.85	1.15	1.45	1.65	2.05	2.50	3.10
Non-hydraulic lime putty with pozzolanic/cement additive – 0:1:3	14	0.21	0.49	0.70	0.95	1.15	1.40	1.70	2.05	2.40
Hydraulic lime – 0:1:2	14	0.21	0.49	0.70	0.95	1.15	1.40	1.70	2.05	2.40
Non-hydraulic lime mortar – 0:1:3	28[2]	0.21	0.42	0.55	0.70	0.75	0.85	1.05	1.15	1.40

NOTES:
1. For columns or piers of cross sectional area $A < 0.2\,m^2$, the basic compressive strength should be multiplied by $\gamma = (0.7 + 1.5A)$.
2. Longer may be required if the weather is not warm and dry.

Source: CP111: 1970.

Capacity reduction factors for slenderness and eccentricity

Slenderness ratio	Slenderness reduction factor, β				
	Axially loaded	Eccentricity of loading			
		t/6	t/4	t/3	t/3 to t/2
6	1.000	1.000	1.000	1.000	1.000
8	0.950	0.930	0.920	0.910	0.885
10	0.890	0.850	0.830	0.810	0.771
12	0.840	0.780	0.750	0.720	0.657
14	0.780	0.700	0.660	0.620	0.542
16	0.730	0.630	0.580	0.530	0.428
18	0.670	0.550	0.490	0.430	0.314
20	0.620	0.480	0.410	0.340	0.200
22	0.560	0.400	0.320	0.240	–
24	0.510	0.330	0.240	–	–
26	0.450	0.250	–	–	–
27	0.430	0.220	–	–	–

Concentrated loads

Although CP111 indicates that concentrated compressive stresses up to 1.5 times the permissible compressive stresses are acceptable, it is now thought that this guidance is not conservative as it does not take account of the bearing width or position.

Therefore it is generally accepted that bearing stresses should be kept within the basic permissible stresses. For historic buildings, this typically means maximum bearing stresses of $0.42 \, \text{N/mm}^2$ for stock bricks in traditional lime mortar or $0.7 \, \text{N/mm}^2$ where the structure has been 'engineered', perhaps with flettons in arches or vaults.

Source: CP111:1970.

Lintel design to BS 5977

BS 5977 sets out the method for load assessment of lintels in masonry structures for openings up to 4.5 m in single storey construction or up to 3.6 m in normal domestic two to three storey buildings. The method assumes that the masonry over an opening in a simple wall will arch over the opening. The code guidance must be applied with common sense as building elevations are rarely simple and load will be channelled down piers between openings. Typically there should be not less than 0.6 m or 0.2L of masonry to each side of the opening (where L is the clear span), not less than 0.6L of masonry above the lintel at midspan and not less than 0.6 m of masonry over the lintel supports. When working on existing buildings, the effect of new openings on existing lintels should be considered.

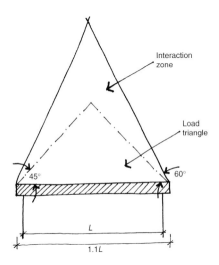

Loading assumptions:

1. The weight of the masonry in the loaded triangle is carried on the lintel – not the masonry in the zone of interaction.
2. Any point load or distributed load applied within the load triangle is dispersed at 45° and carried by the lintel.
3. Half of any point, or distributed, load applied to the masonry within the zone of interaction is carried by the lintel.

Where there are no openings above the lintel, and the loading assumptions apply, no loads outside the interaction zone need to be considered. Openings which are outside the zone of interaction, or which cut across the zone of interaction completely, need not be considered and do not add load to the lintel. However, openings which cut into (rather than across) the zone of interaction can have a significant effect on lintel loading as all the self-weight of the wall and applied loads above the line X–Y are taken into account. As for the other loads applied in the zone of interaction, they are halved and spread out at 45° from the line X–Y to give a line load on the lintel below.

All the loading conditions are illustrated in the following example:

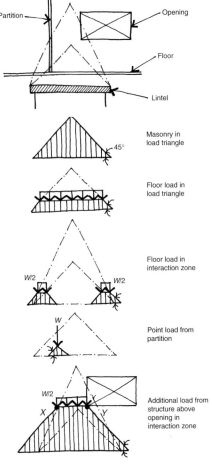

Source: BS 5977: Part 1: 1981.

Masonry accessories

Joist hangers

Joist hangers provide quick, economic and reliable timber to timber, timber to masonry and timber to steel junctions. Joist hangers should comply with BS 6178 and be galvanized for general use or stainless steel for special applications. Normally 600 mm of masonry over the hangers is required to provide restraint and ensure full load carrying capacity. Coursing adjustments should be made in the course below the course carrying the joist hanger to avoid supporting the hangers on cut blocks. The end of the joist should be packed tight to the back of the hanger, have enough bearing on the hanger and be fixed through every provided hole with 3.75 × 30 mm square twist sheradized nails. The back of the hanger must be tight to the wall and should not be underslung from beam supports. If the joist needs to be cut down to fit into the joist hanger, it may exceed the load capacity of the hanger. If the joist hangers are not installed to the manufacturer's instructions, they can be overloaded and cause collapse.

Straps for robustness

Masonry walls must be strapped to floors and roofs for robustness in order to allow for any out of plane forces, accidental loads and wind suction around the roof line. The traditional strap is 30 × 5 mm with a characteristic tensile strength of 10 kN. Straps are typically galvanized mild steel or austenitic stainless steel, fixed to three joists with four fixings and built into the masonry wall at a maximum spacing of 2 m. A typical strap can provide a restraining force of 5 kN/m depending on the security of the fixings. Compressive loads are assumed to be transferred by direct contact between the wall and floor/roof structures. Building Regulations and BS 8103: Part 1 set out recommendations for the fixing and spacing of straps.

Padstones

Used to spread the load at the bearings of steel beams on masonry walls. The plan area of the padstone is determined by the permissible concentrated bearing stress on the masonry. The depth of the padstone is based on a 45° load spread from the edges of the steel beam on the padstone until the padstone area is sufficient that the bearing stresses are within permissible values.

Proprietary pre-stressed concrete beam lintels

The following values are working loads for beam lintels which do not act compositely with the masonry above the opening manufactured by Supreme Concrete. Supreme Concrete stock pre-stressed concrete lintels from 0.6 m to 3.6 m long in 0.15 m increments but can produce special lintels up to about 4.2 m long. The safe loads given below are based on 150 mm end bearing. A separate range of fair faced lintels are also available.

Maximum uniformly distributed service load
kN

Lintel width mm	Lintel depth mm	Reference mm	Clear span between supports m						
			0.6	0.9	1.5	2.1	2.7	3.3	3.9
100	65	P100	8.00	5.71	3.64	2.67	2.11	1.74	–
	100	S10	17.92	12.80	8.15	5.97	4.72	3.90	–
	140	R15A	35.95	25.68	16.34	11.98	9.46	7.81	6.66
	215	R22A	59.42	53.71	34.18	25.07	19.79	16.35	13.93
140	65	P150	10.99	7.85	4.99	3.66	2.89	2.39	–
	100	R15	27.41	19.58	12.46	9.14	7.21	5.96	5.08
	215	R21A	73.90	73.90	52.65	38.61	30.48	25.18	21.45
215	65	P220	26.24	18.74	11.93	8.75	6.91	5.70	–
	100	R22	38.61	27.58	17.55	12.87	10.16	8.39	7.15
	140	R21	73.90	53.26	33.89	24.85	19.62	16.21	13.81

Source: Supreme Concrete Lintels (2008). Note that this information is subject to change at any time. Consult the latest Supreme Concrete literature for up to date information.

Profiled steel lintels

Profiled galvanized steel lintels are particularly useful for cavity wall construction, as they can be formed to support both leaves and incorporate insulation. Profiled steel lintels are supplied to suit cavity widths from 50 mm to 165 mm, single leaf walls, standard and heavy duty loading conditions, wide inner or outer leaves and timber construction. Special lintels are also available for corners and arches. The following lintels are selected from the range produced by I.G. Ltd.

Lintel reference	Ext. leaf mm	Cavity mm	Int. leaf mm	Height mm	Available lengths (in 150 mm increments) m	Gauge mm	Total uniformly distributed load for load ratio 1 kN	Total uniformly distributed load for load ratio 2 kN
L1/S 100	102	90–110	125	88	0.60–1.20	1.6	12	10
				88	1.35–1.50	2.0	16	13
				107	1.65–1.80	2.0	19	16
				125	1.95–2.10	2.0	21	17
				150	2.25–2.40	2.0	23	18
				162	2.55–2.70	2.6	27	22
				171	2.85–3.00	2.6	27	20
				200	3.15–3.60	3.2	27	20
				200	3.75–4.05	3.2	26	19
				200[1]	4.20–4.80	3.2	27	22
L1/HD 100	102	90–110	125	110	0.60–1.20	3.2	30	22
				135[1]	1.35–1.50	3.2	30	22
				163[1]	1.65–2.10	3.2	40	35
				203[1]	2.25–2.55	3.2	40	35
				203[1]	2.70–3.00	3.2	40	35
				203[1]	3.15–3.60	3.2	35	32
				203[1]	3.75–4.20	3.2	33	28
L1/XHD 100	102	90–110	125	163[1]	0.60–1.50	3.2	50	45
				163[1]	1.65–1.80	3.2	50	45
				203[1]	1.95–2.10	3.2	55	45
L1/S WIL 100	102	90–110	150	82	0.60–1.20	2.0	13	11
				107	1.35–1.80	2.0	17	14
				142	1.95–2.40	2.0	23	18
				177	2.55–3.00	2.6	24	18
				188[1]	3.15–3.60	3.2	30	26
				188[1]	3.75–4.20	3.2	27	25

Product								
L1/HD WIL 100 95 88 125	102	90–110	150	113 135[1] 165[1] 165[1] 188[1]	0.60–1.20 1.35–1.50 1.65–1.80 1.95–2.10 2.25–2.70	3.2 3.2 3.2 3.2 3.2	20 25 35 30 36	17 22 27 25 32
L5/100[2] 95 88 100	102	50–110	125 max	229	0.60–1.50 1.65–2.10 2.25–3.00 3.15–4.05 4.20–4.80	2.9 2.9 2.9 3.2 3.2		70 60 50 45 40
L6/100[3] 207 6 250 45	102	50–110	125 max	207	0.60–4.80 5.20 5.40 5.80 6.20 6.60		86 75 70 65 60 55 End bearing 200 mm	
L9 200	200–215			55 55 100	0.60–1.50 1.65–1.80 1.95–2.70	2.5 2.5 3.0	6 6 10	

Profiled steel lintels – continued

Lintel reference	Ext. leaf mm	Cavity mm	Int. leaf mm	Height mm	Available lengths (in 150 mm increments) m	Gauge mm	Total uniformly distributed load for load ratio 1 kN	Total uniformly distributed load for load ratio 2 kN
L10	102			60	0.60–1.20	3.0	4	
				110	1.35–1.80	3.0	8	
				210	1.95–2.70	2.8	10	
				210	2.85–3.00	3.0	6	
L11	102			150	0.60–1.80	2.5	16	
				225	1.95–2.40	2.5	20	
				225	2.55–3.00	3.0	22	
				225	3.15–3.60	3.0	12	

NOTES:
1. Indicates that a continuous bottom plate is added to lintel.
2. L5 and L11 lintels are designed assuming composite action with the masonry over the lintel.
3. L6 lintels are made with 203 × 133UB30 supporting the inner leaf.
4. The L1/S lintel is also available as L1/S 110 for cavities 110–125 mm, L1/S 130 for 130–145 mm and L1/S 150 for 150–165 mm.
5. IG can provide details for wide inner leaf lintels for cavities greater than 100 mm on request.
6. Loads in tables are unfactored. A lintel should not exceed in a max. deflection of L/325 when subject to the safe working load.
7. Load ratio 1 – applies to walls with an inner to outer load ratio of between 1:1 and 3:1. This ratio is normally applicable to lintels that support masonry or masonry and timber floors.
8. Load ratio 2 – applies to walls with an inner to outer load ratio of between 4:1 and 19:1. This ratio is normally applicable to lintels that support concrete floors or are at eaves details.

Source: IG Lintels Ltd (2007).

8
Reinforced Concrete

The Romans are thought to have been the first to use the binding properties of volcanic ash in mass concrete structures. The art of making concrete was then lost until Portland Cement was discovered in 1824 by Joseph Aspedin from Leeds. His work was developed by two Frenchmen, Monier and Lambot, who began experimenting with reinforcement. Deformed bars were developed in America in the 1870s, and the use of reinforced concrete has developed worldwide since 1900–1910. Concrete consists of a paste of aggregate, cement and water which can be reinforced with steel bars, or occasionally fibres, to enhance its flexural strength. Concrete constituents are as follows:

Cement Limestone and clay fired to temperatures of about 1400°C and ground to a powder. Grey is the standard colour but white can be used to change the mix appearance. The cement content of a mix affects the strength and finished surface appearance.

Aggregate Coarse aggregate (10 to 20 mm) and sand make up about 75% of the mix volume. Coarse aggregate can be natural dense stone or lightweight furnace by-products.

Water Water is added to create the cement paste which coats the aggregate. The water/cement ratio must be carefully controlled as the addition of water to a mix will increase workability and shrinkage, but will reduce strength if cement is not added.

Reinforcement Reinforcement normally consists of deformed steel bars. Traditionally the main bars were typically high yield steel ($f_y = 460\,N/mm^2$) and the links mild steel ($f_y = 250\,N/mm^2$). However, the new standards on bar bending now allow small diameter high yield bars to be bent to the same small radii as mild steel bars. This may mean that the use of mild steel links will reduce. The bars can be loose, straight or shaped, or as high yield welded mesh. Less commonly steel, plastic or glass fibres can be added (1 to 2% by volume) instead of bars to improve impact and cracking resistance, but this is generally only used for ground bearing slabs.

Admixtures Workability, durability and setting time can be affected by the use of admixtures.

Formwork Generally designed by the contractor as part of the temporary works, this is the steel, timber or plastic mould used to keep the liquid concrete in place until it has hardened. Formwork can account for up to half the cost of a concrete structure and should be kept simple and standardized where possible.

Summary of material properties

Density   17 to 24 kN/m^3 depending on the density of the chosen aggregate.

Compressive strength   Design strengths have a good range. $F_{cu} = 7$ to 60 N/mm^2.

Tensile strength   Poor at about 8 to 15% of F_{cu}. Reinforcement provides flexural strength.

Modulus of elasticity   This varies with the mix design strength, reinforcement content and age. Typical short-term (28 days) values are: 24 to 32 kN/mm^2. Long-term values are about 30 to 50% of the short-term values.

Linear coefficient of thermal expansion   8 to $12 \times 10^{-6\circ}$ C.

Shrinkage   As water is lost in the chemical hydration reaction with the cement, the concrete section will shrink. The amount of shrinkage depends on the water content, aggregate properties and section geometry. Normally, a long-term shrinkage strain of 0.03% can be assumed, of which 90% occurs in the first year.

Creep   Irreversible plastic flow will occur under sustained compressive loads. The amount depends on the temperature, relative humidity, applied stress, loading period, strength of concrete, allowed curing time and size of element. It can be assumed that about 40% and 80% of the final creep occurs in one month and 30 months respectively. The final (30 year) creep value is estimated from $\sigma\phi/E$, where σ is the applied stress, E is the modulus of elasticity of the concrete at the age of loading and ϕ is the creep factor which varies between about 1.0 and 3.2 for UK concrete loaded at 28 days.

Concrete mixes

Concrete mix design is not an absolute science. The process is generally iterative, based on an initial guess at the optimum mix constituents, followed by testing and mix adjustments on a trial-and-error basis.

There are different ways to specify concrete. A Prescribed Mix is where the purchaser specifies the mix proportions and is responsible for ensuring (by testing) that these proportions produce a suitable mix. A Designed Mix is where the engineer specifies the required performance, the concrete producer selects the mix proportions and concrete cubes are tested in parallel with the construction to check the mix compliance and consistency. Grades of Designed Mixes are prefixed by C. Special Proprietary Mixes such as self-compacting or waterproof concrete can also be specified, such as Pudlo or Caltite.

The majority of the concrete in the UK is specified on the basis of strength, workability and durability as a designated mix to BS EN 206 and BS 8500. This means that the ready mix companies must operate third party accredited quality assurance (to BS EN ISO 9001), which substantially reduces the number of concrete cubes which need to be tested. Grades of Designated mixes are prefixed by GEN, FND or RC depending on their proposed use.

Cementitious content

Cement is a single powder (containing for example OPC and fly ash) supplied to the concrete producer and denoted 'CEM'. Where a concrete producer combines cement, fly ash, ggbs and/or limestone fines at the works, the resulting cement combination is denoted 'C' and must set out the early day strengths as set out in BS 8500-2. As a result mixes can have the same constituents and strength (for example cement combination type IIA), with their labels indicating whether they are as a result of processes by the cement manufacturers (e.g. CEM II/A) or producers (e.g. CIIA). However, use of broad designations is suitable for most purposes, with cement and cement combinations being considered the same for specification purposes.

Cement combination types

CEM 1	Portland cement (OPC)
SRPC	Sulphate resisting portland cement
IIA	OPC + 6–20% fly ash, ggbs or limestone fines
IIB-S	OPC + 21–35% ggbs
IIB-V	OPC + 21–35% fly ash
IIIA	OPC + 36–65% ggbs (low early strength)
IIIB	OPC + 66–80% ggbs (low early strength)
IVB-V	OPC + 36–55% fly ash

Designated concrete mixes to BS 8500 & BS EN 206

Designated mix	Compressive strength N/mm²	Strength class	Typical application	Min cement content kg/m³	Max free water/ cement ratio	Typical slump mm	Consistence class
GEN 0	7.5	C−/17.5	Kerb bedding and backing	120	n/a	nominal 10	S1
GEN 1	10	C8/10	Blinding and mass concrete fill	180	n/a	75	S3
			Drainage works	180	n/a	10–50	S1
			Oversite below suspended slabs	180	n/a	75	S3
GEN 3	20	C16/20	Mass concrete foundations	220	n/a	75	S3
			Trench fill foundations	220	n/a	125	S4
FND 2, 3, 4 or 4A	35	C28/35	Foundations in sulphate conditions: 2, 3, 4 or 4A	320–380	0.5–0.35	75	S3
RC 30	30	C25/30	Reinforced concrete	260	0.65	50–100	S3
RC 35	35	C28/35		280	0.60	50–100	S3
RC 40	40	C32/40		300	0.55	50–100	S3
RC 50	50	C40/50		340	0.45	50–100	S3

Where strength class C28/35 indicates that the minimum characteristic cylinder strength is 28 N/mm² and cube strength 35 N/mm².

Source: BS EN 206: Part 1: 2004: Table A.13 adapted.

Traditional prescribed mix proportions

Concrete was traditionally specified on the basis of prescribed volume proportions of cement to fine and coarse aggregates. This method cannot allow for variability in the mix constituents, and as a result mix strengths can vary widely. This variability means that prescribed mixes batched by volume are rarely used for anything other than small works where the concrete does not need to be of a particularly high quality.

Typical volume batching ratios and the probable strengths achieved (with a slump of 50 mm to 75 mm) are:

Typical prescribed mix volume batching proportions Cement : sand : 20 mm aggregate	Probable characteristic crushing strength, F_{cu} N/mm^2
1:1.5:3	40
1:2:4	30
1:3:6	20

Concrete cube testing for strength of designed mixes

Concrete cube tests should be taken to check compliance of the mix with the design and specification. The amount of testing will depend on how the mix has been designed or specified. If the concrete is a designed mix from a ready mix plant BS 5328 gives the following minimum rates for sampling:

1 sample per	Maximum quantity of concrete at risk under any one decision	Examples of applicable structures
10 m^3 or 10 batches	40 m^3	Masts, columns, cantilevers
20 m^3 or 20 batches	80 m^3	Beams, slabs, bridges, decks
50 m^3 or 50 batches	200 m^3	Solid rafts, breakwaters

At least one 'sample' should be taken, for each type of concrete mix on the day it is placed, prepared to the requirements of BS 1881. If the above table is not used, 60 m^3 should be the maximum quantity of concrete represented by four consecutive test results. Higher rates of sampling should be adopted for critical elements.

A sample consists of two concrete cubes for each test result. Where results are required for 7 and 28 day strengths, four cubes should be prepared. The concrete cubes are normally cured under water at a minimum of 20°C ± 2°C. If the cubes are not cured at this temperature, their crushing strength can be seriously reduced.

Cube results must be assessed for validity using the following rules for 20 N/mm^2 concrete or above:

- A cube test result is said to be the mean of the strength of two cube tests. Any individual test result should not be more than 3 N/mm^2 below the specified characteristic compressive strength.
- When the difference between the two cube tests (i.e. four cubes) divided by their mean is greater than 15% the cubes are said to be too variable in strength to provide a valid result.
- If a group of test results consists of four consecutive cube results (i.e. eight cubes). The mean of the group of test results should be at least 3 N/mm^2 above the specified characteristic compressive strength.

Separate tests are required to establish the conformity of the mix on the basis of workability, durability, etc.

Source: BS 5328: Part 1: 1997.

Durability and fire resistance

Concrete durability

Concrete durability requirements based on BS 8500 (also known as BS EN 206-1) are based on the expected mode of deterioration, therefore all relevant exposure classes should be identified to establish the worst case for design. Concrete cover is also specified on the basis of a minimum cover, with the addition of a margin for fixing tolerance (normally between 5–15 mm) based on the expected quality control and possible consequences of low cover. In severe exposure conditions (e.g. marine structures, foundations) and/or for exposed architectural concrete, stainless steel reinforcement can be used to improve corrosion resistance, and prevent spalling and staining. Austenitic is the most common type of stainless steel used in reinforcing bars.

Carbonation

Carbon dioxide from the atmosphere combines with water in the surface of concrete elements to form carbonic acid. This reacts with the alkaline concrete to form carbonates which reduces the concrete pH and its passive protection of the steel rebar. With further air and water, the steel rebar can corrode (expanding in volume by 2–4 times) causing cracking and spalling. Depth of carbonation is most commonly identified by the destructive phenolphthalein test which turns from clear to magenta where carbonation has not occurred. In practice corrosion due to carbonation is a minor risk compared to chloride related problems, due to the way concrete is typically used and detailed in the UK.

Chlorides

Chlorides present in air or water from de-icing salts or a marine environment can enter the concrete through cracks, diffusion, capillary action or hydrostatic head to attack steel reinforcement. Chloride ions destroy any passive oxide layer on the steel, leaving it at risk of corrosion in the presence of air and water, resulting in cracking and spalling of the concrete. Although lower water/cement ratios and higher cover are thought to be the main way that will reduce the risk of chloride related corrosion, these measures are largely unsuccessful if not also accompanied by good workmanship and detailing to help keep moisture out of the concrete – particularly on horizontal surfaces.

Freeze thaw

Structures such as dams, spillways, tunnel inlets and exposed horizontal surfaces are likely to be saturated for much of the time. In low temperatures, ice crystals can form in the pores of saturated concrete (expanding by up to 8 times compared to the original water volume) and generate internal stresses which break down the concrete structure. Repeated cycles have a cumulative effect. Air-entrainment has been found to provide resistance to free thaw effects as well as reduce the adverse effects of chlorides.

Aggressive chemicals

All concretes are vulnerable to attack by salts present in solution and to attack by acids. Such chemicals can be present in natural ground and groundwater, as well as contaminated land, so precautions are often necessary to protect buried concrete. Although sulphates are relatively common in natural clay soils, sulphate attack is believed to be a relatively rare cause of deterioration (with acid attack being even rarer). Protection is usually provided by using a sulphate resisting cement or use of slag as a cement replacement. Flowing water is also more aggressive than still water.

Exposure classification and recommendations for resisting corrosion of reinforcement

	Class	Exposure Condition	Recommended strength class, max. w/c ratio, min. cement content (kg/m³)[1]							
		Examples from pr EN 206	Nominal Cover to Reinforcement							
			15 + Δc	20 + Δc	25 + Δc	30 + Δc	35 + Δc	40 + Δc	45 + Δc	50 + Δc
No risk	X0	Completely dry	Recommended that this exposure is not applicable to reinforced concrete							
Carbonation induced corrosion or rebar	XC1	Dry or permanently wet / Inside buildings with low humidity	C20/25 0.7 240	<<<	<<<	<<<	<<<	<<<	<<<	<<<
	XC2	Wet, rarely dry / Water retaining structures, many foundations	–	–	C25/30 0.65 260	<<<	<<<	<<<	<<<	<<<
	XC3*	Moderate humidity / Internal concrete or external sheltered from rain	–	C40/50 0.45 340	C32/40 0.55 300	C28/35 0.6 280	C25/30 0.65 260	<<<	<<<	<<<
	XC4*	Cyclic wet and dry / More severe than XC3 exposure								
Chloride induced corrosion of rebar excl. seawater chlorides	XD1	Moderate humidity / Exposure to direct spray	–	–	C40/50 0.45 360	C32/40 0.55 320	C28/35 0.6 300	<<<	<<<	<<<
	XD2	Wet, rarely dry / Swimming pools or industrial water contact	–	–	–	–	–	C28/35 0.55 320	<<<	<<<
	XD3	Cyclic wet and dry / Parts of bridges, pavements, car parks	–	–	–	–	–	C45/55 0.35 380	C40/50 0.4 380	C35/45 0.45 360
Seawater induced corrosion of rebar	XS1	Airborne salts but no direct contact / Near to, or on, the coast	–	–	–	–	–	C35/45 0.5 340	<<<	<<<
	XS2	Wet, rarely dry / Parts of marine structures	–	–	–	–	–	C28/35 0.55 320	<<<	<<<
	XS3	Tidal, splash and spray zones / Parts of marine structures	–	–	–	–	–	–	C45/55 0.35 380	C40/50 0.4 380
Freeze thaw attack on concrete	XF	Damp or saturated surfaces subject to freeze thaw degradation	Refer to BS 8500-1 Table A.8							
Aggressive chemical attack on concrete & rebar	DS	Surfaces exposed to aggressive chemicals present in natural soils or contaminated land	Refer to BS 8500-2 Table 1							

Notes:

1. For all cement/combination types unless noted * indicating except IVB.

2. Design life of 50 years with normal weight concrete and 20 mm aggregate. See BS 8500 for other variations.

3. More economic to follow BS 8500 if specific cement/combination is to be specified.

4. BS 8500 sets a minimum standard for cement content not to be reduced, therefore <<< indicates minimum standard specified in cell to left.

5. Where Δc is fixing tolerance, normally 5 mm to 15 mm.

Source: BS 8500: Part 1.

Minimum dimensions and cover for fire resistance periods

Member	Requirements	Fire rating hours					
		0.5	1.0	1.5	2.0	3.0	4.0
Columns fully exposed to fire	Minimum column width	150	200	250	300	400	450
	Cover*	20	20	20	25	25	25
Walls (0.4 to 1% steel)	Minimum wall thickness	100	120	140	160	200	240
	Cover*	20	20	20	25	25	25
Beams	Minimum beam width	200	200	200	200	240	280
	Cover for simply supported*	20	20	20	40	60	70
	Cover for continuous*	20	20	20	30	40	50
Slabs with plain soffits	Minimum slab thickness	75	95	110	125	150	170
	Cover for simply supported*	20	20	25	35	45	55
	Cover for continuous*	20	20	20	25	35	45
Ribbed slabs (open soffit and no stirrups)	Minimum top slab thickness	75	95	110	125	150	170
	Minimum rib width	125	125	125	125	150	175
	Cover for simply supported*	20	20	35	45	55	65
	Cover for continuous*	20	20	20	35	45	55

*Cover required to all reinforcement including links. If cover >35 mm special detailing is required to reduce the risk of spalling.

Source: BS 8110: Part 1: 1997.

Preliminary sizing of concrete elements

Typical span/depth ratios

Element	Typical spans m	Overall depth or thickness		
		Simply supported	Continuous	Cantilever
One way spanning slabs	5–6	$L/22$–30	$L/28$–36	$L/7$–10
Two way spanning slabs	6–11	$L/24$–35	$L/34$–40	–
Flat slabs	4–8	$L/27$	$L/36$	$L/7$–10
Close centre ribbed slabs (ribs at 600 mm c/c)	6–14	$L/23$	$L/31$	$L/9$
Coffered slabs (ribs at 900–1500 mm c/c)	8–14	$L/15$–20	$L/18$–24	$L/7$
Post tensioned flat slabs	9–10	$L/35$–40	$L/38$–45	$L/10$–12
Rectangular beams (width >250 mm)	3–10	$L/12$	$L/15$	$L/6$
Flanged beams	5–15	$L/10$	$L/12$	$L/6$
Columns	2.5–8	$H/10$–20	$H/10$–20	$H/10$
Walls	2–4	$H/30$–35	$H/45$	$H/15$–18
Retaining walls	2–8	–	–	$H/10$–14

NOTE:
125 mm is normally the minimum concrete floor thickness for fire resistance.

Preliminary sizing

Beams   Although the span/depth ratios are a good indication, beams tend to need more depth to fit sufficient reinforcement into the section in order to satisfy deflection requirements. Check the detailing early – especially for clashes with steel at column/beam junctions. The shear stress should be limited to 2 N/mm^2 for preliminary design.

Solid slabs   Two way spanning slabs are normally about 90% of the thickness of one way spanning slabs.

Profiled slabs   Obtain copies of proprietary mould profiles to minimize shuttering costs. The shear stress in ribs should be limited to 0.6 N/mm^2 for preliminary design.

Columns   A plain concrete section with no reinforcement can take an axial stress of about $0.45F_{cu}$. The minimum column dimensions for a stocky braced column = clear column height/17.5.

A simple allowance for moment transfer in the continuous junction between slab and column can be made by factoring up the load from the floor immediately above the column being considered (by 1.25 for interior, 1.50 for edge and 2.00 for corner columns). The column design load is this factored load plus any other column loads.

For stocky columns, the column area (A_c) can be estimated by: $A_c = N/15$, $N/18$ or $N/21$ for columns in RC35 concrete containing 1%, 2% or 3% high yield steel respectively.

Reinforcement

The ultimate design strength is $f_y = 250\,N/mm^2$ for mild steel and $f_y = 500\,N/mm^2$ high yield reinforcement.

Weight of reinforcement bars by diameter (kg/m)

6 mm	8 mm	10 mm	12 mm	16 mm	20 mm	25 mm	32 mm	40 mm
0.222	0.395	0.616	0.888	1.579	2.466	3.854	6.313	9.864

Reinforcement area (mm²) for groups of bars

Number of bars	Bar diameter mm								
	6	8	10	12	16	20	25	32	40
1	28	50	79	113	201	314	491	804	1257
2	57	101	157	226	402	628	982	1608	2513
3	85	151	236	339	603	942	1473	2413	3770
4	113	201	314	452	804	1257	1963	3217	5027
5	141	251	393	565	1005	1571	2454	4021	6283
6	170	302	471	679	1206	1885	2945	4825	7540
7	198	352	550	792	1407	2199	3436	5630	8796
8	226	402	628	905	1608	2513	3927	6434	10053
9	254	452	707	1018	1810	2827	4418	7238	11310

Reinforcement area (mm²/m) for different bar spacing

Spacing mm	Bar diameter mm								
	6	8	10	12	16	20	25	32	40
50	565	1005	1571	2262	4021	6283	9817	–	–
75	377	670	1047	1508	2681	4189	6545	10723	–
100	283	503	785	1131	2011	3142	4909	8042	12566
125	226	402	628	905	1608	2513	3927	6434	10053
150	188	335	524	754	1340	2094	3272	5362	8378
175	162	287	449	646	1149	1795	2805	4596	7181
200	141	251	393	565	1005	1571	2454	4021	6283
225	126	223	349	503	894	1396	2182	3574	5585
250	113	201	314	452	804	1257	1963	3217	5027

Reinforcement mesh to BS 4483

Fabric reference	Longitudinal wires			Cross wires			Mass kg/m²
	Diameter mm	Pitch mm	Area mm²/m	Diameter mm	Pitch mm	Area mm²/m	
Square mesh – High tensile steel							
A393	10	200	393	10	200	393	6.16
A252	8	200	252	8	200	252	3.95
A193	7	200	193	7	200	193	3.02
A142	6	200	142	6	200	142	2.22
A98	5	200	98	5	200	98	1.54
Structural mesh – High tensile steel							
B131	12	100	1131	8	200	252	10.90
B785	10	100	785	8	200	252	8.14
B503	8	100	503	8	200	252	5.93
B385	7	100	385	7	200	193	4.53
B283	6	100	283	7	200	193	3.73
B196	5	100	196	7	200	193	3.05
Long mesh – High tensile steel							
C785	10	100	785	6	400	70.8	6.72
C636	9	100	636	6	400	70.8	5.55
C503	8	100	503	5	400	49	4.34
C385	7	100	385	5	400	49	3.41
C283	6	100	283	5	400	49	2.61
Wrapping mesh – Mild steel							
D98	5	200	98	5	200	98	1.54
D49	2.5	100	49	2.5	100	49	0.77

Stock sheet size	Longitudinal wires	Cross wires	Sheet area
	Length 4.8 m	Width 2.4 m	11.52 m²

Source: BS 4486: 1985.

Shear link reinforcement areas

Shear link area, A_{sv} mm^2					Shear link area/link bar spacing, A_{sv}/S_v mm^2/mm								
No. of link legs	Link diameter mm				Link spacing, S_v mm								
	6	8	10	12	100	125	150	175	200	225	250	275	300
2	56				0.560	0.448	0.373	0.320	0.280	0.249	0.224	0.204	0.187
		100			1.000	0.800	0.667	0.571	0.500	0.444	0.400	0.364	0.333
			158		1.580	1.264	1.053	0.903	0.790	0.702	0.632	0.575	0.527
				226	2.260	1.808	1.507	1.291	1.130	1.004	0.904	0.822	0.753
3	84				0.840	0.672	0.560	0.480	0.420	0.373	0.336	0.305	0.280
		150			1.500	1.200	1.000	0.857	0.750	0.667	0.600	0.545	0.500
			237		2.370	1.896	1.580	1.354	1.185	1.053	0.948	0.862	0.790
				339	3.390	2.712	2.260	1.937	1.695	1.507	1.356	1.233	1.130
4	112				1.120	0.896	0.747	0.640	0.560	0.498	0.448	0.407	0.373
		200			2.000	1.600	1.333	1.143	1.000	0.889	0.800	0.727	0.667
			316		3.160	2.528	2.107	1.806	1.580	1.404	1.264	1.149	1.053
				452	4.520	3.616	3.013	2.583	2.260	2.009	1.808	1.644	1.507
6	168				1.680	1.344	1.120	0.960	0.840	0.747	0.672	0.611	0.560
		300			3.000	2.400	2.000	1.714	1.500	1.333	1.200	1.091	1.000
			474		4.740	3.792	3.160	2.709	2.370	2.107	1.896	1.724	1.580
				678	6.780	5.424	4.520	3.874	3.390	3.013	2.712	2.465	2.260

Concrete design to BS 8110

Partial safety factors for ultimate limit state

Load combination	Load type					
	Dead		Live		Earth	Wind
	Adverse	Beneficial	Adverse	Beneficial	and water pressures	
Dead and imposed (and earth and water pressure)	1.4	1.0	1.6	0.0	1.2	–
Dead and wind (and earth and water pressure)	1.4	1.0	–	–	1.2	1.4
Dead and wind and imposed (and earth and water pressure)	1.2	1.2	1.2	1.2	1.2	1.2

Effective depth

Effective depth, d, is the depth from compression face of section to the centre of area of the main reinforcement group allowing for layering, links and concrete cover.

Design of beams
Design moments and shears in beams with more than three spans

	At outer support	Near middle of end span	At first interior support	At middle of interior span	At interior supports
Moment	0	$\frac{WL}{11}$	$\frac{-WL}{9}$	$\frac{WL}{14}$	$\frac{-2WL}{25}$
Shear	$\frac{W}{2}$	–	$\frac{2W}{3}$	–	$\frac{5W}{9}$

Source: BS 8110: Part 1: 1997.

Ultimate moment capacity of beam section

$M_u = 0.156F_{cu}bd^2$ where there is less than 10% moment redistribution.

Factors for lever arm (z/d) and neutral axis (x/d) depth

$k = \dfrac{M}{F_{cu}bd^2}$	0.043	0.050	0.070	0.090	0.110	0.130	0.145	0.156
z/d	0.950	0.941	0.915	0.887	0.857	0.825	0.798	0.777
x/d	0.13	0.15	0.19	0.25	0.32	0.39	0.45	0.50

Where z = lever arm and x = neutral axis depth.

$$\frac{z}{d} = 0.5 + \sqrt{0.25 - \frac{k}{0.9}} \le 0.95 \quad \text{and} \quad x = \frac{(d - z)}{0.45}$$

Area of tension reinforcement for rectangular beams

If the applied moment is less than M_u, then the area of tension reinforcement,
$A_{s_{required}} = M/[0.87\left(\frac{z}{d}\right)f_y d]$

If the applied moment is greater than M_u, then the area of compression steel is
$A'_{s_{required}} = (K - 0.156)\,F_{cu}bd^2/[0.87f_y(d - d')]$ and the area of tension reinforcement is,
$A'_{s_{required}} = 0.156F_{cu}bd^2/[0.87f_y z] + A'_s$ if redistribution is less than 10%.

Equivalent breadth and depth of neutral axis for flanged beams

Flanged beams	Simply supported	Continuous	Cantilever
T beam	$b_w + L/5$	$b_w + L/7$	b_w
L beam	$b_w + L/10$	$b_w + L/13$	b_w

Where b_w = breadth of web, L = actual flange width or beam spacing, h_f is the depth of the flange.

Calculate k using b_w. From k, calculate $0.9x$ from the tabulated values of the neutral axis depth, x/d.

If $0.9x \le h_f$, the neutral axis is in the beam flange and steel areas can be calculated as rectangular beams.

If $0.9x > h_f$, the neutral axis is in the beam web and steel areas can be calculated as BS 8110: clause 3.4.4.5.

Source: BS 8110: Part 1: 1997.

Shear stresses in beams

The applied shear stress is $v = V/b_v d$.

Shear capacity of concrete

The shear capacity of concrete, V_c, relates to the section size, effective depth and percentage reinforcement.

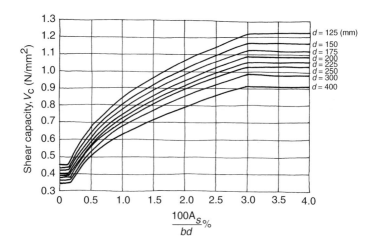

Form and area of shear reinforcement in beams

Value of applied shear stress v (N/mm²)	Form of shear reinforcement to be provided	Area of shear reinforcement to be provided (mm²)
$v < 0.5v_c$ throughout beam	Minimum links should normally be provided other than in elements of minor importance such as lintels, etc.	Suggested minimum: $A_{sv} > \dfrac{0.2b_v S_v}{0.87f_{yv}}$
$0.5v_c < v < (0.4 + v_c)$	Minimum links for whole length of beam to provide a shear resistance of 0.4 N/mm²	$A_{sv} > \dfrac{0.4b S_v}{0.87f_{yv}}$
$(0.4 + v_c) < v < 0.8\sqrt{F_{cu}}$ or 5 N/mm²	Links provided for whole length of beam at no more than 0.75d spacing along the span. No tension bar should be more than 150 mm from a vertical shear link leg[*]	$A_{sv} > \dfrac{b_v S_v (v - v_c)}{0.87f_{yv}}$

[*]Bent-up bars can be used in combination with links as long as no more than 50% of the shear resistance comes from the bent-up bars as set out in BS 8110: clause 3.4.5.6.

Source: BS 8110: Part 1: 1997.

Design of solid slabs

Solid slabs are supported on walls or beams.

With simple supports the applied moment is about $M = Wl_xl_y/24$ allowing for bending in two directions, where l_x and l_y can be different span lengths.

Design moments and shear forces for a one way spanning continuous solid slab

	End support/slab connection				At first interior support	At middle of interior span	At interior supports
	Simple support		Continuous				
	At outer support	Near middle of end span	At outer support	Near middle of end span			
Moment	0	$\dfrac{WL}{11.5}$	$\dfrac{-WL}{25}$	$\dfrac{WL}{13}$	$\dfrac{-WL}{11.5}$	$\dfrac{WL}{15.5}$	$\dfrac{-WL}{15.5}$
Shear	$\dfrac{W}{2.5}$	–	$\dfrac{6W}{13}$	–	$\dfrac{3W}{5}$	–	$\dfrac{W}{2}$

Where W is the load on one span and L is the length of one span.

Design moments for a two way spanning continuous solid slab
Where $l_y / l_x \le 1.5$ the following formulae and coefficients can be used to calculate moments in orthogonal directions $M_x = \beta_x Wl_x$ and $M_y = \beta_y Wl_y$ for the given edge conditions:

Type of panel	Moments considered*	Coefficient β_x for short span l_x				Coefficient β_y for long span l_y
		$\dfrac{l_y}{l_x} = 1.0$	$\dfrac{l_y}{l_x} = 1.2$	$\dfrac{l_y}{l_x} = 1.5$	$\dfrac{l_y}{l_x} = 2.0$	
Interior panel	Continuous edge	$\dfrac{-1}{32}$	$\dfrac{-1}{23}$	$\dfrac{-1}{18}$	$\dfrac{-1}{15}$	$\dfrac{-1}{31}$
	Midspan	$\dfrac{1}{41}$	$\dfrac{1}{31}$	$\dfrac{1}{25}$	$\dfrac{1}{20}$	$\dfrac{1}{41}$
One short edge discontinuous	Continuous edge	$\dfrac{-1}{25}$	$\dfrac{-1}{20}$	$\dfrac{-1}{17}$	$\dfrac{-1}{14}$	$\dfrac{-1}{27}$
	Midspan	$\dfrac{1}{34}$	$\dfrac{1}{27}$	$\dfrac{1}{23}$	$\dfrac{1}{20}$	$\dfrac{1}{35}$
One long edge discontinuous	Continuous edge	$\dfrac{-1}{25}$	$\dfrac{-1}{17}$	$\dfrac{-1}{13}$	$\dfrac{-1}{11}$	$\dfrac{-1}{27}$
	Midspan	$\dfrac{1}{33}$	$\dfrac{1}{23}$	$\dfrac{1}{18}$	$\dfrac{1}{14}$	$\dfrac{1}{35}$
Two adjacent edges discontinuous	Continuous edge	$\dfrac{-1}{21}$	$\dfrac{-1}{15}$	$\dfrac{-1}{12}$	$\dfrac{-1}{10}$	$\dfrac{-1}{22}$
	Midspan	$\dfrac{1}{27}$	$\dfrac{1}{21}$	$\dfrac{1}{16}$	$\dfrac{1}{14}$	$\dfrac{1}{29}$

*These moments apply to the full width of the slab in each direction. The area of reinforcement to be provided top and bottom, both ways, at corners where the slab is not continuous = 75% of the reinforcement for the short span, across a width $l_x/5$ both ways.

Form and area of shear reinforcement in solid slabs
The allowable shear stress, v_c, is the same as that calculated for beams, but the slab section should be sized to avoid shear reinforcement. If required, Table 3.16 in BS 8110 sets out the reinforcement requirements.

Source: BS 8110: Part 1: 1997.

Design of flat slabs

Flat slabs are solid slabs on concrete which sit on points or columns instead of linear wall or beam supports. Slab depth should be selected to satisfy deflection requirements and to resist shear around the column supports. Any recognized method of elastic analysis can be used, but BS 8110 suggests that the slabs be split into bay-wide subframes with columns or sections of columns projecting above and below the slab.

Simplified bending moment analysis in flat slabs

A simplified approach is permitted by BS 8110 which allows moments to be calculated on the basis of the values for one way spanning solid slabs on continuous supports less the value of $0.15Wh_c$ where $h_c = \sqrt{4A_{col}/\pi}$ and A_{col} = column area. Alternatively, the following preliminary moments for regular grid with a minimum of three bays can be used for feasibility or preliminary design purposes only:

Preliminary target moments and forces for flat slab design

	End support/slab connection				At first interior support	At middle of interior span	At interior supports
	Simple support		Continuous				
	At outer support	Near middle of end span	At outer support	Near middle of end span			
Column strip moments	0	$\dfrac{WL^2}{11}$	$\dfrac{-WL^2}{20}$	$\dfrac{WL^2}{10}$	$\dfrac{-2WL^2}{13}$	$\dfrac{WL^2}{11}$	$\dfrac{-2WL^2}{15}$
Middle strip moments	0	$\dfrac{WL^2}{11}$	$\dfrac{-WL^2}{20}$	$\dfrac{WL^2}{10}$	$\dfrac{-WL^2}{20}$	$\dfrac{WL^2}{11}$	$\dfrac{-WL^2}{20}$

W is a UDL in kN/m^2, L is the length of one span and M is in kNm/m width of slab.

Moment transfer between the slab and exterior columns is limited to $M_{t\,max.} = 0.15F_{cu}b_e d^2$ where b_e depends on the slab to column connection as given in Figure 3.13 in BS 8110. Subframe moments may need to be adjusted to keep the assumed moment transfer within the value of $M_{t\,max.}$.

Distribution of bending moments in flat slabs

The subframes used in the analysis are further split into middle and column strips. Loads are more concentrated on the column strips. Typically, for hogging (negative) moments, 75% of the total subframe design moment will be distributed to the column strip. For sagging (positive) moments, 55% of the total subframe design moment will be distributed to the column strip. Special provision must be made for holes in panels and panels with marginal beams or supporting walls. BS 8110 suggests that where $l_y/l_x \leq 2.0$, column strips are normally $l_x/2$ wide centred on the grid. The slab should be detailed so that 66 per cent of the support reinforcement is located in the width $l_x/4$ centred over the column.

Punching shear forces in flat slabs

The critical shear case for flat slabs is punching shear around the column heads. The basic shear, V, is equal to the full design load over the area supported by the column which must be converted to effective shear forces to account for moment transfer between the slab and columns.

For slabs with equal spans, the effective shears are: $V_{eff} = 1.15V$ for internal columns, $V_{eff} = 1.25V$ for corner columns and $V_{eff} = 1.25V$ for edge columns for moments parallel to the slab edge or $V_{eff} = 1.4V$ edge columns for moments perpendicular to the slab edge.

Punching shear checks in flat slabs

The shear stress at the column face should be checked: $v_o = V_{eff}/U_o d$ (where U_o is the column perimeter in contact with the slab). This should be less than the lesser of $0.8\sqrt{F_{cu}}$ or $5\,\text{N/mm}^2$.

Perimeters radiating out from the column should then be checked: $v_i = V_{eff}/U_i d$ where U_i is the perimeter of solid slab spaced off the column. The first perimeter checked ($i = 1$) is spaced $1.5d$ from the column face with subsequent shear perimeters spaced at $0.75d$ intervals. Successive perimeters are checked until the applied shear stress is less than the allowable stress, v_c. BS 8110: clause 3.7.6 sets out the detailing procedure and gives rules for the sharing of shear reinforcement between perimeters.

The position of the column relative to holes and free edges must be taken into account when calculating the perimeter of the slab/column junction available to resist the shear force.

Stiffness and deflection

BS 8110 gives basic span/depth ratio which limit the total deflection to span/250 and live load and creep deflections to the lesser of span/500 or 20 mm, for spans up to 10 m.

Basic span/depth ratios for beams

Support conditions	Rectangular sections $\dfrac{b_w}{b} = 1.0$	Flanged section $\dfrac{b_w}{b} \leq 0.3$
Cantilever	7	5.6
Simply supported	20	16.0
Continuous	26	20.8

For values of $b_w/b > 0.3$ linear interpolation between the flanged and rectangular values is permitted.

Allowable span/depth ratio

Allowable span/depth $= F_1 \times F_2 \times F_3 \times F_4 \times$ Basic span/depth ratio

Where:

F_1 modification factor to reduce deflections in beams with spans over 10 m. $F_1 =$ 10/span, where $F_1 < 1.0$

F_2 modification for tension reinforcement

F_3 modification for compression reinforcement

F_4 modification for stair waists where the staircase occupies at least 60% of the span and there is no stringer beam, $F_4 = 1.15$

The service stress in the bars, $f_s = \dfrac{2f_y A_{s\,required}}{3A_{s\,provided}}$

Source: BS 8110: Part 1: 1997.

198

Modification factor for tension reinforcement

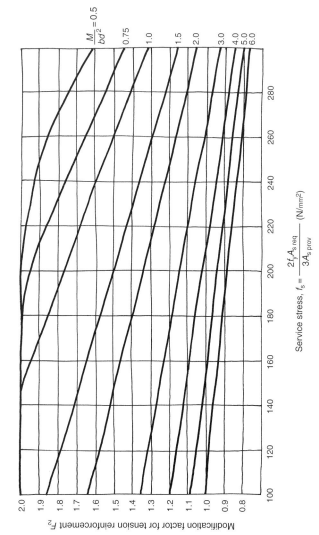

Modification factor for compression reinforcement

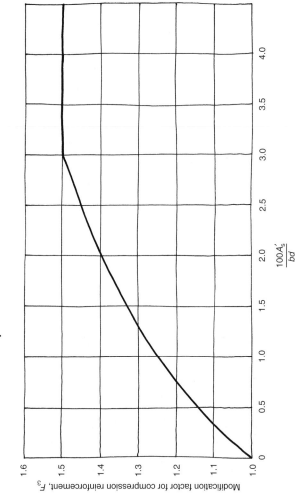

Columns

Vertical elements (of clear height, l, and dimensions, $b \times h$) are considered as columns if $h > 4b$, otherwise they should be considered as walls. Generally the clear column height between restraints should be less than $60b$. It must be established early in the design whether the columns will be in a braced frame where stability is to be provided by shear walls or cores, or whether the columns will be unbraced, meaning that they will maintain the overall stability for the structure. This has a huge effect on the effective length of columns, $l_e = \beta l$, as the design method for columns depends on their slenderness, l_{ex}/b or l_{ey}/h. A column is considered 'stocky' if the slenderness is less than 15 for braced columns or 10 for unbraced columns. Columns exceeding these limits are considered to be 'slender'.

Effective length coefficient (β) for columns

End condition at top of column		End condition at base of column					
		Condition 1		Condition 2		Condition 3	
		Braced	Unbraced	Braced	Unbraced	Braced	Unbraced
Condition 1	'Moment' connection to a beam or foundation which is at least as deep as the column dimension*	0.75	1.20	0.80	1.30	0.90	1.60
Condition 2	'Moment' connection to a beam or foundation which is shallower than the column dimension*	0.80	1.30	0.85	1.50	0.95	1.80
Condition 3	'Pinned' connection	0.90	1.60	0.95	1.80	1.00	n/a
Condition 4	'Free' end	n/a	2.20	n/a	n/a	n/a	n/a

*Column dimensions measured in the direction under consideration.

Source: BS 8110: Part 1: 1997.

Framing moments transferred to columns

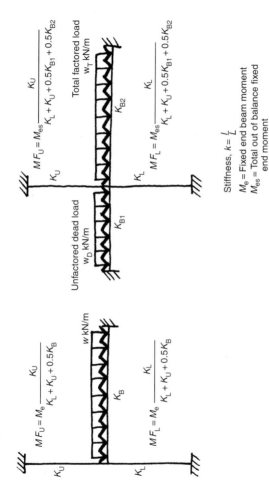

$$K_U$$

$$MF_U = M_e \frac{K_U}{K_L + K_U + 0.5K_B}$$

$$K_B$$

$$w \text{ kN/m}$$

$$MF_L = M_e \frac{K_L}{K_L + K_U + 0.5K_B}$$

$$K_L$$

$$MF_U = M_{es} \frac{K_U}{K_L + K_U + 0.5K_{B1} + 0.5K_{B2}}$$

Total factored load w_T kN/m

$$K_U$$

$$K_{B2}$$

$$MF_L = M_{es} \frac{K_L}{K_L + K_U + 0.5K_{B1} + 0.5K_{B2}}$$

$$K_L$$

Unfactored dead load w_D kN/m

$$K_{B1}$$

Stiffness, $k = \frac{I}{L}$

M_e = Fixed end beam moment
M_{es} = Total out of balance fixed
 end moment

Column design methods

Column design charts must be used where the column has to resist axial and bending stresses. Stocky columns need only normally be designed for the maximum design moment about one axis. The minimum design moment is the axial load multiplied by the greater of the eccentricity or $h/20$ in the plane being considered.

If a full frame analysis has not been carried out, the effect of moment transfer can be approximated by using column subframes or by using increasing axial loads by 10% for symmetrical simply supported loads.

Where only a nominal eccentricity moment applies, stocky columns carrying axial load can be designed for: $N = 0.4F_{cu}A_c + 0.75A_sf_y$.

Slender columns can be designed in the same way as short columns, but must resist an additional moment due to eccentricity caused by the deflection of the column as set out in clause 3.8.3 of BS 8110.

Biaxial bending in columns

When it is necessary to consider biaxial moments, the design moment about one axis is enhanced to allow for the biaxial bending effects and the column is designed about the enhanced axis. Where M is the applied moment, d_x is the effective depth across the x–x axis and d_y is the effective depth across the y–y axis:

If $\dfrac{M_x}{M_y} \geq \dfrac{d_x}{d_y}$ the increased moment about the x–x axis is $M_{x_{enhanced}} = M_x + \dfrac{\beta d_x M_y}{d_y}$.

If $\dfrac{M_y}{M_x} < \dfrac{d_y}{d_x}$ the increased moment about the y–y axis is $M_{y_{enhanced}} = M_y + \dfrac{\beta d_y M_x}{d_x}$ where β is:

$\dfrac{N}{bhF_{cu}}$	0.00	0.10	0.20	0.30	0.40	0.50	$\geq$0.60
β	1.00	0.88	0.77	0.65	0.53	0.42	0.30

Source: BS 8110: Part 1: 1997.

203

Reinforced concrete column design charts

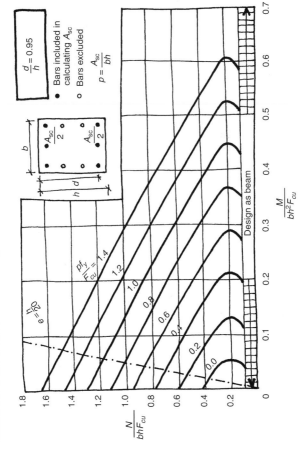

Source: IStructE (2002).

204

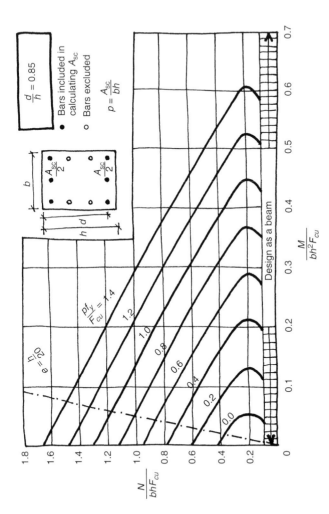

Source: IStructE (2002).

205

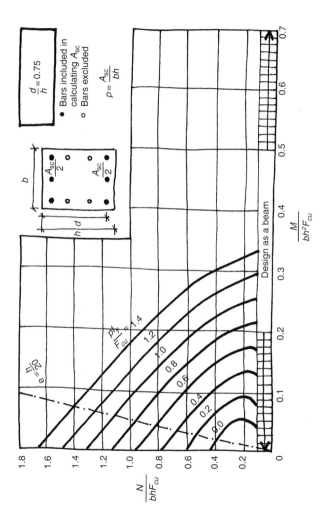

Source: IStructE (2002).

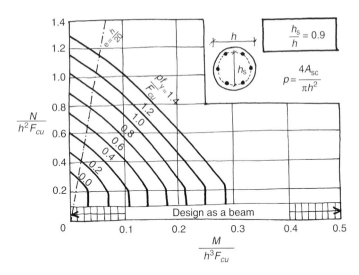

Source: IStructE (2002).

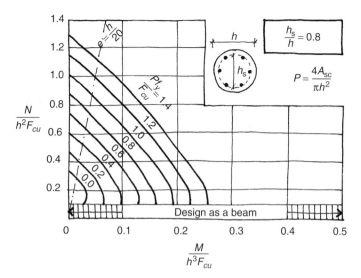

Source: IStructE (2002).

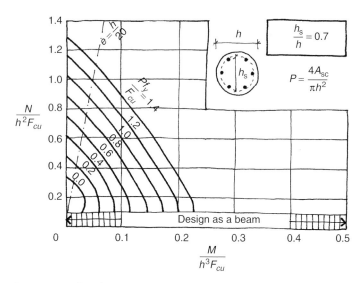

Source: IStructE (2002).

Selected detailing rules for high yield reinforcement to BS 8110

Generally no more than four bars should be arranged in contact at any point. Minimum percentages of reinforcement are intended to control cracking and maximum percentages are intended to ensure that concrete can be placed and adequately compacted around the reinforcement. A_c is the area of the concrete section.

Minimum percentages of reinforcement

For tension reinforcement in rectangular beams/slabs in bending	$A_{s\ min} = 0.13\%\ A_c$
For compression reinforcement (if required) in rectangular beams in bending	$A_{s\ min} = 0.2\%\ A_c$
For compression reinforcement in columns	$A_{s\ min} = 0.4\%\ A_c$

Maximum percentages of reinforcement

For beams	$A_{s\ max} = 4\%\ A_c$
For vertically cast columns	$A_{s\ max} = 6\%\ A_c$
For horizontally cast columns	$A_{s\ max} = 8\%\ A_c$
At lap positions in vertically or horizontally cast columns	$A_{s\ max} = 10\%\ A_c$

Selected rules for maximum distance between bars in tension

The maximum bar spacings as set out in clause 3.12.11.2, BS 8110 will limit the crack widths to 0.3 mm. The clear spacing of high yield bars in beams should be less than 135 mm at supports and 160 mm at midspan. In no case should the clear spacing between bars in slabs exceed $3d$ or 750 mm. Reinforcement to resist shrinkage cracking in walls should be at least 0.25% of the concrete cross sectional area for high yield bars, using small diameter bars at relatively close centres.

Typical bond lengths

Bond is the friction and adhesion between the concrete and the steel reinforcement. It depends on the properties of the concrete and steel, as well as the position of the bars within the concrete. Bond forces are transferred through the concrete rather than relying on contact between steel bars. Deformed Type 2 high yield bars are the most commonly used. For a bar diameter, ϕ, basic bond lengths for tension and compression laps are 40ϕ, 38ϕ and 35ϕ, for $30\,\text{N/mm}^2$, $35\,\text{N/mm}^2$ or $40\,\text{N/mm}^2$ concrete respectively. Tension lap lengths need to be multiplied by 1.4 if the surface concrete cover is less than 2ϕ. If the surface concrete cover to a lap in a corner $<2\phi$, or the distance between adjacent laps is less than 75 mm or 6ϕ, the bond length should be multiplied by 1.4. If both of these situations occur the bond length should be multiplied by 2.0.

Reinforcement bar bending to BS 8666

BS 8666 sets down the specification for the scheduling, dimensioning, bending and cutting of steel reinforcement for concrete.

Minimum scheduling radius, diameter and bending allowances for reinforcement bars (mm)

Nominal bar diameter, D	Minimum radius for schedule R	Minimum diameter of bending former M	Minimum end dimension P	
			Bend $\geq$ 150° (min 5d straight)	Bend $<$ 150° (min 10d straight)
6	12	24	110	110
8	16	32	115	115
10	20	40	120	130
12	24	48	125	160
16	32	64	130	210
20	70	140	190	290
25	87	175	240	365
32	112	224	305	465
40	140	280	380	580
50	175	350	475	725

NOTES:
1. Grade 250 bars are no longer commonly used.
2. Grade H bars (formerly known as T) denote high yield Type 2 deformed bars $f_y = 500 \, \text{N/mm}^2$. Ductility grades A, B and C are available within the classification, with B being most common, C being used where extra ductility is required (e.g. earthquake design) and A for bars (12 mm diameter and less) being bent to tight radii where accuracy is particularly important.
3. Due to 'spring back' the actual bend radius will be slightly greater than half of the bending former diameter.

Source: BS 8666: 2005.

Bar bending shape codes to BS 8666

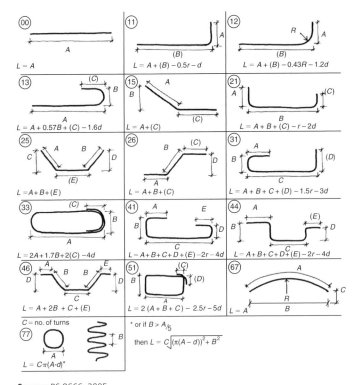

Source: BS 8666: 2005.

Reinforcement estimates

'Like fountain pens, motor cars and wives, steel estimates have some personal features. It is difficult to lay down hard and fast rules and one can only provide a guide to the uninitiated.' This marvellous (but now rather dated) quote was the introduction to an unpublished guide to better reinforcement estimates. These estimates are difficult to get right and the best estimate is based on a proper design and calculations.

DO NOT: Give a reinforcement estimate to anyone without an independent check by another engineer.

DO: Remember that you use more steel than you think and that although you may remember to be generous, you will inevitably omit more than you overestimate. Compare estimates with similar previous projects. Try to keep the QS happy by differentiating between mild and high tensile steel, straight and bent bars, and bars of different sizes. Apply a factor of safety to the final estimate. Keep a running total of the steel scheduled during preparation of the reinforcement drawings so that if the original estimate starts to look tight, it may be possible to make the ongoing steel detailing more economical.

As a useful check on a detailed estimate, the following are typical reinforcement quantities found in different structural elements:

Slabs	80–110 kg/m^3
RC pad footings	70–90 kg/m^3
Transfer slabs	150 kg/m^3
Pile caps/rafts	115 kg/m^3
Columns	150–450 kg/m^3
Ground beams	230 kg/m^3
Beams	220 kg/m^3
Retaining walls	110 kg/m^3
Stairs	135 kg/m^3
Walls	65 kg/m^3

'All up' estimates for different building types:

Heavy industrial	125 kg/m^3
Commercial	95 kg/m^3
Institutional	85 kg/m^3

Source: Price & Myers (2001).

9
Structural Steel

The method of heating iron ore in a charcoal fire determines the amount of carbon in the iron alloy. The following three iron ore products contain differing amounts of carbon: cast iron, wrought iron and steel.

Cast iron involves the heat treatment of iron castings and was developed as part of the industrial revolution between 1800 and 1900. It has a high carbon content and is therefore quite brittle which means that it has a much greater strength in compression than in tension. Typical allowable working stresses were 23 N/mm^2 tension, 123 N/mm^2 compression and 30 N/mm^2 shear.

Wrought iron has relatively uniform properties and, between the 1840s and 1900, wrought iron took over from cast iron for structural use, until it was in turn superseded by mild steel. Typical allowable working stresses were 81 N/mm^2 tension, 61 N/mm^2 compression and 77 N/mm^2 shear.

'Steel' can cover many different alloys of iron, carbon and other alloying elements to alter the properties of the alloys. The steel can be formed into structural sections by casting, hot rolling or cold rolling. Mild steel which is now mostly used for structural work was first introduced in the mid-nineteenth century.

Types of steel products

Cast steel
Castings are generally used for complex or non-standard structural components. The casting shape and moulding process must be carefully controlled to limit residual stresses. Sand casting is a very common method, but the lost wax method is generally used where a very fine surface finish is required.

Cold rolled
Cold rolling is commonly used for lightweight sections, such as purlins and wind posts, etc. Work hardening and residual stresses caused by the cold working cause an increase in the yield strength but this is at the expense of ductility and toughness. Cold rolled steel cannot be designed using the same method as hot rolled steel and design methods are given in BS 5950: Part 5.

Hot rolled steel
Most steel in the UK is produced by continuous casting where ingots or slabs are preheated to about 1300°C and the working temperatures fall as processing continues through the intermediate stages. The total amount of rolling work and the finishing temperatures are controlled to keep the steel grain size fine – which gives a good combination of strength and toughness. Although hollow sections (RHS, CHS and SHS) are often cold bent into shape, they tend to be hot finished and are considered 'hot rolled' for design purposes. This pocket book deals only with hot rolled steel.

Summary of hot rolled steel material properties

Density	$78.5 \, kN/m^3$
Tensile strength	$275–460 \, N/mm^2$ yield stress and $430–550 \, N/mm^2$ ultimate strength
Poisson's ratio	0.3
Modulus of elasticity, E	$205 \, kN/mm^2$
Modulus of rigidity, G	$80 \, kN/mm^2$
Linear coefficient of thermal expansion	$12 \times 10^{-6}/°C$

Mild steel section sizes and tolerances

Fabrication tolerances

BS 4 covers the dimensions of many of the hot rolled sections produced by Corus. Selected rolling tolerances for different sections are covered by the following standards:

UB and UC sections: BS EN 10034

Section height (mm)	$h \leq 180$	$180 < h \leq 400$	$400 < h \leq 700$	$700 < h$
Tolerance (mm)	$+3/-2$	$+4/-2$	$+5/-3$	± 5

Flange width (mm)	$b \leq 110$	$110 < b \leq 210$	$210 < b \leq 325$	$325 < b$
Tolerance (mm)	$+4/-1$	$+4/-2$	± 4	$+6/-5$

Out of squareness for flange width (mm)		$b \leq 110$	$110 < b$
Tolerance (mm)		1.5	2% of b up to max 6.5 mm

Straightness for section height (mm)	$80 < h \leq 180$	$180 < h \leq 360$	$360 < h$
Tolerance on section length (mm)	$0.003L$	$0.0015L$	$0.001L$

RSA sections: BS EN 10056–2

Leg length (mm)	$h \geq 50$	$50 < h \leq 100$	$100 < h \leq 150$	$150 < h \leq 200$	$200 \leq h$
Tolerance (mm)	± 1	± 2	± 3	± 4	$+6/-4$

Straightness for section height		$h \leq 150$	$h \leq 200$	$200 < h$
Tolerance along section length (mm)		$0.004L$	$0.002L$	$0.001L$

PFC sections: BS EN 10279

Section height (mm)	$h \leq 65$	$65 < h \leq 200$	$200 < h \leq 400$	$400 < h$
Tolerance (mm)	± 1.5	± 2	± 3	± 4

Out of squareness for flange width		$b \leq 100$	$100 < b$
Tolerance (mm)		2.0	2.5% of b

Straightness		$h \leq 150$	$150 < h \leq 300$	$300 < h$
Tolerance across flanges (mm)		$0.005L$	$0.003L$	$0.002L$
Tolerance parallel to web (mm)		$0.003L$	$0.002L$	$0.0015L$

Hot finished RHS, SHS and CHS sections: BS EN 10210–2

Straightness	0.2%L and 3 mm over any 1 m
Depth, breadth of diameter:	± 1% (min ± 0.5 mm and max ± 10 mm)
Squareness of side for SHS and RHS:	$90° \pm 1°$
Twist for SHS and RHS:	2 mm + 0.5 mm per m maximum
Twist for EHS	4 mm + 1.0 mm per m maximum

Examples of minimum bend radii for selected steel sections

The minimum radius to which any section can be curved depends on its metallurgical properties, particularly its ductility, cross sectional geometry and end use (the latter determines the standard required for the appearance of the work). It is therefore not realistic to provide a definitive list of the radii to which every section can be curved due to the wide number of end uses, but a selection of examples is possible. Normal bending tolerances are about 8 mm on the radius. In cold rolling the steel is deformed in the yield stress range and therefore becomes work hardened and displays different mechanical properties (notably a loss of ductility). However, if the section is designed to be working in the elastic range there is generally no significant difference to its performance.

Section	Typical bend radius for S275 steel m
610 × 305 UB 238	40.0
533 × 210 UB 122	30.0
305 × 165 UB 40	15.0
250 × 150 × 12.5 RHS	9.0
305 × 305 UC 118	5.5
300 × 100 PFC 46	4.6
150 × 150 × 12.5 SHS	3.0
254 × 203 RSJ 82	2.4
191 × 229 TEE 49	1.5
152 × 152 UC 37	1.5
125 × 65 PFC 15	1.0
152 × 127 RSJ 37	0.8

Source: Angle Ring Company Limited (2002).

Hot rolled section tables

UK beams – dimensions and properties

UKB designation	Mass per metre	Depth of section	Width of section	Thickness		Root radius	Depth between fillets	Ratios for local buckling		Second moment of area		Lateral torsional buckling ratio
				Web	Flange			Flange	Web	Axis x–x	Axis y–y	
		D	B	t	T	r	d	$B/2T$	d/t	I_x	I_y	D/T
	kg/m	mm	mm	mm	mm	mm	mm			cm⁴	cm⁴	
1016 × 305 × 487 †	486.7	1036.3	308.5	30.0	54.1	30.0	868.1	2.85	28.9	1022000	26700	19
1016 × 305 × 437 †	437.0	1026.1	305.4	26.9	49.0	30.0	868.1	3.12	32.3	910000	23400	21
1016 × 305 × 393 †	392.7	1015.9	303.0	24.4	43.9	30.0	868.1	3.45	35.6	808000	20500	23
1016 × 305 × 349 †	349.4	1008.1	302.0	21.1	40.0	30.0	868.1	3.78	41.1	723000	18500	25
1016 × 305 × 314 †	314.3	999.9	300.0	19.1	35.9	30.0	868.1	4.18	45.5	644000	16200	28
1016 × 305 × 272 †	272.3	990.1	300.0	16.5	31.0	30.0	868.1	4.84	52.6	554000	14000	32
1016 × 305 × 249 †	248.7	980.1	300.0	16.5	26.0	30.0	868.1	5.77	52.6	481000	11800	38
1016 × 305 × 222 †	222.0	970.3	300.0	16.0	21.1	30.0	868.1	7.11	54.3	408000	9550	46
914 × 419 × 388	388.0	921.0	420.5	21.4	36.6	24.1	799.6	5.74	37.4	720000	45400	25
914 × 419 × 343	343.3	911.8	418.5	19.4	32.0	24.1	799.6	6.54	41.2	626000	39200	28
914 × 305 × 289	289.1	926.6	307.7	19.5	32.0	19.1	824.4	4.81	42.3	504000	15600	29
914 × 305 × 253	253.4	918.4	305.5	17.3	27.9	19.1	824.4	5.47	47.7	436000	13300	33
914 × 305 × 224	224.2	910.4	304.1	15.9	23.9	19.1	824.4	6.36	51.6	376000	11200	38
914 × 305 × 201	200.9	903.0	303.3	15.1	20.2	19.1	824.4	7.51	54.6	325000	9420	45
838 × 292 × 226	226.5	850.9	293.8	16.1	26.8	17.8	761.7	5.48	47.3	340000	11400	32
838 × 292 × 194	193.8	840.7	292.4	14.7	21.7	17.8	761.7	6.74	51.8	279000	9070	39
838 × 292 × 176	175.9	834.9	291.7	14.0	18.8	17.8	761.7	7.76	54.4	246000	7800	44
762 × 267 × 197	196.8	769.8	268.0	15.6	25.4	16.5	686.0	5.28	44.0	240000	8170	30
762 × 267 × 173	173.0	762.2	266.7	14.3	21.6	16.5	686.0	6.17	48.0	205000	6850	35
762 × 267 × 147	146.9	754.0	265.2	12.8	17.5	16.5	686.0	7.58	53.6	169000	5460	43
762 × 267 × 134	133.9	750.0	264.4	12.0	15.5	16.5	686.0	8.53	57.2	151000	4790	48
686 × 254 × 170	170.2	692.9	255.8	14.5	23.7	15.2	615.1	5.40	42.4	170000	6630	29
686 × 254 × 152	152.4	687.5	254.5	13.2	21.0	15.2	615.1	6.06	46.6	150000	5780	33
686 × 254 × 140	140.1	683.5	253.7	12.4	19.0	15.2	615.1	6.68	49.6	136000	5180	36
686 × 254 × 125	125.2	677.9	253.0	11.7	16.2	15.2	615.1	7.81	52.6	118000	4380	42
610 × 305 × 238	238.1	635.8	311.4	18.4	31.4	16.5	540.0	4.96	29.3	209000	15800	20
610 × 305 × 179	179.0	620.2	307.1	14.1	23.6	16.5	540.0	6.51	38.3	153000	11400	26
610 × 305 × 149	149.2	612.4	304.8	11.8	19.7	16.5	540.0	7.74	45.8	126000	9310	31
610 × 229 × 140	139.9	617.2	230.2	13.1	22.1	12.7	547.6	5.21	41.8	112000	4510	28
610 × 229 × 125	125.1	612.2	229.0	11.9	19.6	12.7	547.6	5.84	46.0	98600	3930	31
610 × 229 × 113	113.0	607.6	228.2	11.1	17.3	12.7	547.6	6.60	49.3	87300	3430	35
610 × 229 × 101	101.2	602.6	227.6	10.5	14.8	12.7	547.6	7.69	52.2	75800	2910	41
610 × 178 × 100 †	100.3	607.4	179.2	11.3	17.2	12.7	547.6	5.21	48.5	72500	1660	35
610 × 178 × 92 †	92.2	603.0	178.8	10.9	15.0	12.7	547.6	5.96	50.2	64600	1440	40
610 × 178 × 82 †	81.8	598.6	177.9	10.0	12.8	12.7	547.6	6.95	54.8	55900	1210	47
533 × 312 × 273 †	273.3	577.1	320.2	21.1	37.6	12.7	476.5	4.26	22.6	199000	20600	15
533 × 312 × 219 †	218.8	560.3	317.4	18.3	29.2	12.7	476.5	5.43	26.0	151000	15600	19
533 × 312 × 182 †	181.5	550.7	314.5	15.2	24.4	12.7	476.5	6.44	31.3	123000	12700	23
533 × 312 × 151 †	150.6	542.5	312.0	12.7	20.3	12.7	476.5	7.68	37.5	101000	10300	27

†Additional sizes to BS4 available in UK.

Radius of gyration		Elastic modulus		Plastic modulus		Buckling parameter	Torsional index	Warping constant	Torsional constant	Area of section
Axis x–x	Axis y–y	Axis x–x	Axis y–y	Axis x–x	Axis y–y					
r_x	r_y	Z_x	Z_y	S_x	S_y	u	x	H	J	A
cm	cm	cm³	cm³	cm³	cm³			dm⁶	cm⁴	cm²
40.6	6.57	19700	1730	23200	2800	0.867	21.1	64.4	4300	620
40.4	6.49	17700	1540	20800	2470	0.868	23.1	56.0	3190	557
40.2	6.40	15900	1350	18500	2170	0.868	25.5	48.4	2330	500
40.3	6.44	14300	1220	16600	1940	0.872	27.9	43.3	1720	445
40.1	6.37	12900	1080	14800	1710	0.872	30.7	37.7	1260	400
40.0	6.35	11200	934	12800	1470	0.873	35.0	32.2	835	347
39.0	6.09	9820	784	11300	1240	0.861	39.8	26.8	582	317
38.0	5.81	8410	636	9810	1020	0.850	45.7	21.5	390	283
38.2	9.59	15600	2160	17700	3340	0.885	26.7	88.9	1730	494
37.8	9.46	13700	1870	15500	2890	0.883	30.1	75.8	1190	437
37.0	6.51	10900	1010	12600	1600	0.867	31.9	31.2	926	368
36.8	6.42	9500	871	10900	1370	0.866	36.2	26.4	626	323
36.3	6.27	8270	739	9530	1160	0.861	41.3	22.1	422	286
35.7	6.07	7200	621	8350	982	0.854	46.8	18.4	291	256
34.3	6.27	7980	773	9160	1210	0.870	35.0	19.3	514	289
33.6	6.06	6640	620	7640	974	0.862	41.6	15.2	306	247
33.1	5.90	5890	535	6810	842	0.856	46.5	13.0	221	224
30.9	5.71	6230	610	7170	958	0.869	33.2	11.3	404	251
30.5	5.58	5390	514	6200	807	0.864	38.1	9.39	267	220
30.0	5.40	4470	411	5160	647	0.858	45.2	7.40	159	187
29.7	5.30	4020	362	4640	570	0.854	49.8	6.46	119	171
28.0	5.53	4920	518	5630	811	0.872	31.8	7.42	308	217
27.8	5.46	4370	455	5000	710	0.871	35.5	6.42	220	194
27.6	5.39	3990	409	4560	638	0.868	38.7	5.72	169	178
27.2	5.24	3480	346	3990	542	0.862	43.9	4.80	116	159
26.3	7.23	6590	1020	7490	1570	0.886	21.3	14.5	785	303
25.9	7.07	4930	743	5550	1140	0.886	27.7	10.2	340	228
25.7	7.00	4110	611	4590	937	0.886	32.7	8.17	200	190
25.0	5.03	3620	391	4140	611	0.875	30.6	3.99	216	178
24.9	4.97	3220	343	3680	535	0.873	34.1	3.45	154	159
24.6	4.88	2870	301	3280	469	0.870	38.0	2.99	111	144
24.2	4.75	2520	256	2880	400	0.864	43.1	2.52	77.0	129
23.8	3.60	2390	185	2790	296	0.855	38.7	1.44	95.0	128
23.4	3.50	2140	161	2510	258	0.848	42.8	1.24	71.0	117
23.2	3.40	1870	136	2190	218	0.843	48.5	1.04	48.8	104
23.9	7.69	6890	1290	7870	1990	0.890	15.9	15.0	1290	348
23.3	7.48	5400	982	6120	1510	0.884	19.8	11.0	642	279
23.1	7.40	4480	806	5040	1240	0.885	23.5	8.77	373	231
22.9	7.32	3710	659	4150	1010	0.885	27.9	7.01	216	192

Source: Copyright Corus 2008 – reproduced with the kind permission of Corus.

UK beams – dimensions and properties – continued

UKB designation	Mass per metre	Depth of section	Width of section	Thickness Web	Thickness Flange	Root radius	Depth between fillets	Ratios for local buckling Flange	Ratios for local buckling Web	Second moment of area Axis x–x	Second moment of area Axis y–y	Lateral torsional buckling ratio
		D	B	t	T	r	d	$B/2T$	d/t	I_x	I_y	D/T
	kg/m	mm	mm	mm	mm	mm	mm			cm⁴	cm⁴	
533 × 210 × 138 †	138.3	549.1	213.9	14.7	23.6	12.7	476.5	4.53	32.4	86100	3860	23
533 × 210 × 122	122.0	544.5	211.9	12.7	21.3	12.7	476.5	4.97	37.5	76000	3390	26
533 × 210 × 109	109.0	539.5	210.8	11.6	18.8	12.7	476.5	5.61	41.1	66800	2940	29
533 × 210 × 101	101.0	536.7	210.0	10.8	17.4	12.7	476.5	6.03	44.1	61500	2690	31
533 × 210 × 92	92.1	533.1	209.3	10.1	15.6	12.7	476.5	6.71	47.2	55200	2390	34
533 × 210 × 82	82.2	528.3	208.8	9.6	13.2	12.7	476.5	7.91	49.6	47500	2010	40
533 × 165 × 85 †	84.8	534.9	166.5	10.3	16.5	12.7	476.5	5.05	46.3	48500	1270	32
533 × 165 × 75 †	74.7	529.1	165.9	9.7	13.6	12.7	476.5	6.10	49.1	41100	1040	39
533 × 165 × 66 †	65.7	524.7	165.1	8.9	11.4	12.7	476.5	7.24	53.5	35000	859	46
457 × 191 × 161 †	161.4	492.0	199.4	18.0	32.0	10.2	407.6	3.12	22.6	79800	4250	15
457 × 191 × 133 †	133.3	480.6	196.7	15.3	26.3	10.2	407.6	3.74	26.6	63800	3350	18
457 × 191 × 106 †	105.8	469.2	194.0	12.6	20.6	10.2	407.6	4.71	32.3	48900	2510	23
457 × 191 × 98	98.3	467.2	192.8	11.4	19.6	10.2	407.6	4.92	35.8	45700	2350	24
457 × 191 × 89	89.3	463.4	191.9	10.5	17.7	10.2	407.6	5.42	38.8	41000	2090	26
457 × 191 × 82	82.0	460.0	191.3	9.9	16.0	10.2	407.6	5.98	41.2	37100	1870	29
457 × 191 × 74	74.3	457.0	190.4	9.0	14.5	10.2	407.6	6.57	45.3	33300	1670	32
457 × 191 × 67	67.1	453.4	189.9	8.5	12.7	10.2	407.6	7.48	48.0	29400	1450	36
457 × 152 × 82	82.1	465.8	155.3	10.5	18.9	10.2	407.6	4.11	38.8	36600	1180	25
457 × 152 × 74	74.2	462.0	154.4	9.6	17.0	10.2	407.6	4.54	42.5	32700	1050	27
457 × 152 × 67	67.2	458.0	153.8	9.0	15.0	10.2	407.6	5.13	45.3	28900	913	31
457 × 152 × 60	59.8	454.6	152.9	8.1	13.3	10.2	407.6	5.75	50.3	25500	795	34
457 × 152 × 52	52.3	449.8	152.4	7.6	10.9	10.2	407.6	6.99	53.6	21400	645	41
406 × 178 × 85 †	85.3	417.2	181.9	10.9	18.2	10.2	360.4	5.00	33.1	31700	1830	23
406 × 178 × 74	74.2	412.8	179.5	9.5	16.0	10.2	360.4	5.61	37.9	27300	1550	26
406 × 178 × 67	67.1	409.4	178.8	8.8	14.3	10.2	360.4	6.25	41.0	24300	1360	29
406 × 178 × 60	60.1	406.4	177.9	7.9	12.8	10.2	360.4	6.95	45.6	21600	1200	32
406 × 178 × 54	54.1	402.6	177.7	7.7	10.9	10.2	360.4	8.15	46.8	18700	1020	37
406 × 140 × 53 †	53.3	406.6	143.3	7.9	12.9	10.2	360.4	5.55	45.6	18300	635	32
406 × 140 × 46	46.0	403.2	142.2	6.8	11.2	10.2	360.4	6.35	53.0	15700	538	36
406 × 140 × 39	39.0	398.0	141.8	6.4	8.6	10.2	360.4	8.24	56.3	12500	410	46
356 × 171 × 67	67.1	363.4	173.2	9.1	15.7	10.2	311.6	5.52	34.2	19500	1360	23
356 × 171 × 57	57.0	358.0	172.2	8.1	13.0	10.2	311.6	6.62	38.5	16000	1110	28
356 × 171 × 51	51.0	355.0	171.5	7.4	11.5	10.2	311.6	7.46	42.1	14100	968	31
356 × 171 × 45	45.0	351.4	171.1	7.0	9.7	10.2	311.6	8.82	44.5	12100	811	36
356 × 127 × 39	39.1	353.4	126.0	6.6	10.7	10.2	311.6	5.89	47.2	10200	358	33
356 × 127 × 33	33.1	349.0	125.4	6.0	8.5	10.2	311.6	7.38	51.9	8250	280	41

†Additional sizes to BS4 available in UK.

Radius of gyration		Elastic modulus		Plastic modulus		Buckling parameter	Torsional index	Warping constant	Torsional constant	Area of section
Axis x–x	Axis y–y	Axis x–x	Axis y–y	Axis x–x	Axis y–y					
r_x	r_y	Z_x	Z_y	S_x	S_y	u	x	H	J	A
cm	cm	cm^3	cm^3	cm^3	cm^3			dm^6	cm^4	cm^2
22.1	4.68	3140	361	3610	568	0.873	25.0	2.67	250	176
22.1	4.67	2790	320	3200	500	0.877	27.6	2.32	178	155
21.9	4.60	2480	279	2830	436	0.875	30.9	1.99	126	139
21.9	4.57	2290	256	2610	399	0.874	33.2	1.81	101	129
21.7	4.51	2070	228	2360	355	0.872	36.5	1.60	75.7	117
21.3	4.38	1800	192	2060	300	0.864	41.6	1.33	51.5	105
21.2	3.44	1820	153	2100	243	0.862	35.5	0.857	73.8	108
20.8	3.30	1550	125	1810	200	0.853	41.1	0.691	47.9	95.2
20.5	3.20	1340	104	1560	166	0.847	47.0	0.566	32.0	83.7
19.7	4.55	3240	426	3780	672	0.882	16.4	2.25	515	206
19.4	4.44	2660	341	3070	535	0.880	19.6	1.73	292	170
19.0	4.32	2080	259	2390	405	0.877	24.4	1.27	146	135
19.1	4.33	1960	243	2230	379	0.881	25.7	1.18	121	125
19.0	4.29	1770	218	2010	338	0.880	28.3	1.04	90.7	114
18.8	4.23	1610	196	1830	304	0.877	30.9	0.922	69.2	104
18.8	4.20	1460	176	1650	272	0.877	33.9	0.818	51.8	94.6
18.5	4.12	1300	153	1470	237	0.872	37.9	0.705	37.1	85.5
18.7	3.37	1570	153	1810	240	0.873	27.4	0.591	89.2	105
18.6	3.33	1410	136	1630	213	0.873	30.1	0.518	65.9	94.5
18.4	3.27	1260	119	1450	187	0.869	33.6	0.448	47.7	85.6
18.3	3.23	1120	104	1290	163	0.868	37.5	0.387	33.8	76.2
17.9	3.11	950	84.6	1100	133	0.859	43.9	0.311	21.4	66.6
17.1	4.11	1520	201	1730	313	0.881	24.4	0.728	93.0	109
17.0	4.04	1320	172	1500	267	0.882	27.6	0.608	62.8	94.5
16.9	3.99	1190	153	1350	237	0.880	30.5	0.533	46.1	85.5
16.8	3.97	1060	135	1200	209	0.880	33.8	0.466	33.3	76.5
16.5	3.85	930	115	1050	178	0.871	38.3	0.392	23.1	69.0
16.4	3.06	899	88.6	1030	139	0.870	34.1	0.246	29.0	67.9
16.4	3.03	778	75.7	888	118	0.871	38.9	0.207	19.0	58.6
15.9	2.87	629	57.8	724	90.8	0.858	47.5	0.155	10.7	49.7
15.1	3.99	1070	157	1210	243	0.886	24.4	0.412	55.7	85.5
14.9	3.91	896	129	1010	199	0.882	28.8	0.330	33.4	72.6
14.8	3.86	796	113	896	174	0.881	32.1	0.286	23.8	64.9
14.5	3.76	687	94.8	775	147	0.874	36.8	0.237	15.8	57.3
14.3	2.68	576	56.8	659	89.0	0.871	35.2	0.105	15.1	49.8
14.0	2.58	473	44.7	543	70.2	0.863	42.2	0.081	8.79	42.1

Source: Copyright Corus 2008 – reproduced with the kind permission of Corus.

UK beams – dimensions and properties – continued

UKB designation	Mass per metre	Depth of section	Width of section	Thickness		Root radius	Depth between fillets	Ratios for local buckling		Second moment of area		Lateral torsional buckling ratio
				Web	Flange			Flange	Web	Axis x–x	Axis y–y	
		D	B	t	T	r	d	$B/2T$	d/t	I_x	I_y	D/T
	kg/m	mm	mm	mm	mm	mm	mm			cm⁴	cm⁴	
305 × 165 × 54	54.0	310.4	166.9	7.9	13.7	8.9	265.2	6.09	33.6	11700	1060	23
305 × 165 × 46	46.1	306.6	165.7	6.7	11.8	8.9	265.2	7.02	39.6	9900	896	26
305 × 165 × 40	40.3	303.4	165.0	6.0	10.2	8.9	265.2	8.09	44.2	8500	764	30
305 × 127 × 48	48.1	311.0	125.3	9.0	14.0	8.9	265.2	4.48	29.5	9570	461	22
305 × 127 × 42	41.9	307.2	124.3	8.0	12.1	8.9	265.2	5.14	33.2	8200	389	25
305 × 127 × 37	37.0	304.4	123.4	7.1	10.7	8.9	265.2	5.77	37.4	7170	336	28
305 × 102 × 33	32.8	312.7	102.4	6.6	10.8	7.6	275.9	4.74	41.8	6500	194	29
305 × 102 × 28	28.2	308.7	101.8	6.0	8.8	7.6	275.9	5.78	46.0	5370	155	35
305 × 102 × 25	24.8	305.1	101.6	5.8	7.0	7.6	275.9	7.26	47.6	4460	123	44
254 × 146 × 43	43.0	259.6	147.3	7.2	12.7	7.6	219.0	5.80	30.4	6540	677	20
254 × 146 × 37	37.0	256.0	146.4	6.3	10.9	7.6	219.0	6.72	34.8	5540	571	23
254 × 146 × 31	31.1	251.4	146.1	6.0	8.6	7.6	219.0	8.49	36.5	4410	448	29
254 × 102 × 28	28.3	260.4	102.2	6.3	10.0	7.6	225.2	5.11	35.7	4000	179	26
254 × 102 × 25	25.2	257.2	101.9	6.0	8.4	7.6	225.2	6.07	37.5	3410	149	31
254 × 102 × 22	22.0	254.0	101.6	5.7	6.8	7.6	225.2	7.47	39.5	2840	119	37
203 × 133 × 30	30.0	206.8	133.9	6.4	9.6	7.6	172.4	6.97	26.9	2900	385	22
203 × 133 × 25	25.1	203.2	133.2	5.7	7.8	7.6	172.4	8.54	30.2	2340	308	26
203 × 102 × 23	23.1	203.2	101.8	5.4	9.3	7.6	169.4	5.47	31.4	2100	164	22
178 × 102 × 19	19.0	177.8	101.2	4.8	7.9	7.6	146.8	6.41	30.6	1360	137	23
152 × 89 × 16	16.0	152.4	88.7	4.5	7.7	7.6	121.8	5.76	27.1	834	89.8	20
127 × 76 × 13	13.0	127.0	76.0	4.0	7.6	7.6	96.6	5.00	24.2	473	55.7	17

Radius of gyration		Elastic modulus		Plastic modulus		Buckling parameter	Torsional index	Warping constant	Torsional constant	Area of section
Axis $x–x$	Axis $y–y$	Axis $x–x$	Axis $y–y$	Axis $x–x$	Axis $y–y$					
r_x	r_y	Z_x	Z_y	S_x	S_y	u	x	H	J	A
cm	cm	cm³	cm³	cm³	cm³			dm⁶	cm⁴	cm²
13.0	3.93	754	127	846	196	0.889	23.6	0.234	34.8	68.8
13.0	3.90	646	108	720	166	0.891	27.1	0.195	22.2	58.7
12.9	3.86	560	92.6	623	142	0.889	31.0	0.164	14.7	51.3
12.5	2.74	616	73.6	711	116	0.873	23.3	0.102	31.8	61.2
12.4	2.70	534	62.6	614	98.4	0.872	26.5	0.0846	21.1	53.4
12.3	2.67	471	54.5	539	85.4	0.872	29.7	0.0725	14.8	47.2
12.5	2.15	416	37.9	481	60.0	0.866	31.6	0.0442	12.2	41.8
12.2	2.08	348	30.5	403	48.4	0.859	37.4	0.0349	7.40	35.9
11.9	1.97	292	24.2	342	38.8	0.846	43.4	0.027	4.77	31.6
10.9	3.52	504	92.0	566	141	0.891	21.2	0.103	23.9	54.8
10.8	3.48	433	78.0	483	119	0.890	24.3	0.0857	15.3	47.2
10.5	3.36	351	61.3	393	94.1	0.880	29.6	0.0660	8.55	39.7
10.5	2.22	308	34.9	353	54.8	0.874	27.5	0.0280	9.57	36.1
10.3	2.15	266	29.2	306	46.0	0.866	31.5	0.0230	6.42	32.0
10.1	2.06	224	23.5	259	37.3	0.856	36.4	0.0182	4.15	28.0
8.71	3.17	280	57.5	314	88.2	0.881	21.5	0.0374	10.3	38.2
8.56	3.10	230	46.2	258	70.9	0.877	25.6	0.0294	5.96	32.0
8.46	2.36	207	32.2	234	49.7	0.888	22.5	0.0154	7.02	29.4
7.48	2.37	153	27.0	171	41.6	0.888	22.6	0.0099	4.41	24.3
6.41	2.10	109	20.2	123	31.2	0.890	19.6	0.00470	3.56	20.3
5.35	1.84	74.6	14.7	84.2	22.6	0.895	16.3	0.00200	2.85	16.5

Source: Copyright Corus 2008 – reproduced with the kind permission of Corus.

UK columns – dimensions and properties

UKC designation	Mass per metre	Depth of section	Width of section	Thickness		Root radius	Depth between fillets	Ratios for local buckling		Second moment of area		Lateral torsional buckling ratio
				Web	Flange			Flange	Web	Axis x–x	Axis y–y	
		D	B	t	T	r	d	$B/2T$	d/t	I_x	I_y	D/T
	kg/m	mm	mm	mm	mm	mm	mm			cm⁴	cm⁴	
356 × 406 × 634	633.9	474.6	424.0	47.6	77.0	15.2	290.2	2.75	6.10	275000	98100	6
356 × 406 × 551	551.0	455.6	418.5	42.1	67.5	15.2	290.2	3.10	6.89	227000	82700	7
356 × 406 × 467	467.0	436.6	412.2	35.8	58.0	15.2	290.2	3.55	8.11	183000	67800	8
356 × 406 × 393	393.0	419.0	407.0	30.6	49.2	15.2	290.2	4.14	9.48	147000	55400	9
356 × 406 × 340	339.9	406.4	403.0	26.6	42.9	15.2	290.2	4.70	10.9	123000	46900	9
356 × 406 × 287	287.1	393.6	399.0	22.6	36.5	15.2	290.2	5.47	12.8	99900	38700	11
356 × 406 × 235	235.1	381.0	394.8	18.4	30.2	15.2	290.2	6.54	15.8	79100	31000	13
356 × 368 × 202	201.9	374.6	374.7	16.5	27.0	15.2	290.2	6.94	17.6	66300	23700	14
356 × 368 × 177	177.0	368.2	372.6	14.4	23.8	15.2	290.2	7.83	20.2	57100	20500	15
356 × 368 × 153	152.9	362.0	370.5	12.3	20.7	15.2	290.2	8.95	23.6	48600	17600	17
356 × 368 × 129	129.0	355.6	368.6	10.4	17.5	15.2	290.2	10.5	27.9	40200	14600	20
305 × 305 × 283	282.9	365.3	322.2	26.8	44.1	15.2	246.7	3.65	9.21	78900	24600	8
305 × 305 × 240	240.0	352.5	318.4	23.0	37.7	15.2	246.7	4.22	10.7	64200	20300	9
305 × 305 × 198	198.1	339.9	314.5	19.1	31.4	15.2	246.7	5.01	12.9	50900	16300	11
305 × 305 × 158	158.1	327.1	311.2	15.8	25.0	15.2	246.7	6.22	15.6	38700	12600	13
305 × 305 × 137	136.9	320.5	309.2	13.8	21.7	15.2	246.7	7.12	17.90	32800	10700	15
305 × 305 × 118	117.9	314.5	307.4	12.0	18.7	15.2	246.7	8.22	20.6	27700	9060	17
305 × 305 × 97	96.9	307.9	305.3	9.9	15.4	15.2	246.7	9.91	24.9	22200	7310	20
254 × 254 × 167	167.1	289.1	265.2	19.2	31.7	12.7	200.3	4.18	10.4	30000	9870	9
254 × 254 × 132	132.0	276.3	261.3	15.3	25.3	12.7	200.3	5.16	13.1	22500	7530	11
254 × 254 × 107	107.1	266.7	258.8	12.8	20.5	12.7	200.3	6.31	15.6	17500	5930	13
254 × 254 × 89	88.9	260.3	256.3	10.3	17.3	12.7	200.3	7.41	19.4	14300	4860	15
254 × 254 × 73	73.1	254.1	254.6	8.6	14.2	12.7	200.3	8.96	23.3	11400	3910	18
203 × 203 × 127 †	127.5	241.4	213.9	18.1	30.1	10.2	160.8	3.55	8.88	15400	4920	8
203 × 203 × 113 †	113.5	235.0	212.1	16.3	26.9	10.2	160.8	3.94	9.87	13300	4290	9
203 × 203 × 100 †	99.6	228.6	210.3	14.5	23.7	10.2	160.8	4.44	11.1	11300	3680	10
203 × 203 × 86	86.1	222.2	209.1	12.7	20.5	10.2	160.8	5.10	12.7	9450	3130	11
203 × 203 × 71	71.0	215.8	206.4	10.0	17.3	10.2	160.8	5.97	16.1	7620	2540	12
203 × 203 × 60	60.0	209.6	205.8	9.4	14.2	10.2	160.8	7.25	17.1	6120	2060	15
203 × 203 × 52	52.0	206.2	204.3	7.9	12.5	10.2	160.8	8.17	20.4	5260	1780	16
203 × 203 × 46	46.1	203.2	203.6	7.2	11.0	10.2	160.8	9.25	22.3	4570	1550	18
152 × 152 × 51 †	51.2	170.2	157.4	11.0	15.7	7.6	123.6	5.01	11.2	3230	1020	11
152 × 152 × 44 †	44.0	166.0	155.9	9.5	13.6	7.6	123.6	5.73	13.0	2700	860	12
152 × 152 × 37	37.0	161.8	154.4	8.0	11.5	7.6	123.6	6.71	15.5	2210	706	14
152 × 152 × 30	30.0	157.6	152.9	6.5	9.4	7.6	123.6	8.13	19.0	1750	560	17
152 × 152 × 23	23.0	152.4	152.2	5.8	6.8	7.6	123.6	11.20	21.3	1250	400	22

†Additional sizes to BS4 available in the UK.

Radius of gyration		Elastic modulus		Plastic modulus		Buckling parameter	Torsional index	Warping constant	Torsional constant	Area of section
Axis x–x	Axis y–y	Axis x–x	Axis y–y	Axis x–x	Axis y–y					
r_x	r_y	Z_x	Z_y	S_x	S_y	u	x	H	J	A
cm	cm	cm³	cm³	cm³	cm³			dm⁶	cm⁴	cm²
18.4	11.0	11600	4630	14200	7110	0.843	5.46	38.8	13700	808
18.0	10.9	9960	3950	12100	6060	0.841	6.05	31.1	9240	702
17.5	10.7	8380	3290	10000	5030	0.839	6.86	24.3	5810	595
17.1	10.5	7000	2720	8220	4150	0.837	7.86	18.9	3550	501
16.8	10.4	6030	2330	7000	3540	0.836	8.85	15.5	2340	433
16.5	10.3	5070	1940	5810	2950	0.835	10.2	12.3	1440	366
16.3	10.2	4150	1570	4690	2380	0.834	12.1	9.54	812	299
16.1	9.60	3540	1260	3970	1920	0.844	13.4	7.16	558	257
15.9	9.54	3100	1100	3460	1670	0.844	15.0	6.09	381	226
15.8	9.49	2680	948	2960	1430	0.844	17.0	5.11	251	195
15.6	9.43	2260	793	2480	1200	0.844	19.9	4.18	153	164
14.8	8.27	4320	1530	5110	2340	0.855	7.65	6.35	2030	360
14.5	8.15	3640	1280	4250	1950	0.854	8.74	5.03	1270	306
14.2	8.04	3000	1040	3440	1580	0.854	10.2	3.88	734	252
13.9	7.90	2370	808	2680	1230	0.851	12.5	2.87	378	201
13.7	7.83	2050	692	2300	1050	0.851	14.2	2.39	249	174
13.6	7.77	1760	589	1960	895	0.850	16.2	1.98	161	150
13.4	7.69	1450	479	1590	726	0.850	19.3	1.56	91.2	123
11.9	6.81	2080	744	2420	1140	0.851	8.49	1.63	626	213
11.6	6.69	1630	576	1870	878	0.850	10.3	1.19	319	168
11.3	6.59	1310	458	1480	697	0.848	12.4	0.898	172	136
11.2	6.55	1100	379	1220	575	0.850	14.5	0.717	102	113
11.1	6.48	898	307	992	465	0.849	17.3	0.562	57.6	93.1
9.75	5.50	1280	460	1520	704	0.854	7.38	0.549	427	162
9.59	5.45	1130	404	1330	618	0.853	8.11	0.464	305	145
9.44	5.39	988	350	1150	534	0.852	9.02	0.386	210	127
9.28	5.34	850	299	977	456	0.850	10.2	0.318	137	110
9.18	5.30	706	246	799	374	0.853	11.9	0.250	80.2	90.4
8.96	5.20	584	201	656	305	0.846	14.1	0.197	47.2	76.4
8.91	5.18	510	174	567	264	0.848	15.8	0.167	31.8	66.3
8.82	5.13	450	152	497	231	0.847	17.7	0.143	22.2	58.7
7.04	3.96	379	130	438	199	0.848	10.1	0.061	48.8	65.2
6.94	3.92	326	110	372	169	0.848	11.5	0.050	31.7	56.1
6.85	3.87	273	91.5	309	140	0.848	13.3	0.040	19.2	47.1
6.76	3.83	222	73.3	248	112	0.849	16.0	0.031	10.5	38.3
6.54	3.70	164	52.6	182	80.1	0.840	20.7	0.021	4.63	29.2

Source: Copyright Corus 2008 – reproduced with the kind permission of Corus.

Rolled joists – dimensions and properties

Inside slope = 8°

RSJ designation	Mass per metre	Depth of section	Width of section	Thickness		Radius		Depth between fillets	Ratios for local buckling		Second moment of area		Lateral torsional buckling ratio
				Web	Flange	Root	Toe		Flange	Web	Axis x–x	Axis y–y	
		D	T	t	T	r_1	r_2	d	$B/2T$	d/t	I_x	I_y	D/T
	kg/m	mm	mm	mm	mm	mm		mm			cm⁴	cm⁴	
254 × 203 × 82	82.0	254.0	203.2	10.2	19.9	19.6	9.7	166.6	5.11	16.3	12000	2280	13
254 × 114 × 37	37.2	254.0	114.3	7.6	12.8	12.4	6.1	199.3	4.46	26.2	5080	269	20
203 × 152 × 52	52.3	203.2	152.4	8.9	16.5	15.5	7.6	133.2	4.62	15.0	4800	816	12
152 × 127 × 37	37.3	152.4	127.0	10.4	13.2	13.5	6.6	94.3	4.81	9.07	1820	378	12
127 × 114 × 29	29.3	127.0	114.3	10.2	11.5	9.9	4.8	79.5	4.97	7.79	979	242	11
127 × 114 × 27	26.9	127.0	114.3	7.4	11.4	9.9	5.0	79.5	5.01	10.7	946	236	11
127 × 76 × 16	16.5	127.0	76.2	5.6	9.6	9.4	4.6	86.5	3.97	15.4	571	60.8	13
114 × 114 × 27	27.1	114.3	114.3	9.5	10.7	14.2	3.2	60.8	5.34	6.40	736	224	11
102 × 102 × 23	23.0	101.6	101.6	9.5	10.3	11.1	3.2	55.2	4.93	5.81	486	154	10
102 × 44 × 7	7.5	101.6	44.5	4.3	6.1	6.9	3.3	74.6	3.65	17.3	153	7.82	17
89 × 89 × 19	19.5	88.9	88.9	9.5	9.9	11.1	3.2	44.2	4.49	4.65	307	101	9
76 × 76 × 15	15.0	76.2	80.0	8.9	8.4	9.4	4.6	38.1	4.76	4.28	172	60.9	9
76 × 76 × 13	12.8	76.2	76.2	5.1	8.4	9.4	4.6	38.1	4.54	7.47	158	51.8	9

Radius of gyration		Elastic modulus		Plastic modulus		Buckling parameter	Torsional index	Warping constant	Torsional constant	Area of section
Axis x–x	Axis y–y	Axis x–x	Axis y–y	Axis x–x	Axis y–y					
r_x	r_y	Z_x	Z_y	S_x	S_y	u	x	H	J	A
cm	cm	cm³	cm³	cm³	cm³			dm⁶	cm⁴	cm²
10.7	4.67	947	224	1080	371	0.888	11.0	0.312	152	105
10.4	2.39	400	47.1	459	79.1	0.885	18.7	0.0392	25.2	47.3
8.49	3.50	472	107	541	176	0.890	10.7	0.0711	64.8	66.6
6.19	2.82	239	59.6	279	99.8	0.867	9.33	0.0183	33.9	47.5
5.12	2.54	154	42.3	181	70.8	0.853	8.77	0.00807	20.8	37.4
5.26	2.63	149	41.3	172	68.2	0.868	9.31	0.00788	16.9	34.2
5.21	1.70	90.0	16.0	104	26.4	0.891	11.8	0.00210	6.72	21.1
4.62	2.55	129	39.2	151	65.8	0.839	7.92	0.00601	18.9	34.5
4.07	2.29	95.6	30.3	113	50.6	0.836	7.42	0.00321	14.2	29.3
4.01	0.907	30.1	3.51	35.4	6.03	0.872	14.9	0.000178	1.25	9.50
3.51	2.02	69.0	22.8	82.7	38.0	0.830	6.58	0.00158	11.5	24.9
3.00	1.78	45.2	15.2	54.2	25.8	0.820	6.42	0.000700	6.83	19.1
3.12	1.79	41.5	13.6	48.7	22.4	0.853	7.21	0.000595	4.59	16.2

Source: Copyright Corus 2008 – reproduced with the kind permission of Corus.

UK parallel flange channels – dimensions and properties

PFC designation	Mass per metre	Depth of section	Width of section	Thickness		Root radius	Depth between fillets	Ratios for local buckling		Second moment of area		Lateral torsional buckling ratio
				Web	Flange			Flange	Web	Axis x–x	Axis y–y	
		D	B	t	T	r	nd	b/t	d/t			D/T
	kg/m	mm	mm	mm	mm	mm	mm			cm⁴	cm⁴	
430 × 100 × 64	64.4	430	100	11.0	19.0	15	362	5.26	32.9	21900	722	23
380 × 100 × 54	54.0	380	100	9.5	17.5	15	315	5.71	33.2	15000	643	22
300 × 100 × 46	45.5	300	100	9.0	16.5	15	237	6.06	26.3	8230	568	18
300 × 90 × 41	41.4	300	90	9.0	15.5	12	245	5.81	27.2	7220	404	19
260 × 90 × 35	34.8	260	90	8.0	14.0	12	208	6.43	26.0	4730	353	19
260 × 75 × 28	27.6	260	75	7.0	12.0	12	212	6.25	30.3	3620	185	22
230 × 90 × 32	32.2	230	90	7.5	14.0	12	178	6.43	23.7	3520	334	16
230 × 75 × 26	25.7	230	75	6.5	12.5	12	181	6.00	27.8	2750	181	18
200 × 90 × 30	29.7	200	90	7.0	14.0	12	148	6.43	21.1	2520	314	14
200 × 75 × 23	23.4	200	75	6.0	12.5	12	151	6.00	25.2	1960	170	16
180 × 90 × 26	26.1	180	90	6.5	12.5	12	131	7.20	20.2	1820	277	14
180 × 75 × 20	20.3	180	75	6.0	10.5	12	135	7.14	22.5	1370	146	17
150 × 90 × 24	23.9	150	90	6.5	12.0	12	102	7.50	15.7	1160	253	13
150 × 75 × 18	17.9	150	75	5.5	10.0	12	106	7.50	19.3	861	131	15
125 × 65 × 15	14.8	125	65	5.5	9.5	12	82.0	6.84	14.9	483	80.0	13
100 × 50 × 10	10.2	100	50	5.0	8.5	9	65.0	5.88	13.0	208	32.3	12

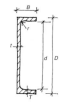

Radius of gyration		Elastic modulus		Elastic NA	Plastic modulus		Plastic NA	Buckling parameter	Torsional index	Warping constant	Torsional constant	Area of section
Axis x–x	Axis y–y	Axis x–x	Axis y–y		Axis x–x	Axis y–y						
				c_y			c_{eq}	u	x	H	J	A
cm	cm	cm³	cm³	cm	cm³	cm³	cm			dm⁶	cm⁴	cm²
16.3	2.97	1020	97.9	2.62	1220	176	0.954	0.917	22.5	0.219	63.0	82.1
14.8	3.06	791	89.2	2.79	933	161	0.904	0.933	21.2	0.150	45.7	68.7
11.9	3.13	549	81.7	3.05	641	148	1.31	0.944	17.0	0.0813	36.8	58.0
11.7	2.77	481	63.1	2.60	568	114	0.879	0.934	18.4	0.0581	28.8	52.7
10.3	2.82	364	56.3	2.74	425	102	1.14	0.943	17.2	0.0379	20.6	44.4
10.1	2.30	278	34.4	2.10	328	62.0	0.676	0.932	20.5	0.0203	11.7	35.1
9.27	2.86	306	55.0	2.92	355	98.9	1.69	0.949	15.1	0.0279	19.3	41.0
9.17	2.35	239	34.8	2.30	278	63.2	1.03	0.945	17.3	0.0153	11.8	32.7
8.16	2.88	252	53.4	3.12	291	94.5	2.24	0.952	12.9	0.0197	18.3	37.9
8.11	2.39	196	33.8	2.48	227	60.6	1.53	0.956	14.7	0.0107	11.1	29.9
7.40	2.89	202	47.4	3.17	232	83.5	2.36	0.950	12.8	0.0141	13.3	33.2
7.27	2.38	152	28.8	2.41	176	51.8	1.34	0.945	15.3	0.00754	7.34	25.9
6.18	2.89	155	44.4	3.30	179	76.9	2.66	0.937	10.8	0.00890	11.8	30.4
6.15	2.40	115	26.6	2.58	132	47.2	1.81	0.945	13.1	0.00467	6.10	22.8
5.07	2.06	77.3	18.8	2.25	89.9	33.2	1.55	0.942	11.1	0.00194	4.72	18.8
4.00	1.58	41.5	9.89	1.73	48.9	17.5	1.18	0.942	10.0	0.000491	2.53	13.0

Source: Copyright Corus 2008 – reproduced with the kind permission of Corus.

UK equal angles – dimensions and properties

RSA designation	Mass per metre	Root radius	Toe radius	Distance of centre of gravity	Second moment of area			Radius of gyration			Elastic modulus	Lateral torsional buckling ratio	Area of section
A × A × T		r_1	r_2	C_x & C_y	Axis x-x, y-y	Axis u-u	Axis v-v	Axis x-x, y-y	Axis u-u	Axis v-v	Axis x-x, y-y	D/T	A
mm × mm × mm	kg/m	mm	mm	cm	cm⁴	cm⁴	cm⁴	cm	cm	cm	cm³		cm²
200 × 200 × 24	71.1	18.0	9.00	5.84	3330	5280	1380	6.06	7.64	3.90	235	8	90.6
200 × 200 × 20	59.9	18.0	9.00	5.68	2850	4530	1170	6.11	7.70	3.92	199	10	76.3
200 × 200 × 18	54.3	18.0	9.00	5.60	2600	4150	1050	6.13	7.75	3.90	181	11	69.1
200 × 200 × 16	48.5	18.0	9.00	5.52	2340	3720	960	6.16	7.76	3.94	162	13	61.8
150 × 150 × 18*	40.1	16.0	8.00	4.38	1060	1680	440	4.55	5.73	2.93	99.8	8	51.2
150 × 150 × 15	33.8	16.0	8.00	4.25	898	1430	370	4.57	5.76	2.93	83.5	10	43.0
150 × 150 × 12	27.3	16.0	8.00	4.12	737	1170	303	4.60	5.80	2.95	67.7	13	34.8
150 × 150 × 10	23.0	16.0	8.00	4.03	624	990	258	4.62	5.82	2.97	56.9	15	29.3
120 × 120 × 15*	26.6	13.0	6.50	3.52	448	710	186	3.63	4.57	2.34	52.8	8	34.0
120 × 120 × 12	21.6	13.0	6.50	3.40	368	584	152	3.65	4.60	2.35	42.7	10	27.5
120 × 120 × 10	18.2	13.0	6.50	3.31	313	497	129	3.67	4.63	2.36	36.0	12	23.2
120 × 120 × 8	14.7	13.0	6.50	3.24	259	411	107	3.71	4.67	2.38	29.5	15	18.8
100 × 100 × 15*	21.9	12.0	6.00	3.02	250	395	105	2.99	3.76	1.94	35.8	7	28.0
100 × 100 × 12	17.8	12.0	6.00	2.90	207	328	85.7	3.02	3.80	1.94	29.1	8	22.7
100 × 100 × 10	15.0	12.0	6.00	2.82	177	280	73.0	3.04	3.83	1.95	24.6	10	19.2
100 × 100 × 8	12.2	12.0	6.00	2.74	145	230	59.9	3.06	3.85	1.96	19.9	13	15.5
90 × 90 × 12*	15.9	11.0	5.50	2.66	149	235	62.0	2.71	3.40	1.75	23.5	8	20.3
90 × 90 × 10	13.4	11.0	5.50	2.58	127	201	52.6	2.72	3.42	1.75	19.8	9	17.1
90 × 90 × 8	10.9	11.0	5.50	2.50	104	166	43.1	2.74	3.45	1.76	16.1	11	13.9
90 × 90 × 7	9.61	11.0	5.50	2.45	92.6	147	38.3	2.75	3.46	1.77	14.1	13	12.2
80 × 80 × 10*	11.9	10.0	5.00	2.34	87.5	139	36.4	2.41	3.03	1.55	15.4	8	15.1
80 × 80 × 8	9.63	10.0	5.00	2.26	72.2	115	29.9	2.43	3.06	1.56	12.6	10	12.3

75 × 75 × 8†	8.99	9.00	4.50	2.14	59.1	93.8	24.5	2.27	2.86	1.46	11.0	9	11.4
75 × 75 × 6†	6.85	9.00	4.50	2.05	45.8	72.7	18.9	2.29	2.89	1.47	8.41	13	8.73
70 × 70 × 7†	7.38	9.00	4.50	1.97	42.3	67.1	17.5	2.12	2.67	1.36	8.41	10	9.40
70 × 70 × 6†	6.38	9.00	4.50	1.93	36.9	58.5	15.3	2.13	2.68	1.37	7.27	12	8.13
65 × 65 × 7†	6.83	9.00	4.50	2.05	33.4	53.0	13.8	1.96	2.47	1.26	7.18	9	8.73
60 × 60 × 8†	7.09	8.00	4.00	1.77	29.2	46.1	12.2	1.80	2.26	1.16	6.89	8	9.03
60 × 60 × 6†	5.42	8.00	4.00	1.69	22.8	36.1	9.44	1.82	2.29	1.17	5.29	10	6.91
60 × 60 × 5†	4.57	8.00	4.00	1.64	19.4	30.7	8.03	1.82	2.30	1.17	4.45	12	5.82
50 × 50 × 6†	4.47	7.00	3.50	1.45	12.8	20.3	5.34	1.50	1.89	0.968	3.61	8	5.69
50 × 50 × 5†	3.77	7.00	3.50	1.40	11.0	17.4	4.55	1.51	1.90	0.973	3.05	10	4.80
50 × 50 × 4†	3.06	7.00	3.50	1.36	8.97	14.2	3.73	1.52	1.91	0.979	2.46	13	3.89
45 × 45 × 4.5†	3.06	7.00	3.50	1.25	7.14	11.4	2.94	1.35	1.71	0.870	2.20	9	3.90
40 × 40 × 5†	2.97	6.00	3.00	1.16	5.43	8.60	2.26	1.20	1.51	0.773	1.91	8	3.79
40 × 40 × 4†	2.42	6.00	3.00	1.12	4.47	7.09	1.86	1.21	1.52	0.777	1.55	10	3.08
35 × 35 × 4†	2.09	5.00	2.50	1.00	2.95	4.68	1.23	1.05	1.32	0.678	1.18	9	2.67
30 × 30 × 4†	1.78	5.00	2.50	0.878	1.80	2.85	0.754	0.892	1.12	0.577	0.850	8	2.27
30 × 30 × 3†	1.36	5.00	2.50	0.835	1.40	2.22	0.585	0.899	1.13	0.581	0.649	10	1.74
25 × 25 × 4†	1.45	3.50	1.75	0.762	1.02	1.61	0.430	0.741	0.931	0.482	0.586	6	1.85
25 × 25 × 3†	1.12	3.50	1.75	0.723	0.803	1.27	0.334	0.751	0.945	0.484	0.452	8	1.42
20 × 20 × 3†	0.882	3.50	1.75	0.598	0.392	0.618	0.165	0.590	0.742	0.383	0.279	7	1.12

*British Standard sections not produced by Corus.

†Addition to British Standard range.

Source: Copyright Corus 2008 – reproduced with the kind permission of Corus.

UKA unequal angles – dimensions and properties

RSA designation	Mass per metre	Root radius	Toe radius	Distance of centre of gravity	Distance of centre of gravity	Angle x–x to u–u axis	Second moment of area	
$D \times B \times T$		r_1	r_2	C_x	C_y	Tan a	Axis x–x	Axis y–y
mm × mm × mm	kg/m	mm	mm	cm	cm		cm⁴	cm⁴
200 × 150 × 18†	47.1	15.0	7.50	6.33	3.85	0.549	2380	1150
200 × 150 × 15	39.6	15.0	7.50	6.21	3.73	0.000	2020	979
200 × 150 × 12	32.0	15.0	7.50	6.08	3.61	0.000	1650	803
200 × 100 × 15	33.8	15.0	7.50	7.16	2.22	0.000	1760	299
200 × 100 × 12	27.3	15.0	7.50	7.03	2.10	0.000	1440	247
200 × 100 × 10	23.0	15.0	7.50	6.93	2.01	0.000	1220	210
150 × 90 × 15	33.9	12.0	6.00	5.21	2.23	0.000	761	205
150 × 90 × 12	21.6	12.0	6.00	5.08	2.12	0.000	627	171
150 × 90 × 10	18.2	12.0	6.00	5.00	2.04	0.000	533	146
150 × 75 × 15	24.8	12.0	6.00	5.52	1.81	0.000	713	119
150 × 75 × 12	20.2	12.0	6.00	5.40	1.69	0.000	588	99.6
150 × 75 × 10	17.0	12.0	6.00	5.31	1.61	0.000	501	85.6
125 × 75 × 12	17.8	11.0	5.50	4.31	1.84	0.000	354	95.5
125 × 75 × 10	15.0	11.0	5.50	4.23	1.76	0.000	302	82.1
125 × 75 × 8	12.2	11.0	5.50	4.14	1.68	0.000	247	67.6
100 × 75 × 12	15.4	10.0	5.00	3.27	2.03	0.000	189	90.2
100 × 75 × 10	13.0	10.0	5.00	3.19	1.95	0.000	162	77.6
100 × 75 × 8	10.6	10.0	5.00	3.10	1.87	0.000	133	64.1
100 × 65 × 10†	12.3	10.0	5.00	3.36	1.63	0.000	154	51.0
100 × 65 × 8†	9.94	10.0	5.00	3.27	1.55	0.000	127	42.2
100 × 65 × 7†	8.77	10.0	5.00	3.23	1.51	0.000	113	37.6
100 × 50 × 8†	8.97	8.00	4.00	3.60	1.13	0.000	116	19.7
100 × 50 × 6†	6.84	8.00	4.00	3.51	1.05	0.000	89.9	15.4
80 × 60 × 7†	7.36	8.00	4.00	2.51	1.52	0.000	59.0	28.4
80 × 40 × 8†	7.07	7.00	3.50	2.94	0.963	0.000	57.6	9.61
80 × 40 × 6†	5.41	7.00	3.50	2.85	0.884	0.000	44.9	7.59
75 × 50 × 8†	7.39	7.00	3.50	2.52	1.29	0.000	52.0	18.4
75 × 50 × 6†	5.65	7.00	3.50	2.44	1.21	0.000	40.5	14.4
70 × 50 × 6†	5.41	7.00	3.50	2.23	1.25	0.000	33.4	14.2
65 × 50 × 5†	4.35	6.00	3.00	1.99	1.25	0.000	23.2	11.9
60 × 40 × 6†	4.46	6.00	3.00	2.00	1.01	0.000	20.1	7.12
60 × 40 × 5†	3.76	6.00	3.00	1.96	0.972	0.000	17.2	6.11
60 × 30 × 5†	3.36	5.00	2.50	2.17	0.684	0.000	15.6	2.63
50 × 30 × 5†	2.96	5.00	2.50	1.73	0.741	0.000	9.36	2.51
45 × 30 × 4†	2.25	4.50	2.25	1.48	0.740	0.000	5.78	2.05
40 × 25 × 4†	1.93	4.00	2.00	1.36	0.623	0.000	3.89	1.16
40 × 20 × 4†	1.77	4.00	2.00	1.47	0.480	0.000	3.59	0.600
30 × 20 × 4†	1.46	4.00	2.00	1.03	0.541	0.000	1.59	0.553
30 × 20 × 3†	1.12	4.00	2.00	0.990	0.502	0.000	1.25	0.437

Second moment of area		Radius of gyration						Elastic modulus		Lateral torsional buckling ratio	Area of section
Axis u–u	Axis v–v	Axis x–x	Axis y–y	Axis u–u	Axis v–v			Axis x–x	Axis y–y	D/T	A
cm⁴	cm⁴	cm	cm	cm	cm			cm³	cm³		cm²
2920	623	6.29	4.37	6.97	3.22			174	103	11	60.0
2480	526	6.33	4.40	7.00	3.23			147	86.9	13	50.5
2030	430	6.36	4.44	7.04	3.25			119	70.5	17	40.8
1860	193	6.40	2.64	6.59	2.12			137	38.5	13	43.0
1530	159	6.43	2.67	6.63	2.14			111	31.3	17	34.8
1290	135	6.46	2.68	6.65	2.15			93.2	26.3	20	29.2
841	126	4.74	2.46	4.98	1.93			77.7	30.4	10	33.9
694	104	4.77	2.49	5.02	1.94			63.3	24.8	13	27.5
591	88.3	4.80	2.51	5.05	1.95			53.3	21.0	15	23.2
753	78.6	4.75	1.94	4.88	1.58			75.2	21.0	10	31.7
623	64.7	4.78	1.97	4.92	1.59			61.3	17.1	13	25.7
531	55.1	4.81	1.99	4.95	1.60			51.6	14.5	15	21.7
391	58.5	3.95	2.05	4.15	1.61			43.2	16.9	10	22.7
334	49.9	3.97	2.07	4.18	1.61			36.5	14.3	13	19.1
274	40.9	4.00	2.09	4.21	1.63			29.6	11.6	16	15.5
230	49.5	3.10	2.14	3.42	1.59			28.0	16.5	8	19.7
197	42.2	3.12	2.16	3.45	1.59			23.8	14.0	10	16.6
162	34.6	3.14	2.18	3.47	1.60			19.3	11.4	13	13.5
175	30.1	3.14	1.81	3.35	1.39			23.2	10.5	10	15.6
144	24.8	3.16	1.83	3.37	1.40			18.9	8.54	13	12.7
128	22.0	3.17	1.83	3.39	1.40			16.6	7.53	14	11.2
123	12.8	3.19	1.31	3.28	1.06			18.2	5.08	13	11.4
95.4	9.92	3.21	1.33	3.31	1.07			13.8	3.89	17	8.71
72.0	15.4	2.51	1.74	2.77	1.28			10.7	6.34	11	9.38
60.9	6.34	2.53	1.03	2.60	0.838			11.4	3.16	10	9.01
47.6	4.93	2.55	1.05	2.63	0.845			8.73	2.44	13	6.89
59.6	10.8	2.35	1.40	2.52	1.07			10.4	4.95	9	9.41
46.6	8.36	2.37	1.42	2.55	1.08			8.01	3.81	13	7.19
39.7	7.92	2.20	1.43	2.40	1.07			7.01	3.78	12	6.89
28.8	6.32	2.05	1.47	2.28	1.07			5.14	3.19	13	5.54
23.1	4.16	1.88	1.12	2.02	0.855			5.03	2.38	10	5.68
19.7	3.54	1.89	1.13	2.03	0.860			4.25	2.02	12	4.79
16.5	1.71	1.91	0.784	1.97	0.633			4.07	1.14	12	4.28
10.3	1.54	1.57	0.816	1.65	0.639			2.86	1.11	10	3.78
6.65	1.18	1.42	0.850	1.52	0.640			1.91	0.910	11	2.87
4.35	0.700	1.26	0.687	1.33	0.534			1.47	0.619	10	2.46
3.80	0.393	1.26	0.514	1.30	0.417			1.42	0.393	10	2.26
1.81	0.330	0.925	0.546	0.988	0.421			0.807	0.379	8	1.86
1.43	0.256	0.935	0.553	1.00	0.424			0.621	0.292	10	1.43

+British Standard sections not produced by Corus.

†Addition to British Standard range.

Source: Copyright Corus 2008 – reproduced with the kind permission of Corus.

Hot finished rectangular hollow sections – dimensions and properties

RHS designation		Mass per metre	Area of section	Ratios for local buckling		Second moment of area		Radius of gyration		Elastic modulus		Plastic modulus		Torsional constants		Surface area of section
Size	Thickness					Axis x-x	Axis y-y	Axis x-x	Axis y-y	Axis x-x	Axis y-y	Axis x-x	Axis y-y			
$D \times B$	t		A	B/t	D/t	I_x	I_y	r_x	r_y	Z_x	Z_y	S_x	S_y	J	C	
mm × mm	mm	kg/m	cm²			cm⁴	cm⁴	cm	cm	cm³	cm³	cm³	cm³	cm⁴	cm³	m²/m
50 × 30	3.2	3.61	4.60	12.6	6.38	14.2	6.20	1.76	1.16	5.68	4.13	7.25	5.00	14.2	6.80	0.152
60 × 40	3.0	4.35	5.54	17.0	10.3	26.5	13.9	2.18	1.58	8.82	6.95	10.9	8.19	29.2	11.2	0.192
	4.0	5.64	7.19	12.0	7.00	32.8	17.0	2.14	1.54	10.9	8.52	13.8	10.3	36.7	13.7	0.190
	5.0	6.85	8.73	9.00	5.00	38.1	19.5	2.09	1.50	12.7	9.77	16.4	12.2	43.0	15.7	0.187
80 × 40	3.2	5.62	7.16	22.0	9.50	57.2	18.9	2.83	1.63	14.3	9.46	18.0	11.0	46.2	16.1	0.232
	4.0	6.90	8.79	17.0	7.00	68.2	22.2	2.79	1.59	17.1	11.1	21.8	13.2	55.2	18.9	0.230
	5.0	8.42	10.7	13.0	5.00	80.3	25.7	2.74	1.55	20.1	12.9	26.1	15.7	65.1	21.9	0.227
	6.3	10.3	13.1	9.70	3.35	93.3	29.2	2.67	1.49	23.3	14.6	31.1	18.4	75.6	24.8	0.224
	8.0	12.5	16.0	7.00	2.00	106	32.1	2.58	1.42	26.5	16.1	36.5	21.2	85.8	27.4	0.219
90 × 50	3.6	7.40	9.42	22.0	10.9	98.3	38.7	3.23	2.03	21.8	15.5	27.2	18.0	89.4	25.9	0.271
	5.0	9.99	12.7	15.0	7.00	127	49.2	3.16	1.97	28.3	19.7	36.0	23.5	116	32.9	0.267
	6.3	12.3	15.6	11.3	4.94	150	57.0	3.10	1.91	33.3	22.8	43.2	28.0	138	38.1	0.264
100 × 50	3.0	6.71	8.54	30.3	13.7	110	36.8	3.58	2.08	21.9	14.7	27.3	16.8	88.4	25.0	0.292
	3.2	7.13	9.08	28.3	12.6	116	38.8	3.57	2.07	23.2	15.5	28.9	17.7	93.4	26.4	0.292
	4.0	8.78	11.2	22.0	9.50	140	46.2	3.53	2.03	27.9	18.5	35.2	21.5	113	31.4	0.290
	5.0	10.8	13.7	17.0	7.00	167	54.3	3.48	1.99	33.3	21.7	42.6	25.8	135	36.9	0.287
	6.3	13.3	16.9	12.9	4.94	197	63.0	3.42	1.93	39.4	25.2	51.3	30.8	160	42.9	0.284
	8.0	16.3	20.8	9.50	3.25	230	71.7	3.33	1.86	46.0	28.7	61.4	36.3	186	48.9	0.279
100 × 60	3.6	8.53	10.9	24.8	13.7	145	64.8	3.65	2.44	28.9	21.6	35.6	24.9	142	35.6	0.311
	5.0	11.6	14.7	17.0	9.00	189	83.6	3.58	2.38	37.8	27.9	47.4	32.9	189	45.9	0.307
	6.3	14.2	18.1	12.9	6.52	225	98.1	3.52	2.33	45.0	32.7	57.3	39.5	224	55.5	0.304
	8.0	17.5	22.4	9.50	4.50	264	113	3.44	2.25	52.8	37.8	68.7	47.1	265	62.2	0.299

120 × 60	3.6	9.66	12.3	30.3	13.7	227	76.3	4.30	2.49	37.9	25.4	47.2	28.9	183	43.3	0.351
	5.0	13.1	16.7	21.0	9.00	299	98.8	4.23	2.43	49.9	32.9	63.1	38.4	242	56.0	0.347
	6.3	16.2	20.7	16.0	6.52	358	116	4.16	2.37	59.7	38.8	76.7	46.3	290	65.9	0.344
	8.0	20.1	25.6	12.0	4.50	425	135	4.08	2.30	70.8	45.0	92.7	55.4	344	76.6	0.339
120 × 80	5.0	14.7	18.7	21.0	13.0	365	193	4.42	3.21	60.9	48.2	74.6	56.1	401	77.9	0.387
	6.3	18.2	23.2	16.0	9.70	440	230	4.36	3.15	73.3	57.6	91.0	68.2	487	92.9	0.384
	8.0	22.6	28.8	12.0	7.00	525	273	4.27	3.08	87.5	68.1	111	82.6	587	110	0.379
	10.0	27.4	34.9	9.00	5.00	609	313	4.18	2.99	102	78.1	131	97.3	688	126	0.374
150 × 100	5.0	18.6	23.7	27.0	17.0	739	392	5.58	4.07	98.5	78.5	119	90.1	807	127	0.487
	6.3	23.1	29.5	20.8	12.9	898	474	5.52	4.01	120	94.8	147	110	986	153	0.484
	8.0	28.9	36.8	15.8	9.50	1090	569	5.44	3.94	145	114	180	135	1200	183	0.479
	10.0	35.3	44.9	12.0	7.00	1280	665	5.34	3.85	171	133	216	161	1430	214	0.474
	12.5	42.8	54.6	9.00	5.00	1490	763	5.22	3.74	198	153	256	190	1680	246	0.468
160 × 80	4.0	14.4	18.4	37.0	17.0	612	207	5.77	3.35	76.5	51.7	94.7	58.3	493	88.1	0.470
	5.0	17.8	22.7	29.0	13.0	744	249	5.72	3.31	93.0	62.3	116	71.1	600	106	0.467
	6.3	22.2	28.2	22.4	9.70	903	299	5.66	3.26	113	74.8	142	86.8	730	127	0.464
	8.0	27.6	35.2	17.0	7.00	1090	356	5.57	3.18	136	89.0	175	106	883	151	0.459
	10.0	33.7	42.9	13.0	5.00	1280	411	5.47	3.10	161	103	209	125	1040	175	0.454
200 × 100	5.0	22.6	28.7	37.0	17.0	1500	505	7.21	4.19	149	101	185	114	1200	172	0.587
	6.3	28.1	35.8	28.7	12.9	1830	613	7.15	4.14	183	123	228	140	1480	208	0.584
	8.0	35.1	44.8	22.0	9.50	2230	739	7.06	4.06	223	148	282	172	1800	251	0.579
	10.0	43.1	54.9	17.0	7.00	2660	869	6.96	3.98	266	174	341	206	2160	295	0.574
	12.5	52.7	67.1	13.0	5.00	3140	1000	6.84	3.87	314	201	408	245	2540	341	0.568

Source: Copyright Corus 2008 – reproduced with the kind permission of Corus.

Hot finished rectangular hollow sections – dimensions and properties – continued

RHS designation		Mass per metre	Area of section	Ratios for local buckling		Second moment of area		Radius of gyration		Elastic modulus		Plastic modulus		Torsional constants		Surface area of section
Size	Thickness					Axis x-x	Axis y-y	Axis x-x	Axis y-y	Axis x-x	Axis y-y	Axis x-x	Axis y-y			
$D \times B$	t		A	B/t	D/t	I_x	I_y	r_x	r_y	Z_x	Z_y	S_x	S_y	J	C	
mm × mm	mm	kg/m	cm²			cm⁴	cm⁴	cm	cm	cm³	cm³	cm³	cm³	cm⁴	cm³	m²/m
200 × 150	8.0	41.4	52.8	22.0	15.8	2970	1890	7.50	5.99	297	253	359	294	3640	398	0.679
	10.0	51.0	64.9	17.0	12.0	3570	2260	7.41	5.91	357	302	436	356	4410	475	0.674
250 × 120	10.0	54.1	68.9	22.0	9.00	5310	1640	8.78	4.88	425	273	539	318	4090	468	0.714
	12.5	66.4	84.6	17.0	6.60	6330	1930	8.65	4.77	506	321	651	381	4880	549	0.708
250 × 150	5.0	30.4	38.7	47.0	27.0	3360	1530	9.31	6.28	269	204	324	228	3280	337	0.787
	6.3	38.0	48.4	36.7	20.8	4140	1870	9.25	6.22	331	250	402	283	4050	413	0.784
	8.0	47.7	60.8	28.3	15.8	5110	2300	9.17	6.15	409	306	501	350	5020	506	0.779
	10.0	58.8	74.9	22.0	12.0	6170	2760	9.08	6.06	494	367	611	426	6090	605	0.774
	12.5	72.3	92.1	17.0	9.00	7390	3270	8.96	5.96	591	435	740	514	7330	717	0.768
	16.0	90.3	115	12.6	6.38	8880	3870	8.79	5.80	710	516	906	625	8870	849	0.759
260 × 140*	5.0	30.4	38.7	49.0	25.0	3530	1350	9.55	5.91	272	193	331	216	3080	326	0.787
	6.3	38.0	48.4	38.3	19.2	4360	1660	9.49	5.86	335	237	411	267	3800	399	0.784
	8.0	47.7	60.8	29.5	14.5	5370	2030	9.40	5.78	413	290	511	331	4700	488	0.779
	10.0	58.8	74.9	23.0	11.0	6490	2430	9.31	5.70	499	347	624	402	5700	584	0.774
	12.5	72.3	92.1	17.8	8.20	7770	2880	9.18	5.59	597	411	756	485	6840	690	0.768
	16.0	90.3	115	13.3	5.75	9340	3400	9.01	5.44	718	486	925	588	8260	815	0.759
300 × 100	8.0	47.7	60.8	34.5	9.50	6310	1080	10.2	4.21	420	216	546	245	3070	387	0.779
	10.0	58.8	74.9	27.0	7.00	7610	1280	10.1	4.13	508	255	666	296	3680	458	0.774
300 × 150	8.0	54.0	68.8	34.5	15.8	8010	2700	10.8	6.27	534	360	663	407	6450	613	0.879
	10.0	66.7	84.9	27.0	12.0	9720	3250	10.7	6.18	648	433	811	496	7840	736	0.874
	12.5	82.1	105	21.0	9.00	11700	3860	10.6	6.07	779	514	986	600	9450	874	0.868
	16.0	103	131	15.8	6.38	14200	4600	10.4	5.92	944	613	1210	732	11500	1040	0.859
300 × 200	10.0	74.5	94.9	27.0	17.0	11800	6280	11.2	8.13	788	628	956	721	12900	1020	0.974
	12.5	91.9	117	21.0	13.0	14300	7540	11.0	8.02	952	754	1170	877	15700	1220	0.968

300 × 250	5.0	42.2	53.7	57.0	47.0	7410	5610	11.7	10.2	494	449	575	508	9770	697	1.09
	6.3	52.8	67.3	44.6	36.7	9190	6950	11.7	10.2	613	556	716	633	12200	862	1.08
	8.0	66.5	84.8	34.5	28.3	11400	8630	11.6	10.1	761	690	896	791	15200	1070	1.08
	10.0	82.4	105	27.0	22.0	13900	10500	11.5	10.0	928	840	1100	971	18600	1300	1.07
	12.5	102	130	21.0	17.0	16900	12700	11.4	9.89	1120	1010	1350	1190	22700	1560	1.07
	16.0	128	163	15.8	12.6	20600	15500	11.2	9.74	1380	1240	1670	1470	28100	1900	1.06
350 × 150	5.0	38.3	48.7	67.0	27.0	7660	2050	12.5	6.49	437	274	543	301	5160	477	0.987
	6.3	47.9	61.0	52.6	20.8	9480	2530	12.5	6.43	542	337	676	373	6390	586	0.984
	8.0	60.3	76.8	40.8	15.8	11800	3110	12.4	6.36	673	414	844	464	7930	721	0.979
	10.0	74.5	94.9	32.0	12.0	14300	3740	12.3	6.27	818	498	1040	566	9630	867	0.974
	12.5	91.9	117	25.0	9.00	17300	4450	12.2	6.17	988	593	1260	686	11600	1030	0.968
	16.0	115	147	18.9	6.38	21100	5320	12.0	6.01	1210	709	1560	840	14100	1230	0.959

*Special order only. Minimum tonnage applies.

Source: Copyright Corus 2008 – reproduced with the kind permission of Corus.

Hot finished rectangular hollow sections – dimensions and properties – continued

Size	Mass per metre	Thickness	Area of section	Ratios for local buckling		Second moment of area		Radius of gyration		Elastic modulus		Plastic modulus		Torsional constants		Surface area of section
D × B		t	A	D/t	B/t	Axis x-x Ix	Axis y-y Iy	Axis x-x rx	Axis y-y ry	Axis x-x Zx	Axis y-y Zy	Axis x-x Sx	Axis y-y Sy	J	C	
mm × mm	kg/m	mm	cm²			cm⁴	cm⁴	cm	cm	cm³	cm³	cm³	cm³	cm⁴	cm³	m²/m
350 × 250	46.1	5.0	58.7	47.0	67.0	10600	6360	13.5	10.4	607	509	716	569	12200	817	1.19
	57.8	6.3	73.6	36.7	52.6	13200	7890	13.4	10.4	754	631	892	709	15200	1010	1.18
	72.8	8.0	92.8	28.3	40.8	16400	9800	13.3	10.3	940	784	1120	888	19000	1250	1.18
	90.2	10.0	115	22.0	32.0	20100	11900	13.2	10.2	1150	955	1380	1090	23400	1530	1.17
	112	12.5	142	17.0	25.0	24400	14400	13.1	10.1	1400	1160	1690	1330	28500	1840	1.17
	126	14.2	160	14.6	21.6	27200	16000	13.0	10.0	1550	1280	1890	1490	31900	2040	1.16
	141	16.0	179	12.6	18.9	30000	17700	12.9	9.93	1720	1410	2100	1660	35300	2250	1.16
	120	16.0	153	4.50	22.0	26300	3560	13.1	4.82	1320	593	1760	709	10800	1080	0.999
400 × 150	42.2	5.0	53.7	27.0	77.0	10700	2320	14.1	6.57	534	309	671	337	6130	547	1.09
	52.8	6.3	67.3	20.8	60.5	13300	2850	14.0	6.51	663	380	836	418	7600	673	1.08
	66.5	8.0	84.8	15.8	47.0	16500	3510	13.9	6.43	824	468	1050	521	9420	828	1.08
	82.4	10.0	105	12.0	37.0	20100	4230	13.8	6.35	1010	564	1290	636	11500	998	1.07
	102	12.5	130	9.00	29.0	24400	5040	13.7	6.24	1220	672	1570	772	13800	1190	1.07
	115	14.2	146	7.56	25.2	27100	5550	13.6	6.16	1360	740	1760	859	15300	1310	1.06
	128	16.0	163	6.38	22.0	29800	6040	13.5	6.09	1490	805	1950	947	16800	1430	1.06
400 × 200	72.8	8.0	92.8	22.0	47.0	19600	6660	14.5	8.47	978	666	1200	743	15700	1140	1.18
	90.2	10.0	115	17.0	37.0	23900	8080	14.4	8.39	1200	808	1480	911	19300	1380	1.17
	112	12.5	142	13.0	29.0	29100	9740	14.3	8.28	1450	974	1810	1110	23400	1660	1.17
	126	14.2	160	11.1	25.2	32400	10800	14.2	8.21	1620	1080	2030	1240	26100	1830	1.16
	141	16.0	179	9.50	22.0	35700	11800	14.1	8.13	1790	1180	2260	1370	28900	2010	1.16

400 × 300	8.0	85.4	109	47.0	34.5	25700	16500	15.4	12.3	1290	1100	1520	1250	31000	1750	1.38
	10.0	106	135	37.0	27.0	31500	20200	15.3	12.2	1580	1350	1870	1540	38200	2140	1.37
	12.5	131	167	29.0	21.0	38500	24600	15.2	12.1	1920	1640	2300	1880	46800	2590	1.37
	14.2	148	189	25.2	18.1	43000	27400	15.1	12.1	2150	1830	2580	2110	52500	2890	1.36
	16.0	166	211	22.0	15.8	47500	30300	15.0	12.0	2380	2020	2870	2350	58300	3180	1.36
450 × 250	8.0	85.4	109	53.3	28.3	30100	12100	16.6	10.6	1340	971	1620	1080	27100	1630	1.38
	10.0	106	135	42.0	22.0	36900	14800	16.5	10.5	1640	1190	2000	1330	33300	1990	1.37
	12.5	131	167	33.0	17.0	45000	18000	16.4	10.4	2000	1440	2460	1630	40700	2410	1.37
	14.2	148	189	28.7	14.6	50300	20000	16.3	10.3	2240	1600	2760	1830	45600	2680	1.36
	16.0	166	211	25.1	12.6	55700	22000	16.2	10.2	2480	1760	3070	2030	50500	2950	1.36
500 × 200	8.0	85.4	109	59.5	22.0	34000	8140	17.7	8.65	1360	814	1710	896	21100	1430	1.38
	10.0	106	135	47.0	17.0	41800	9890	17.6	8.56	1670	989	2110	1100	25900	1740	1.37
	12.5	131	167	37.0	13.0	51000	11900	17.5	8.45	2040	1190	2590	1350	31500	2100	1.37
	14.2	148	189	32.2	11.1	56900	13200	17.4	8.38	2280	1320	2900	1510	35200	2320	1.36
	16.0	166	211	28.3	9.50	63000	14500	17.3	8.30	2520	1450	3230	1670	38900	2550	1.36
500 × 300	8.0	97.9	125	59.5	34.5	43700	20000	18.7	12.6	1750	1330	2100	1480	42600	2200	1.58
	10.0	122	155	47.0	27.0	53800	24400	18.6	12.6	2150	1630	2600	1830	52500	2700	1.57
	12.5	151	192	37.0	21.0	65800	29800	18.5	12.5	2630	1990	3200	2240	64400	3280	1.57
	14.2	170	217	32.2	18.1	73700	33200	18.4	12.4	2950	2220	3590	2520	72200	3660	1.56
	16.0	191	243	28.3	15.8	81800	36800	18.3	12.3	3270	2450	4010	2800	80300	4040	1.56
	20.0	235	300	22.0	12.0	98800	44100	18.2	12.1	3950	2940	4890	3410	97400	4840	1.55

Source: Copyright Corus 2008 – reproduced with the kind permission of Corus.

Hot finished square hollow sections – dimensions and properties

SHS designation		Mass per metre	Area of section	Ratio for local buckling	Second moment of area	Radius of gyration	Elastic modulus	Plastic modulus	Torsional constants		Surface area of section
Size $D \times D$ mm × mm	Thickness t mm	kg/m	A cm²	D/t	I cm⁴	r cm	Z cm³	S cm³	J cm⁴	C cm³	m²/m
40 × 40	3.0	3.41	4.34	10.3	9.78	1.50	4.89	5.97	15.7	7.10	0.152
	3.2	3.61	4.60	9.50	10.2	1.49	5.11	6.28	16.5	7.42	0.152
	4.0	4.39	5.59	7.00	11.8	1.45	5.91	7.44	19.5	8.54	0.150
	5.0	5.28	6.73	5.00	13.4	1.41	6.68	8.66	22.5	9.60	0.147
50 × 50	3.0	4.35	5.54	13.7	20.2	1.91	8.08	9.70	32.1	11.8	0.192
	3.2	4.62	5.88	12.6	21.2	1.90	8.49	10.2	33.8	12.4	0.192
	4.0	5.64	7.19	9.50	25.0	1.86	9.99	12.3	40.4	14.5	0.190
	5.0	6.85	8.73	7.00	28.9	1.82	11.6	14.5	47.6	16.7	0.187
	6.3	8.31	10.6	4.94	32.8	1.76	13.1	17.0	55.2	18.8	0.184
60 × 60	3.0	5.29	6.74	17.0	36.2	2.32	12.1	14.3	56.9	17.7	0.232
	3.2	5.62	7.16	15.8	38.2	2.31	12.7	15.2	60.2	18.6	0.232
	4.0	6.90	8.79	12.0	45.4	2.27	15.1	18.3	72.5	22.0	0.230
	5.0	8.42	10.7	9.00	53.3	2.23	17.8	21.9	86.4	25.7	0.227
	6.3	10.3	13.1	6.52	61.6	2.17	20.5	26.0	102	29.6	0.224
	8.0	12.5	16.0	4.50	69.7	2.09	23.2	30.4	118	33.4	0.219
70 × 70	3.6	7.40	9.42	16.4	68.6	2.70	19.6	23.3	108	28.7	0.271
	5.0	9.99	12.7	11.0	88.5	2.64	25.3	30.8	142	36.8	0.267
	6.3	12.3	15.6	8.11	104	2.58	29.7	36.9	169	42.9	0.264
	8.0	15.0	19.2	5.75	120	2.50	34.2	43.8	200	49.2	0.259
80 × 80	3.6	8.53	10.9	19.2	105	3.11	26.2	31.0	164	38.5	0.311
	4.0	9.41	12.0	17.0	114	3.09	28.6	34.0	180	41.9	0.310
	5.0	11.6	14.7	13.0	137	3.05	34.2	41.1	217	49.8	0.307
	6.3	14.2	18.1	9.70	162	2.99	40.5	49.7	262	58.7	0.304
	8.0	17.5	22.4	7.00	189	2.91	47.3	59.5	312	68.3	0.299
90 × 90	3.6	9.66	12.3	22.0	152	3.52	33.8	39.7	237	49.7	0.351
	4.0	10.7	13.6	19.5	166	3.50	37.0	43.6	260	54.2	0.350
	5.0	13.1	16.7	15.0	200	3.45	44.4	53.0	316	64.8	0.347
	6.3	16.2	20.7	11.3	238	3.40	53.0	64.3	382	77.0	0.344
	8.0	20.1	25.6	8.25	281	3.32	62.6	77.6	459	90.5	0.339
100 × 100	4.0	11.9	15.2	22.0	232	3.91	46.4	54.4	361	68.2	0.390
	5.0	14.7	18.7	17.0	279	3.86	55.9	66.4	439	81.8	0.387
	6.3	18.2	23.2	12.9	336	3.80	67.1	80.9	534	97.8	0.384
	8.0	22.6	28.8	9.50	400	3.73	79.9	98.2	646	116	0.379
	10.0	27.4	34.9	7.00	462	3.64	92.4	116	761	133	0.374
120 × 120	5.0	17.8	22.7	21.0	498	4.68	83.0	97.6	777	122	0.467
	6.3	22.2	28.2	16.0	603	4.62	100	120	950	147	0.464
	8.0	27.6	35.2	12.0	726	4.55	121	146	1160	176	0.459
	10.0	33.7	42.9	9.00	852	4.46	142	175	1380	206	0.454
	12.5	40.9	52.1	6.60	982	4.34	164	207	1620	236	0.448
140 × 140	5.0	21.0	26.7	25.0	807	5.50	115	135	1250	170	0.547
	6.3	26.1	33.3	19.2	984	5.44	141	166	1540	206	0.544
	8.0	32.6	41.6	14.5	1200	5.36	171	204	1890	249	0.539
	10.0	40.0	50.9	11.0	1420	5.27	202	246	2270	294	0.534
	12.5	48.7	62.1	8.20	1650	5.16	236	293	2700	342	0.528
150 × 150	5.0	22.6	28.7	27.0	1000	5.90	134	156	1550	197	0.587
	6.3	28.1	35.8	20.8	1220	5.85	163	192	1910	240	0.584
	8.0	35.1	44.8	15.8	1490	5.77	199	237	2350	291	0.579
	10.0	43.1	54.9	12.0	1770	5.68	236	286	2830	344	0.574
	12.5	52.7	67.1	9.00	2080	5.57	277	342	3380	402	0.568

SHS designation		Mass per metre	Area of section	Ratio for local buckling	Second moment of area	Radius of gyration	Elastic modulus	Plastic modulus	Torsional constants		Surface area of section
Size $D \times D$ mm × mm	Thickness t mm	kg/m	A cm²	D/t	I cm⁴	r cm	Z cm³	S cm³	J cm⁴	C cm³	m²/m
160 × 160	5.0	24.1	30.7	29.0	1230	6.31	153	178	1890	226	0.627
	6.3	30.1	38.3	22.4	1500	6.26	187	220	2330	275	0.624
	8.0	37.6	48.0	17.0	1830	6.18	229	272	2880	335	0.619
	10.0	46.3	58.9	13.0	2190	6.09	273	329	3480	398	0.614
	12.5	56.6	72.1	9.80	2580	5.98	322	395	4160	467	0.608
180 × 180	6.3	34.0	43.3	25.6	2170	7.07	241	281	3360	355	0.704
	8.0	42.7	54.4	19.5	2660	7.00	296	349	4160	434	0.699
	10.0	52.5	66.9	15.0	3190	6.91	355	424	5050	518	0.694
	12.5	64.4	82.1	11.4	3790	6.80	421	511	6070	613	0.688
	16.0	80.2	102	8.25	4500	6.64	500	621	7340	724	0.679
200 × 200	5.0	30.4	38.7	37.0	2450	7.95	245	283	3760	362	0.787
	6.3	38.0	48.4	28.7	3010	7.89	301	350	4650	444	0.784
	8.0	47.7	60.8	22.0	3710	7.81	371	436	5780	545	0.779
	10.0	58.8	74.9	17.0	4470	7.72	447	531	7030	655	0.774
	12.5	72.3	92.1	13.0	5340	7.61	534	643	8490	778	0.768
	16.0	90.3	115	9.50	6390	7.46	639	785	10300	927	0.759
250 × 250	6.3	47.9	61.0	36.7	6010	9.93	481	556	9240	712	0.984
	8.0	60.3	76.8	28.3	7460	9.86	596	694	11500	880	0.979
	10.0	74.5	94.9	22.0	9060	9.77	724	851	14100	1070	0.974
	12.5	91.9	117	17.0	10900	9.66	873	1040	17200	1280	0.968
	16.0	115	147	12.6	13300	9.50	1060	1280	21100	1550	0.959
300 × 300	6.3	57.8	73.6	44.6	10500	12.0	703	809	16100	1040	1.18
	8.0	72.8	92.8	34.5	13100	11.9	875	1010	20200	1290	1.18
	10.0	90.2	115	27.0	16000	11.8	1070	1250	24800	1580	1.17
	12.5	112	142	21.0	19400	11.7	1300	1530	30300	1900	1.17
	16.0	141	179	15.8	23900	11.5	1590	1900	37600	2330	1.16
350 × 350	8.0	85.4	109	40.8	21100	13.9	1210	1390	32400	1790	1.38
	10.0	106	135	32.0	25900	13.9	1480	1720	39900	2190	1.37
	12.5	131	167	25.0	31500	13.7	1800	2110	48900	2650	1.37
	16.0	166	211	18.9	38900	13.6	2230	2630	61000	3260	1.36
400 × 400	10.0	122	155	37.0	39100	15.9	1960	2260	60100	2900	1.57
	12.5	151	192	29.0	47800	15.8	2390	2780	73900	3530	1.57
	16.0	191	243	22.0	59300	15.6	2970	3480	92400	4360	1.56
	20.0*	235	300	17.0	71500	15.4	3580	4250	112000	5240	1.55

*SAW process (single longitudinal seam weld, slightly proud).

Source: Copyright Corus 2008 – reproduced with the kind permission of Corus.

Hot finished circular hollow sections – dimensions and properties

CHS designation		Mass per metre	Area of section	Ratio for local buckling	Second moment of area	Radius of gyration	Elastic modulus	Plastic modulus	Torsional constants		Surface area of section
Outside diameter D	Thickness t		A	D/t	I	r	Z	S	J	C	
mm	mm	kg/m	cm²		cm⁴	cm	cm³	cm³	cm⁴	cm³	m²/m
26.9	3.2	1.87	2.38	8.41	1.70	0.846	1.27	1.81	3.41	2.53	0.085
33.7	2.6	1.99	2.54	13.0	3.09	1.10	1.84	2.52	6.19	3.67	0.106
	3.2	2.41	3.07	10.5	3.60	1.08	2.14	2.99	7.21	4.28	0.106
	4.0	2.93	3.73	8.43	4.19	1.06	2.49	3.55	8.38	4.97	0.106
42.4	2.6	2.55	3.25	16.3	6.46	1.41	3.05	4.12	12.9	6.10	0.133
	3.2	3.09	3.94	13.3	7.62	1.39	3.59	4.93	15.2	7.19	0.133
	4.0	3.79	4.83	10.6	8.99	1.36	4.24	5.92	18.0	8.48	0.133
	5.0	4.61	5.87	8.48	10.5	1.33	4.93	7.04	20.9	9.86	0.133
48.3	3.2	3.56	4.53	15.1	11.6	1.60	4.80	6.52	23.2	9.59	0.152
	4.0	4.37	5.57	12.1	13.8	1.57	5.70	7.87	27.5	11.4	0.152
	5.0	5.34	6.80	9.66	16.2	1.54	6.69	9.42	32.3	13.4	0.152
60.3	3.2	4.51	5.74	18.8	23.5	2.02	7.78	10.4	46.9	15.6	0.189
	4.0	5.55	7.07	15.1	28.2	2.00	9.34	12.7	56.3	18.7	0.189
	5.0	6.82	8.69	12.1	33.5	1.96	11.1	15.3	67.0	22.2	0.189
76.1	2.9	5.24	6.67	26.2	44.7	2.59	11.8	15.5	89.5	23.5	0.239
	3.2	5.75	7.33	23.8	48.8	2.58	12.8	17.0	97.6	25.6	0.239
	4.0	7.11	9.06	19.0	59.1	2.55	15.5	20.8	118	31.0	0.239
	5.0	8.77	11.2	15.2	70.9	2.52	18.6	25.3	142	37.3	0.239
88.9	3.2	6.76	8.62	27.8	79.2	3.03	17.8	23.5	158	35.6	0.279
	4.0	8.38	10.7	22.2	96.3	3.00	21.7	28.9	193	43.3	0.279
	5.0	10.3	13.2	17.8	116	2.97	26.2	35.2	233	52.4	0.279
	6.3	12.8	16.3	14.1	140	2.93	31.5	43.1	280	63.1	0.279
114.3	3.2	8.77	11.2	35.7	172	3.93	30.2	39.5	345	60.4	0.359
	3.6	9.83	12.5	31.8	192	3.92	33.6	44.1	384	67.2	0.359
	4.0	10.9	13.9	28.6	211	3.90	36.9	48.7	422	73.9	0.359
	5.0	13.5	17.2	22.9	257	3.87	45.0	59.8	514	89.9	0.359
	6.3	16.8	21.4	18.1	313	3.82	54.7	73.6	625	109	0.359
139.7	5.0	16.6	21.2	27.9	481	4.77	68.8	90.8	961	138	0.439
	6.3	20.7	26.4	22.2	589	4.72	84.3	112	1177	169	0.439
	8.0	26.0	33.1	17.5	720	4.66	103	139	1441	206	0.439
	10.0	32.0	40.7	14.0	862	4.60	123	169	1724	247	0.439
168.3	5.0	20.1	25.7	33.7	856	5.78	102	133	1710	203	0.529
	6.3	25.2	32.1	26.7	1050	5.73	125	165	2110	250	0.529
	8.0	31.6	40.3	21.0	1300	5.67	154	206	2600	308	0.529
	10.0	39.0	49.7	16.8	1560	5.61	186	251	3130	372	0.529
	12.5	48.0	61.2	13.5	1870	5.53	222	304	3740	444	0.529
193.7	5.0	23.3	29.6	38.7	1320	6.67	136	178	2640	273	0.609
	6.3	29.1	37.1	30.7	1630	6.63	168	221	3260	337	0.609
	8.0	36.6	46.7	24.2	2020	6.57	208	276	4030	416	0.609
	10.0	45.3	57.7	19.4	2440	6.50	252	338	4880	504	0.609
	12.5	55.9	71.2	15.5	2930	6.42	303	411	5870	606	0.609

CHS designation		Mass per metre	Area of section	Ratio for local buckling	Second moment of area	Radius of gyration	Elastic modulus	Plastic modulus	Torsional constants		Surface area of section
Outside diameter D	Thickness t		A	D/t	I	r	Z	S	J	C	
mm	mm	kg/m	cm²		cm⁴	cm	cm³	cm³	cm⁴	cm³	m²/m
219.1	5.0	26.4	33.6	43.8	1930	7.57	176	229	3860	352	0.688
	6.3	33.1	42.1	34.8	2390	7.53	218	285	4770	436	0.688
	8.0	41.6	53.1	27.4	2960	7.47	270	357	5920	540	0.688
	10.0	51.6	65.7	21.9	3600	7.40	328	438	7200	657	0.688
	12.5	63.7	81.1	17.5	4350	7.32	397	534	8690	793	0.688
	16.0	80.1	102	13.7	5300	7.20	483	661	10600	967	0.688
244.5	8.0	46.7	59.4	30.6	4160	8.37	340	448	8320	681	0.768
	10.0	57.8	73.7	24.5	5070	8.30	415	550	10100	830	0.768
	12.5	71.5	91.1	19.6	6150	8.21	503	673	12300	1010	0.768
	16.0	90.2	115	15.3	7530	8.10	616	837	15100	1230	0.768
273.0	6.3	41.4	52.8	43.3	4700	9.43	344	448	9390	688	0.858
	8.0	52.3	66.6	34.1	5850	9.37	429	562	11700	857	0.858
	10.0	64.9	82.6	27.3	7150	9.31	524	692	14300	1050	0.858
	12.5	80.3	102	21.8	8700	9.22	637	849	17400	1270	0.858
	16.0	101	129	17.1	10700	9.10	784	1060	21400	1570	0.858
323.9	6.3	49.3	62.9	51.4	7930	11.2	490	636	15900	979	1.02
	8.0	62.3	79.4	40.5	9910	11.2	612	799	19800	1220	1.02
	10.0	77.4	98.6	32.4	12200	11.1	751	986	24300	1500	1.02
	12.5	96.0	122	25.9	14800	11.0	917	1210	29700	1830	1.02
	16.0	121	155	20.2	18400	10.9	1140	1520	36800	2270	1.02
355.6	16.0	134	171	22.2	24700	12.0	1390	1850	49300	2770	1.12
406.4	6.3	62.2	79.2	64.5	15800	14.1	780	1010	31700	1560	1.28
	8.0	78.6	100	50.8	19900	14.1	978	1270	39700	1960	1.28
	10.0	97.8	125	40.6	24500	14.0	1210	1570	49000	2410	1.28
	12.5	121	155	32.5	30000	13.9	1480	1940	60100	2960	1.28
	16.0	154	196	25.4	37400	13.8	1840	2440	74900	3690	1.28
457.0	8.0	88.6	113	57.1	28400	15.9	1250	1610	56900	2490	1.44
	10.0	110	140	45.7	35100	15.8	1540	2000	70200	3070	1.44
	12.5	137	175	36.6	43100	15.7	1890	2470	86300	3780	1.44
	16.0	174	222	28.6	54000	15.6	2360	3110	108000	4720	1.44
508.0	10.0	123	156	50.8	48500	17.6	1910	2480	97000	3820	1.60
	12.5	153	195	40.6	59800	17.5	2350	3070	120000	4710	1.60
	16.0	194	247	31.8	74900	17.4	2950	3870	150000	5900	1.60

Source: Copyright Corus 2008 – reproduced with the kind permission of Corus.

Hot finished elliptical hollow sections – dimensions and properties

EHS designation		Mass per metre	Area of section	Second moment of area		Radius of gyration		Elastic modulus		Plastic modulus		Torsional constants		Surface area
Size H × B	Thickness t		A	I x–x	I y–y	r x–x	r y–y	Z x–x	Z y–y	S x–x	S y–y	J	C	
mm	mm	kg/m	cm²	cm⁴	cm⁴	cm	cm	cm³	cm³	cm³	cm³	cm⁴	cm³	m²/m
150 × 75	4.0	10.7	13.6	301	101	4.70	2.72	40.1	26.9	56.1	34.4	303	60.1	0.363
150 × 75	5.0	13.3	16.9	367	122	4.66	2.69	48.9	32.5	68.9	42.0	367	72.2	0.363
150 × 75	6.3	16.5	21.0	448	147	4.62	2.64	59.7	39.1	84.9	51.5	443	86.3	0.363
200 × 100	5.0	17.9	22.8	897	302	6.27	3.64	89.7	60.4	125	76.8	905	135	0.484
200 × 100	6.3	22.3	28.4	1100	368	6.23	3.60	110	73.5	155	94.7	1110	163	0.484
200 × 100	8.0	28.0	35.7	1360	446	6.17	3.54	136	89.3	193	117	1350	197	0.484
200 × 100	10.0	34.5	44.0	1640	529	6.10	3.47	164	106	235	141	1610	232	0.484
250 × 125	6.3	28.2	35.9	2210	742	7.84	4.55	176	119	246	151	2220	265	0.605
250 × 125	8.0	35.4	45.1	2730	909	7.78	4.49	219	145	307	188	2730	323	0.605
250 × 125	10.0	43.8	55.8	3320	1090	7.71	4.42	265	174	376	228	3290	385	0.605
250 × 125	12.5	53.9	68.7	4000	1290	7.63	4.34	320	207	458	276	3920	453	0.605
300 × 150	8.0	42.8	54.5	4810	1620	9.39	5.44	321	215	449	275	4850	481	0.726
300 × 150	10.0	53.0	67.5	5870	1950	9.32	5.37	391	260	551	336	5870	577	0.726
300 × 150	12.5	65.5	83.4	7120	2330	9.24	5.29	475	311	674	409	7050	686	0.726
300 × 150	16.0	82.5	105	8730	2810	9.12	5.17	582	374	837	503	8530	818	0.726
400 × 200	8.0	57.6	73.4	11700	3970	12.6	7.35	584	397	811	500	11900	890	0.969
400 × 200	10.0	71.5	91.1	14300	4830	12.5	7.28	717	483	1000	615	14500	1080	0.969
400 × 200	12.5	88.6	113	17500	5840	12.5	7.19	877	584	1230	753	17600	1300	0.969
400 × 200	16.0	112	143	21700	7140	12.3	7.07	1090	714	1540	936	21600	1580	0.969
500 × 250	10.0	90	115	28539	9682	15.8	9.2	1142	775	1585	976	28950	1739	1.21
500 × 250	12.5	112	142	35000	11800	15.7	9.10	1400	943	1960	1200	35300	2110	1.21
500 × 250	16.0	142	180	43700	14500	15.6	8.98	1750	1160	2460	1500	43700	2590	1.21

Source: Copyright Corus 2008 – reproduced with the kind permission of Corus.

Mild steel rounds typically available

Bar diameter mm	Weight kg/m	Bar diameter mm	Weight kg/m	Bar diameter mm	Weight kg/m	Bar diameter mm	Weight kg/m
6	0.22	16	1.58	40	9.86	65	26.0
8	0.39	20	2.47	45	12.5	75	34.7
10	0.62	25	3.85	50	15.4	90	49.9
12	0.89	32	6.31	60	22.2	100	61.6

Mild steel square bars typically available

Bar size mm	Weight kg/m	Bar size mm	Weight kg/m	Bar size mm	Weight kg/m
8	0.50	25	4.91	50	19.60
10	0.79	30	7.07	60	28.30
12.5	1.22	32	8.04	75	44.20
16	2.01	40	12.60	90	63.60
20	3.14	45	15.90	100	78.50

Mild steel flats typically available

Bar size mm	Weight kg/m	Bar size mm	Weight kg/m	Bar size mm	Weight kg/m	Bar size mm	Weight kg/m	Bar size mm	Weight kg/m
13 × 3	0.307	45 × 6	2.120	65 × 40	20.40	100 × 15	11.80	160 × 10	12.60
13 × 6	0.611	45 × 8	2.830	70 × 8	4.40	100 × 20	15.70	160 × 12	15.10
16 × 3	0.378	45 × 10	3.530	70 × 10	5.50	100 × 25	19.60	160 × 15	18.80
20 × 3	0.466	45 × 12	4.240	70 × 12	6.59	100 × 30	23.60	160 × 20	25.20
20 × 5	0.785	45 × 15	5.295	70 × 20	11.0	100 × 40	31.40	180 × 6	8.50
20 × 6	0.940	45 × 20	7.070	70 × 25	13.70	100 × 50	39.30	180 × 10	14.14
20 × 10	1.570	45 × 25	8.830	75 × 6	3.54	110 × 6	5.18	180 × 12	17.00
25 × 3	0.589	50 × 3	1.180	75 × 8	4.71	110 × 10	8.64	180 × 15	21.20
25 × 5	0.981	50 × 5	1.960	75 × 10	5.90	110 × 12	10.40	180 × 20	28.30
25 × 6	1.18	50 × 6	2.360	75 × 12	7.07	110 × 20	17.30	180 × 25	35.30
25 × 8	1.570	50 × 8	3.140	75 × 15	8.84	110 × 50	43.20	200 × 6	9.90
25 × 10	1.960	50 × 10	3.93	75 × 20	11.78	120 × 6	5.65	200 × 10	15.70
25 × 12	2.360	50 × 12	4.71	75 × 25	14.72	120 × 10	9.42	200 × 12	18.80
30 × 3	0.707	50 × 15	5.89	75 × 30	17.68	120 × 15	14.10	200 × 15	23.60
30 × 5	1.180	50 × 20	7.85	80 × 6	3.77	120 × 20	18.80	200 × 20	31.40
30 × 6	1.410	50 × 25	9.81	80 × 8	5.02	120 × 25	23.60	200 × 25	39.20
30 × 8	1.880	50 × 30	11.80	80 × 10	6.28	130 × 6	6.10	200 × 30	47.20
30 × 10	2.360	50 × 40	15.70	80 × 12	7.54	130 × 8	8.16	220 × 10	17.25
30 × 12	2.830	55 × 10	4.56	80 × 15	9.42	130 × 10	10.20	220 × 15	25.87
30 × 20	4.710	60 × 8	3.77	80 × 20	12.60	130 × 12	12.20	220 × 20	34.50
35 × 6	1.650	60 × 10	4.71	80 × 25	15.70	130 × 15	15.30	220 × 25	43.20
35 × 10	2.750	60 × 12	5.65	80 × 30	18.80	130 × 20	20.40	250 × 10	19.60
35 × 12	3.300	60 × 15	7.07	80 × 40	25.10	130 × 25	25.50	250 × 12	23.60
35 × 20	5.500	60 × 20	9.42	80 × 50	31.40	140 × 6	6.60	250 × 15	29.40
40 × 3	0.942	60 × 25	11.80	90 × 6	4.24	140 × 10	11.00	250 × 20	39.20
40 × 5	1.570	60 × 30	14.14	90 × 10	7.07	140 × 12	13.20	250 × 25	49.10
40 × 6	1.880	65 × 5	2.55	90 × 12	8.48	140 × 20	22.00	250 × 40	78.40
40 × 8	2.510	65 × 6	3.06	90 × 15	10.60	150 × 6	7.06	250 × 50	98.10
40 × 10	3.140	65 × 8	4.05	90 × 20	14.10	150 × 8	9.42	280 × 12.5	27.48
40 × 12	3.770	65 × 10	5.10	90 × 25	17.70	150 × 10	11.80	300 × 10	23.55
40 × 15	4.710	65 × 12	6.12	100 × 5	3.93	150 × 12	14.10	300 × 12	28.30
40 × 20	6.280	65 × 15	7.65	100 × 6	4.71	150 × 15	17.70	300 × 15	35.30
40 × 25	7.850	65 × 20	10.20	100 × 8	6.28	150 × 20	23.60	300 × 20	47.10
40 × 30	9.420	65 × 25	12.80	100 × 10	7.85	150 × 25	29.40	300 × 25	58.80
45 × 3	1.060	65 × 30	15.30	100 × 12	9.42			300 × 40	94.20

Hot rolled mild steel plates typically available

Thickness mm	Weight kg/m²	Thickness mm	Weight kg/m²	Thickness mm	Weight kg/m²	Thickness mm	Weight kg/m²	Thickness mm	Weight kg/m²
3	23.55	10	78.50	30	235.50	55	431.75	90	706.50
3.2	25.12	12.5	98.12	32	251.20	60	471.00	100	785.00
4	31.40	15	117.75	35	274.75	65	510.25	110	863.50
5	39.25	20	157.00	40	314.00	70	549.50	120	942.00
6	47.10	22.5	176.62	45	353.25	75	588.75	130	1050.50
8	62.80	25	196.25	50	392.50	80	628.00	150	1177.50

Durbar mild steel floor plates typically available

Basic size mm	Weight kg/m²	Basic size mm	Weight kg/m²
2500 × 1250 × 3 3000 × 1500 × 3	26.19	3000 × 1500 × 8 3700 × 1830 × 8 4000 × 1750 × 8 6100 × 1830 × 8	65.44
2000 × 1000 × 4.5 2500 × 1250 × 4.5 3000 × 1250 × 4.5 3700 × 1830 × 4.5 4000 × 1750 × 4.5	37.97	2000 × 1000 × 10 2500 × 1250 × 10 3000 × 1500 × 10 3700 × 1830 × 10	81.14
2000 × 1000 × 6 2500 × 1250 × 6 3000 × 1500 × 6 3700 × 1830 × 6 4000 × 1750 × 6	49.74	2000 × 1000 × 12.5 2500 × 1250 × 12.5 3000 × 1500 × 12.5 3700 × 1830 × 12.5 4000 × 1750 × 12.5	100.77
2000 × 1000 × 8 2500 × 1250 × 8	65.44	The depth of pattern ranges from 1.9 to 2.4 mm.	

Slenderness

Slenderness and elastic buckling

The slenderness (λ) of a structural element indicates how much load the element can carry in compression. Short stocky elements have low values of slenderness and are likely to fail by crushing, while elements with high slenderness values will fail by elastic (reversible) buckling. Slender columns will buckle when the axial compression reaches the critical load. Slender beams will buckle when the compressive stress causes the compression flange to buckle and twist sideways. This is called Lateral Torsional Buckling and it can be avoided (and the load capacity of the beam increased) by restraining the compression flange at intervals or over its full length. Full lateral restraint can be assumed if the construction fixed to the compression flange is capable of resisting a force of not less than 2.5% of the maximum force in that flange distributed uniformly along its length.

Slenderness limits

Slenderness, $\lambda = L_e/r$ where L_e is the effective length and r is the radius of gyration – generally about the weaker axis.

For robustness, members should be selected so that their slenderness does not exceed the following limits:

Members resisting load other than wind	$\lambda \leq 180$
Members resisting self-weight and wind only	$\lambda \leq 250$
Members normally acting as ties but subject to load reversal due to wind	$\lambda \leq 350$

Effective length for different restraint conditions

Effective length of beams – end restraint

Conditions of restraint at the ends of the beams		Effective length	
		Normal loading	Destabilizing loading
Compression flange laterally restrained; beam fully restrained against torsion (rotation about the longitudinal axis)	Both flanges fully restrained against rotation on plan	$0.70L$	$0.85L$
	Compression flange fully restrained against rotation on plan	$0.75L$	$0.90L$
	Both flanges partially restrained against rotation on plan	$0.80L$	$0.95L$
	Compression flange partially restrained against rotation on plan	$0.85L$	$1.00L$
	Both flanges free to rotate on plan	$1.00L$	$1.20L$
Compression flange laterally unrestrained; both flanges free to rotate on plan	Partial torsional restraint against rotation about the longitudinal axis provided by connection of bottom flange to supports	$1.0L + 2D$	$1.2L + 2D$
	Partial torsional restraint against rotation about the longitudinal axis provided only by the pressure of the bottom flange bearing onto the supports	$1.2L + 2D$	$1.4L + 2D$

NOTE:
The illustrated connections are not the only methods of providing the restraints noted in the table.

Source: BS 5950: Part 1: 2000.

Effective length of cantilevers

Conditions of restraint		Effective length*	
Support	Cantilever tip	Normal loading	Destabilizing loading
Continuous with lateral restraint to top flange	Free	3.0L	7.5L
	Top flange laterally restrained	2.7L	7.5L
	Torsional restraint	2.4L	4.5L
	Lateral and torsional restraint	2.1L	3.6L
Continuous with partial torsional restraint	Free	2.0L	5.0L
	Top flange laterally restrained	1.8L	5.0L
	Torsional restraint	1.6L	3.0L
	Lateral and torsional restraint	1.4L	2.4L
Continuous with lateral and torsional restraint	Free	1.0L	2.5L
	Top flange laterally restrained	0.9L	2.5L
	Torsional restraint	0.8L	1.5L
	Lateral and torsional restraint	0.7L	1.2L
Restrained laterally, torsionally and against rotation on plan	Free	0.8L	1.4L
	Top flange laterally restrained	0.7L	1.4L
	Torsional restraint	0.6L	0.6L
	Lateral and torsional restraint	0.5L	0.5L

Cantilever tip restraint conditions

Free	Top flange laterally restrained	Torsional restraint	Lateral and torsional restraint
"Not braced on plan"	"Braced on plan in at least one bay"	"Not braced on plan"	"Braced on plan in at least one bay"

*Increase effective length by 30% for moments applied at cantilever tip.

Source: BS 5950: Part 1: 2000.

Effective length of braced columns – restraint provided by cross bracing or shear wall

Conditions of restraint at the ends of the columns		Effective length
Effectively held in position at both ends	Effectively restrained in direction at both ends	0.70L
	Partially restrained in direction at both ends	0.85L
	Restrained in direction at one end	0.85L
	Not restrained in direction at either end	1.00L

Effective length of unbraced columns – restraint provided by sway of columns

Conditions of restraint at the ends of the columns		Effective length
Effectively held in position and restrained in direction at one end	Other end effectively restrained in direction	1.20L
	Other end partially restrained in direction	1.50L
	Other end not restrained in direction	2.00L

Source: BS 5950: Part 1: 2000.

Durability and fire resistance

Corrosion mechanism and protection

$$4Fe + 3O_2 + 2H_2O = 2Fe_2O_3 \cdot H_2O$$

Iron/Steel + Oxygen + Water = Rust

For corrosion of steel to take place, oxygen and water must both be present. The corrosion rate is affected by the atmospheric pollution and the length of time the steelwork remains wet. Sulphates (typically from industrial pollution) and chlorides (typically in marine environments – coastal is considered to be a 2 km strip around the coast in the UK) can accelerate the corrosion rate. All corrosion occurs at the anode ($-$ve where electrons are lost) and the products of corrosion are deposited at the cathode ($+$ve where the electrons are gained). Both anodic and cathodic areas can be present on a steel surface.

The following factors should be considered in relation to the durability of a structure: the environment, degree of exposure, shape of the members, structural detailing, protective measures and whether inspection and maintenance are possible. Bi-metallic corrosion should also be considered in damp conditions.

Durability exposure conditions

Corrosive environments are classified by BS EN ISO 12944: Part 2 and ISO 9223, and the corrosivity of the environment must be assessed for each project.

Corrosivity category and risk	Examples of typical environments in a temperate climate*	
	Exterior	Interior
C1 – Very low	–	Heated buildings with clean atmospheres, e.g. offices, shops, schools, hotels, etc. (theoretically no protection is needed)
C2 – Low	Atmospheres with low levels of pollution. Mostly rural areas	Unheated buildings where condensation may occur, e.g. depots and sports halls
C3 – Medium	Urban and industrial atmospheres with moderate sulphur dioxide pollution. Coastal areas with low salinity	Production rooms with high humidity and some air pollution, e.g. food processing plants, laundries, breweries, dairies, etc.
C4 – High	Industrial areas and coastal areas with moderate salinity	Chemical plants, swimming pools, coastal ship and boatyards
C5I – Very high (industrial)	Industrial areas with high humidity and aggressive atmosphere	Buildings or areas with almost permanent condensation and high pollution
C5M – Very high (marine)	Coastal and offshore areas with high salinity	Buildings or areas with almost permanent condensation and high pollution

*A hot and humid climate increases the corrosion rate and steel will require additional protection than in a temperate climate.

BS EN ISO 12944: Part 3 gives advice on steelwork detailing to avoid crevices where moisture and dirt can be caught and accelerate corrosion. Some acidic timbers should be isolated from steelwork.

Get advice for each project: Corus can give advice on all steelwork coatings. The Galvanizers' Association, Thermal Spraying and Surface Engineering Association and paint manufacturers also give advice on system specifications.

Methods of corrosion protection

A corrosion protection system should consist of good surface preparation and application of a suitable coating with the required durability and minimum cost.

Mild steel surface preparation to BS EN ISO 8501

Hot rolled structural steelwork (in mild steel) leaves the last rolling process at about 1000°C. As it cools, its surface reacts with the air to form a blue-grey coating called mill scale, which is unstable, will allow rusting of the steel and will cause problems with the adhesion of protective coatings. The steel must be degreased to ensure that any contaminants which might affect the coatings are removed. The mill scale can then be removed by abrasive blast cleaning. Typical blast cleaning surface grades are:

Sa 1   Light blast cleaning

Sa 2   Thorough blast cleaning

Sa 2½   Very thorough blast cleaning

Sa 3   Blast cleaning to visually clean steel

Sa 2½ is used for most structural steel. Sa 3 is often used for surface preparation for metal spray coatings.

Metallic and non-metallic particles can be used to blast clean the steel surface. Chilled angular metallic grit (usually grade G24) provides a rougher surface than round metallic shot, so that the coatings have better adhesion to the steel surface. Acid pickling is often used after blast cleaning to Sa 2½ to remove final traces of mill scale before galvanizing. Coatings must be applied very quickly after the surface preparation to avoid rust reforming and the requirement for reblasting.

Paint coatings for structural steel

Paint provides a barrier coating to prevent corrosion and is made up of pigment (for colour and protection), binder (for formation of the coating film) and solvent (to allow application of the paint before it evaporates and the paint hardens). When first applied, the paint forms a wet film thickness which can be measured and the dry film thickness (DFT – which is normally the specified element) can be predicted when the percentage volume of solids in the paint is known. Primers are normally classified on their protective pigment (e.g. zinc phosphate primer). Intermediate (which build the coating thickness) and finish coats are usually classified on their binders (e.g. epoxies, vinyls, urethanes, etc.). Shop primers (with a DFT of 15–25 μm) can be applied before fabrication but these only provide a couple of weeks' worth of protection. Zinc rich primers generally perform best. Application of paint can be by brush, roller, air spray and airless spray – the latter is the most common in the UK. Application can be done on site or in the shop and where the steel is to be exposed, the method of application should be chosen for practicality and the surface finish. Shop applied coatings tend to need touching up on site if they are damaged in transit.

Metallic coatings for structural steel

Hot dip galvanizing   De-greased, blast cleaned (generally Sa 2½) and then acid pick-led steel is dipped into a flux agent and then into a bath of molten zinc. The zinc reacts with the surface of the steel, forming alloys and as the steel is lifted out a layer of pure zinc is deposited on the outer surface of the alloys. The zinc coating is chemically bonded to the steel and is sacrificial. The Galvanizers' Association can provide details of galvanizing baths around the country, but the average bath size is about 10 m long × 1.2 m wide × 2 m deep. The largest baths available in 2002 in the UK are 21 m × 1.5 m × 2.4 m and 7.6 m × 2.1 m × 3 m. The heat can cause distortions in fabricated, asymmetric or welded elements. Galvanizing is typically 85–140 μm thick and should be carried out to BS EN ISO 1461 and 14713. Paint coatings can be applied on top of the galvanizing for aesthetic or durability reasons and an etch primer is normally required to ensure that the paint properly adheres to the galvanizing.

Thermal spray   Degreased and blast cleaned (generally Sa 3) steel is sprayed with molten particles of aluminium or zinc. The coating is particulate and the pores normally need to be sealed with an organic sealant in order to prevent rust staining. Metal sprayed coatings are mechanically bonded to the steel and work partly by anodic protection and partly by barrier protection. There are no limits on the size of elements which can be coated and there are no distortion problems. Thermal spray is typically 150–200 μm thick in aluminium, 100–150 μm thick in zinc and should be carried out to BS EN 22063 and BS EN ISO 14713. Paint coatings can be applied for aesthetic or durability reasons. Bi-metallic corrosion issues should be considered when selecting fixings for aluminium sprayed elements in damp or external environments.

Weathering steel

Weathering steels are high strength, low alloy, weldable structural steels which form a protective rust coating in air that reaches a critical level within 2–5 years and prevents further corrosion. Cor-ten is the Corus proprietary brand of weathering steel, which has material properties comparable to S355, but the relevant material standard is BS EN 10155. To optimize the use of weathering steel, avoid contact with absorbent surfaces (e.g. concrete), prolonged wetting (e.g. north faces of buildings in the UK), burial in soils, contact with dissimilar metals and exposure to aggressive environments. Even if these conditions are met, rust staining can still affect adjacent materials during the first few years. Weathering bolts (ASTM A325, Type 3 or Cor-ten X) must be used for bolted connections. Standard black bolts should not be used as the zinc coating will be quickly consumed and the fastener corroded. Normal welding techniques can be used.

Stainless steel

Stainless steel is the most corrosion resistant of all the steels due to the presence of chromium in its alloys. The surface of the steel forms a self-healing invisible oxide layer which prevents ongoing corrosion and so the surface must be kept clean and exposed to provide the oxygen required to maintain the corrosion resistance. Stainless steel is resistant to most things, but special precautions should be taken in chlorinated environments. Alloying elements are added in different percentages to alter the durability properties:

SS 304	18% Cr, 10% Ni	Used for general cladding, brick support angles, etc.
SS 409	11% Cr	Sometimes used for lintels
SS 316	17% Cr, 12% Ni, 2.5% Mo	Used in medium marine/ aggressive environments
SS Duplex 2205	22% Cr, 5.5% Ni, 3% Mo	Used in extreme marine and industrial environments

Summary of methods of fire protection

System	Typical thickness[2] for 60 mins protection	Advantages	Disadvantages
Boards Up to 4 hours' protection. Most popular system in the UK	25–30 mm	Clean 'boxed in' appearance; dry application; factory quality boards; needs no steel surface preparation	High cost; complex fitting around details; slow to apply
Vermiculite concrete spray Up to 4 hours' protection. Second most popular system in the UK	20 mm	Cheap; easy on complex junctions; needs no steel surface preparation; often boards used on columns, with spray on the beams	Poor appearance; messy application needs screening; the wet trade will affect following trades; compatibility with corrosion protection needs to be checked
Intumescent paint Maximum 2 hours' protection. Charring starts at 200–250°C	1–4 mm[1]	Good aesthetic; shows off form of steel; easy to cover complex details; can be applied in shop or on site	High cost; not suited to all environments; short periods of resistance; soft, thick, easily damaged coatings; difficult to get a really high quality finish; compatibility with corrosion protection needs to be checked
Flexible blanket Cheap alternative to sprays	20–30 mm	Low cost; dry fixing	Not good aesthetics
Concrete encasement Generally only used when durability is a requirement	25–50 mm	Provides resistance to abrasion, impact, corrosion and weather exposure	Expensive; time consuming; heavy; large thickness required
Concrete filled columns Used for up to 2 hours' protection or to reduce intumescent paint thickness on hollow sections	–	Takes up less plan area; acts as permanent shutter; good durability	No data for CHS posts; minimum section size which can be protected 140 × 140SHS; expensive
Water filled columns Columns interconnected to allow convection cooling. Only used if no other option	–	Long periods of fire resistance	Expensive; lots of maintenance required to control water purity and chemical content
Block filled column webs Up to 30 minutes protection	–	Reduced cost; less plan area; good durability	Limited protection times; not advised for steel in partition walls

NOTES:

1. Coating thickness specified on the basis of the sections' dimensions and the number of sides that will be exposed to fire.
2. Castellated beams need about 20% more fire protection than is calculated for the basic parent material.

Preliminary sizing of steel elements

Typical span/depth ratios

Element	Typical span (L) m	Beam depth
Primary beams/trusses (heavy point loads)	4–12	L/10–15
Secondary beams/trusses (distributed loads)	4–20	L/15–25
Transfer beams/trusses carrying floors	6–30	L/10
Castellated beams	4–12	L/10–15
Plate girders	10–30	L/10–12
Vierendeel girders	6–18	L/8–10
Parallel chord roof trusses	10–100	L/12–20
Pitched roof trusses	8–20	L/5–10
Light roof beams	6–60	L/18–30
Conventional lattice roof girders	5–20	L/12–15
Space frames (allow for l/250 pre-camber)	10–100	L/15–30
Hot rolled universal column	single storey 2–8	L/20–25
	multi-storey 2–4	L/7–18
Hollow section column	single storey 2–8	L/20–35
	multi-storey 2–4	L/7–28
Lattice column	4–10	L/20–25
Portal leg and rafter (haunch depth <0.11)	9–60	L/35–40

Preliminary sizing

Beams
There are no shortcuts. Deflection will tend to govern long spans, while shear will govern short spans with heavy loading. Plate girders or trusses are used when the loading is beyond the capacity of rolled sections.

Columns – typical maximum column section size for braced frames

203 UC	Buildings 2 to 3 storeys high and spans up to 7 m.
254 UC	Buildings up to 5 storeys high.
305 UC	Buildings up to 8 storeys high or supports for low rise buildings with long spans.
354 UC	Buildings from 8 to 12 storeys high.

Columns – enhanced loads for preliminary axial design
An enhanced axial load for columns subject to out of balance loads can be used for preliminary design:

Top storey:	Total axial load $+ 4Y - Y + 2X - X$
Intermediate storey:	Total axial load $+ 2Y - Y + X - X$

Where $X - X$ and $Y - Y$ are the net axial load differences in each direction.

Trusses with parallel chord
Axial force in chord, $F = M_{applied}/d$ where d is the distance between the chord centroids. $I_{truss} = \Sigma (A_c d^2/4)$ where A_c is the area of each chord.

For equal chords this can be simplified to $I_{truss} = A_c d^2/2$.

Portal frames

The *Institution of Structural Engineers' Grey Book* for steel design gives the following preliminary method for sizing plastic portal frames with the following assumptions:

- Plastic hinges are formed at the eaves (in the stanchion) and near the apex, therefore Class 1 sections as defined in BS 5950 should be used.
- Moment at the end of the haunch is $0.87M_p$.
- Wind loading does not control the design.
- Stability of the frame should be checked separately.
- Load, W = vertical rafter load per metre run.

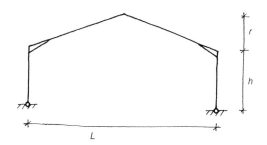

Horizontal base reaction, $H = H_{FR}WL$

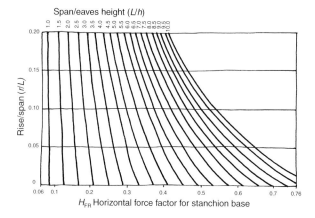

Design moment for rafter, $M_{p\ rafter} = M_{PR}WL^2$

Also consider the high axial force which will be in the rafter and design for combined axial and bending!

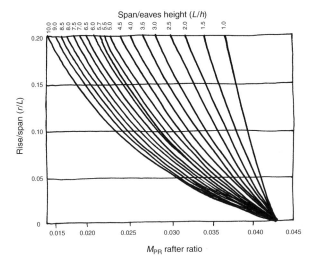

Design moment for stanchion, $M_{p\ stanchion} = M_{PL}WL^2$

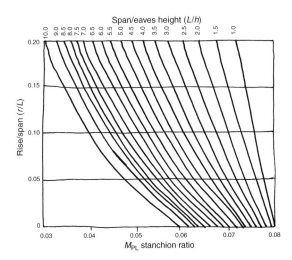

Source: IStructE (2002).

Steel design to BS 5950

BS 5950: Part 1 was written to allow designers to reduce conservatism in steel design. The resulting choice and complication of the available design methods has meant that sections are mainly designed using software or the SCI *Blue Book*. As the code is very detailed, the information about BS 5950 has been significantly summarized – covering only grade S275 steelwork and using the code's conservative design methods.

Corus Construction have an online interactive 'Blue Book' on their website giving capacities for all sections designed to BS 5950.

Partial safety factors

Load combination	Load type				
	Dead	**Imposed**	**Wind**	**Crane loads**	**Earth and water pressures**
Dead and imposed	1.4 or 1.0	1.6	–	–	1.4
Dead and wind	1.4 or 1.0	–	1.4	–	–
Dead and wind and imposed	1.2 or 1.0	1.2	1.2	–	–
Dead and crane loads	1.4	–	–	$V = 1.6$ $H = 1.6$ V and $H = 1.4$	–
Dead and imposed and crane loads	1.2	Crane $V = 1.4$ Crane $H = 1.2$		$V = 1.4$ $H = 1.2$ V and $H = 1.4$	
Dead and wind and crane loads	1.2	–	1.2	1.2	–
Forces due to temperature change	–	1.2	–	–	–
Exceptional snow load due to drifting	–	1.05	–	–	–

Source: BS 5950: Part 1: 2000.

Selected mild steel design strengths

Steel grade	Steel thickness less than or equal to mm	Design strength, p_y N/mm^2
S275	16	275
	40	265
	63	255
S355	16	355
	40	345
	63	335

Generally it is more economic to use S275 where it is required in small quantities (less than 40 tonnes), where deflection instead of strength limits design, or for members such as nominal ties where the extra strength is not required. In other cases it is more economical to consider S355.

Ductility and steel grading

In addition to the strength of the material, steel must be specified for a suitable ductility to avoid brittle fracture, which is controlled by the minimum service temperature, the thickness of steel, the steel grade, the type of detail and the stress and strain levels. Ductility is measured by the Charpy V notch test. In the UK the minimum service temperature expected to occur over the design life of the structure should be taken as −5°C for internal steelwork or −15°C for external steelwork. For steelwork in cold stores or cold climates appropriate lower temperatures should be selected. Tables 4, 5, 6 and 7 in BS 5950 give the detailed method for selection of the appropriate steel grade. Steel grading has become more important now that the UK construction industry is using more imported steel. The latest British Standard has revised the notation used to describe the grades of steel. The equivalent grades are set out below:

Current grading references BS 5950: Part 1: 2000 and BS EN 100 25: 1993				Superseded grading references* BS 5950: Part 1: 1990 and BS 4360: 1990				
Grade	Charpy test temperature °C	Steel use	Max steel thickness mm	Grade	Charpy test temperature °C	Steel use	Max steel thickness mm	
							<100 N/mm^2	>100 N/mm^2
S275	Untested	Internal only	25	43 A	Untested	Internal	50	25
						External	30	15
S275 JR	Room temp. 20°C	Internal only	30	43 B	Room temp. 20°C	Internal	50	25
						External	30	15
S275 J0	0°C	Internal	65	43 C	0°C	Internal	n/a	60
		External	54			External	80	40
S275 J2	−20°C	Internal	94	43 D	−20°C	Internal	n/a	n/a
		External	78			External	n/a	90

*Where the superseded equivalent for grades S355 and S460 are Grades 50 and 55 respectively.

Source: BS 5950: Part 1: 2000.

Section classification and local buckling

Sections are classified by BS 5950 depending on how their cross section behaves under compressive load. Structural sections in thinner plate will tend to buckle locally and this reduces the overall compressive strength of the section and means that the section cannot achieve its full plastic moment capacity. Sections with tall webs tend to be slender under axial compression, while cross sections with wide out-stand flanges tend to be slender in bending. Combined bending and compression can change the classification of a cross section to slender, when that cross section might not be slender under either bending or compression when applied independently.

For plastic design, the designer must therefore establish the classification of a section (for the given loading conditions) in order to select the appropriate design method from those available in BS 5950. For calculations without capacity tables or computer packages, this can mean many design iterations.

BS 5950 has four types of section classification:

Class 1:	Plastic	Cross sections with plastic hinge rotation capacity.
Class 2:	Compact	Cross sections with plastic moment capacity.
Class 3:	Semi-compact	Cross sections in which the stress at the extreme compression fibre can reach the design strength, but the plastic moment capacity cannot be developed.
Class 4:	Slender	Cross sections in which it is necessary to make explicit allowance for the effects of local buckling.

Tables 11 and 12 in BS 5950 classify different hot rolled and fabricated sections based on the limiting width to thickness ratios for each section class. None of the UB, UC, RSJ or PFC sections are slender in pure bending. Under pure axial compression, none of the UC, RSJ or PFC sections are slender, but some UB and hollow sections can be:

UB	Slender if $d/t > 40\varepsilon$
SHS and RHS (hot rolled)	Slender if $d/t > 40\varepsilon$
CHS	Slender if $D/t > 80\varepsilon^2$
Tee stem	Slender if $d/t > 18\varepsilon$

Where D = overall depth, t = plate thickness, d = web depth, p_y = design strength, $\varepsilon = \sqrt{275/P_y}$.

For simplicity only design methods for Class 1 and 2 sections are covered in this book.

Source: BS 5950: Part 1: 2000.

Tension members to BS 5950

Bolted connections: $P_t = (A_e - 0.5a_2)\, p_y$

Welded connections: $P_t = (A_e - 0.3a_2)\, p_y$

If $a_2 = A_g - a_1$ where A_g is the gross section area, A_e is the effective area (which is the net area multiplied by 1.2 for S275 steel, 1.1 for S355 or 1.0 for S460) and a_1 is the area of the connected part (web or flange, etc.).

Flexural members

Shear capacity, P_v

$P_v = 0.6 p_y A_v$

Where A_v is the shear area, which should be taken as:

tD	for rolled I sections (loaded parallel to the web) and rolled T sections
$AD/(D + B)$	for rectangular hollow sections
$t\,(D - T)$	for welded T sections
$0.6A$	for circular hollow sections
$0.9A$	solid bars and plates

t = web thickness, A = cross sectional area, D = overall depth, B = overall breadth, T = flange thickness.

If $d/t > 70$ for a rolled section, or >62 for a welded section, shear buckling must be allowed for (see BS 5950: clause 4.4.5).

Source: BS 5950: Part 1: 2000.

Moment capacity M_C to BS 5950

The basic moment capacity (M_c) depends on the provision of full lateral restraint and the interaction of shear and bending stresses. M_c is limited to $1.2p_yZ$ to avoid irreversible deformation under serviceability loads. Full lateral restraint can be assumed if the construction fixed to the compression flange is capable of resisting not less than 2.5% of the maximum compression force in the flange, distributed uniformly along the length of the flange. Moment capacity (M_c) is generally the controlling capacity for class 1 and 2 sections in the following cases:

● Bending about the minor axis.
● CHS, SHS or small solid circular or square bars.
● RHS in some cases given in clause 4.3.6.1 of BS 5950.
● UB, UC, RSJ, PFC, SHS or RHS if $\lambda < 34$ for S275 steel and $\lambda < 30$ for S355 steel in Class 1 and 2 sections where $\lambda = L_E/r$.

Low shear ($F_v < 0.6P_v$) $M_c = p_yS$

High shear ($F_v > 0.6P_v$) $M_c = p_y (S - \rho S_v)$

Where $\rho = \left(2\dfrac{F_v}{P_v} - 1\right)^2$ and $S_v =$ the plastic modulus of the shear area used to calculate P_v.

Lateral torsional buckling capacity M_b

Lateral torsional buckling (LTB) occurs in tall sections or long beams in bending if not enough restraint is provided to the compression flange. Instability of the compression flange results in buckling of the beam, preventing the section from developing its full plastic capacity, M_c. The reduced bending moment capacity, M_b, depends on the slenderness of the section, λ_{LT}.

A simplified and conservative method of calculating M_b for rolled sections uses D/T and L_e/r_y to determine an ultimate bending stress p_b (from the following graph) where $M_b = p_bS_x$ for Class 1 and 2 sections.

Source: BS 5950: Part 1: 2000.

Ultimate bending strengths for rolled sections, p_b (in S275) BS 5950

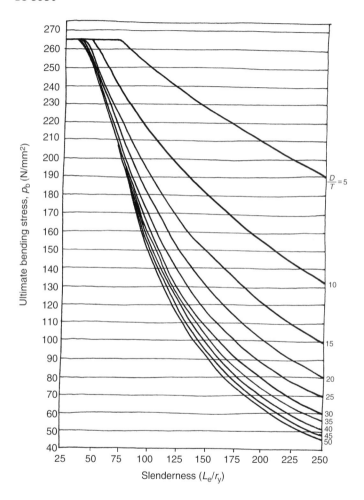

Compression members to BS 5950

The compression capacity of Class 1 and 2 sections can be calculated as $P_c = A_g p_c$, where A_g is the gross area of the section and p_c can be estimated depending on the expected buckling axis and the section type for steel of $\leq 40\,mm$ thickness.

Type of section	Strut curve for value of p_c	
	Axis of buckling	
	x–x	y–y
Hot finished structural hollow section	a	a
Rolled I section	a	b
Rolled H section	b	c
Round, square or flat bar	b	b
Rolled angle, channel or T section/paired rolled sections/compound rolled sections	Any axis: c	

Ultimate compression stresses for rolled sections, p_c

Ultimate compression stresses for rolled sections, p_c (in S275) BS 5950

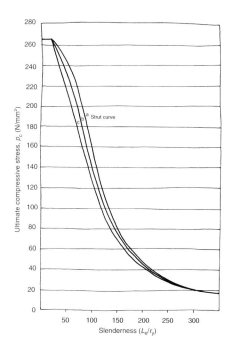

Combined bending and compression to BS 5950

Although each section should have its classification checked for combined bending and axial compression, the capacities from the previous tables can be checked against the following simplified relationship for section Classes 1 and 2:

$$\frac{F_c}{P_c} + \frac{M_x}{M_{cx} \text{ or } M_b} + \frac{M_y}{M_{cy}} < 1.0$$

Section 4.8 in BS 5950 should be referred to in detail for all the relevant checks.

Connections to BS 5950

Welded connections

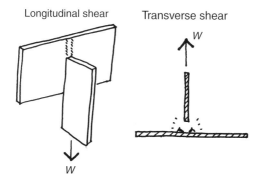

Longitudinal shear   Transverse shear

W

W

The resultant of combined longitudinal and transverse forces should be checked:

$$\left(\frac{F_{\text{L}}}{P_{\text{L}}}\right)^2 + \left(\frac{F_{\text{T}}}{P_{\text{T}}}\right)^2 < 1.0.$$

Ultimate fillet weld capacities for S275 elements joined at 90°

Leg length s mm	Throat thickness $a = 0.7s$ mm	Longitudinal capacity* $P_{\text{L}} = p_{\text{w}}a$ kN/mm	Transverse capacity* $P_{\text{L}} = p_{\text{w}}aK$ kN/mm
4	2.8	0.616	0.770
6	4.2	0.924	1.155
8	5.6	1.232	1.540
12	8.4	1.848	2.310

*Based on values for S275, $p_{\text{w}} = 220\,\text{N/mm}^2$ and $K = 1.25$.

Bolted connections to BS 5950

Limiting bolt spacings

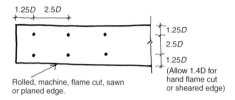

1.25D   2.5D

1.25D
2.5D
1.25D
(Allow 1.4D for
hand flame cut
or sheared edge)

Rolled, machine, flame cut, sawn
or planed edge.

Direct shear

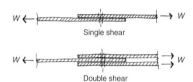

$W \leftarrow$ ⟶ W

Single shear

$W \leftarrow$ ⟶ W

Double shear

Simple moment connection bolt groups

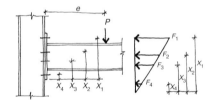

e

P

F_1
F_2
F_3
F_4

x_4 x_3 x_2 x_1

x_1
x_2
x_3
x_4

$$M_{cap} = \left(\frac{\text{no. rows of bolts}}{x_1} \right) P_t \sum x_i^2$$

$$V = \frac{P}{n}$$

$$F_n = P_t \frac{x_n}{x_{n-1}}$$

Where $x_1 = \max x_i$ and x_i = depth from point of rotation to centre of bolt being considered, P_t is the tension capacity of the bolts, n is the number of bolts, V is the shear on
each bolt and F is the tension in each bolt. This is a simplified analysis which assumes
that the bolt furthest from the point of rotation carries the most load. As the connection elements are likely to be flexible, this is unlikely to be the case; however, more complicated analysis requires a computer or standard tables.

Bolt capacity checks For bolts in shear or tension see the following tabulated values.
For bolts in shear and tension check: $(F_v/P_v) + (F_t/P_t) \leq 1.4$ where F indicates the factored design load and P indicates the ultimate bolt capacity.

Selected ultimate bolt capacities for non-pre-loaded ordinary bolts in S275 steel to BS 5950

Diameter of bolt, φ mm	Tensile stress area mm²	Tension capacity kN	Shear capacity		Bearing capacity for end distance = 2φ kN						
			Single kN	Double kN	Thickness of steel passed through mm						
					5	6	8	10	12	15	20
Grade 4.6											
6	20.1	3.9	3.2	6.4	*13.8*	*16.6*	*22.1*	*27.6*	*33.1*	*41.4*	*55.2*
8	36.6	7.0	5.9	11.7	*18.4*	*22.1*	*29.4*	*36.8*	*44.2*	*55.2*	*73.6*
10	58	11.1	9.3	18.6	*23.0*	*27.6*	*36.8*	*46.0*	*55.2*	*69.0*	*92.0*
12	84.3	16.2	13.5	27.0	*27.6*	*33.1*	*44.2*	*55.2*	*66.2*	*82.8*	*110.4*
16	157	30.1	25.1	50.2	*36.8*	*44.2*	*58.9*	*73.6*	*88.3*	*110.4*	*147.2*
20	245	47.0	39.2	78.4	*46.0*	*55.2*	*73.6*	*92.0*	*110.4*	*138.0*	*184.0*
24	353	67.8	56.5	113.0	**55.2**	*66.2*	*88.3*	*110.4*	*132.5*	*165.6*	*220.8*
30	561	107.7	89.8	179.5	**69.0**	**82.8**	*110.4*	*138.0*	*165.6*	*207.0*	*276.0*
Grade 8.8											
6	20.1	9.0	7.5	15.1	13.8	*16.6*	*22.1*	*27.6*	*33.1*	*41.4*	*55.2*
8	36.6	16.4	13.7	27.5	18.4	22.1	*29.4*	*36.8*	*44.2*	*55.2*	*73.6*
10	58	26.0	21.8	43.5	23.0	27.6	*36.8*	*46.0*	*55.2*	*69.0*	*92.0*
12	84.3	37.8	31.6	63.2	**27.6**	33.1	*44.2*	*55.2*	*66.2*	*82.8*	*110.4*
16	157	70.3	58.9	117.8	**36.8**	**44.2**	58.9	*73.6*	*88.3*	*110.4*	*147.2*
20	245	109.8	91.9	183.8	**46.0**	**55.2**	**73.6**	92.0	*110.4*	*138.0*	*184.0*
24	353	158.1	132.4	264.8	**55.2**	**66.2**	**88.3**	110.4	*132.5*	*165.6*	*220.8*
30	561	251.3	210.4	420.8	**69.0**	**82.8**	**110.4**	**138.0**	**165.6**	**207.0**	*276.0*

NOTES:
- 2 mm clearance holes for φ < 24 or 3 mm clearance holes for φ ≥ 24.
- Tabulated tension capacities are nominal tension capacity = $0.8A_t p_t$ which accounts for prying forces.
- Bearing values shown in **bold** are less than the single shear capacity of the bolt.
- Bearing values shown in *italic* are less than the double shear capacity of the bolt.
- Multiply tabulated bearing values by 0.7 if oversized or short slotted holes are used.
- Multiply tabulated bearing values by 0.5 if kidney shaped or long slotted holes are used.
- Shear capacity should be reduced for large packing, grip lengths or long joints.
- Grade 4.6 $p_s = 160 \text{N/mm}^2$, $p_t = 240 \text{N/mm}^2$.
- Grade 8.8 $p_s = 375 \text{N/mm}^2$, $p_t = 560 \text{N/mm}^2$.
- Total packing at a shear plane should not exceed $\dfrac{4\phi}{3}$.

Selected ultimate bolt capacities for non-pre-loaded countersunk bolts in S275 steel to BS 5950

Diameter of bolt, φ mm	Tensile stress area mm²	Tension capacity kN	Shear capacity		Bearing capacity for end distance = 2φ kN						
			Single kN	Double kN	Thickness of steel passed through (mm)						
					5	6	8	10	12	15	20
Grade 4.6											
6	20.1	3.9	3.2	6.4	8.6	11.3	16.8	22.4	27.9	36.2	50.0
8	36.6	7.0	5.9	11.7	–	12.9	20.2	27.6	35.0	46.0	64.4
10	58	11.1	9.3	18.6	–	–	21.9	31.1	40.3	54.1	77.1
12	84.3	16.2	13.5	27.0	–	–	–	34.5	45.5	62.1	89.7
16	157	30.1	25.1	50.2	–	–	–	–	55.2	77.3	114.1
20	245	47.0	39.2	78.4	–	–	–	–	62.1	89.7	135.7
24	353	67.8	56.5	113.0	–	–	–	–	–	85.6	140.8
Grade 8.8											
6	20.1	9.0	7.5	15.1	8.6	11.3	16.8	22.4	27.9	36.2	50.0
8	36.6	16.4	13.7	27.5	–	**12.9**	20.2	27.6	35.0	46.0	64.4
10	58	26.0	21.8	43.5	–	–	21.9	31.1	40.3	54.1	77.1
12	84.3	37.8	31.6	63.2	–	–	–	34.5	45.5	62.1	89.7
16	157	70.3	58.9	117.8	–	–	–	–	**55.2**	77.3	114.1
20	245	109.8	91.9	183.8	–	–	–	–	**62.1**	89.7	135.7
24	353	158.1	132.4	264.8	–	–	–	–	–	**85.6**	140.8

NOTES:

- Values are omitted from the table where the bolt head is too deep to be countersunk into the thickness of the plate.
- 2 mm clearance holes for φ < 24 or 3 mm clearance holes for φ ≥ 24.
- Tabulated tension capacities are nominal tension capacity = $0.8A_t p_t$ which accounts for prying forces.
- Bearing values shown in **bold** are less than the single shear capacity of the bolt.
- Bearing values shown in *italic* are less than the double shear capacity of the bolt.
- Multiply tabulated bearing values by 0.7 if oversized or short slotted holes are used.
- Multiply tabulated bearing values by 0.5 if kidney shaped or long slotted holes are used.
- Shear capacity should be reduced for large packing, grip lengths or long joints.
- Grade 4.6 p_s = 160 N/mm², p_t = 240 N/mm².
- Grade 8.8 p_s = 375 N/mm², p_t = 560 N/mm².
- Total packing at a shear plane should not exceed $\frac{4\phi}{3}$.
- Table based on Unbrako machine screw dimensions.

Steel design to BS 449

BS 449: Part 2 is the 'old' steel design code issued in 1969 but it is (with amendments) still current. The code is based on elastic bending and working stresses and is very simple to use. It is therefore invaluable for preliminary design, for simple steel elements and for checking existing structures. It is normal to compare the applied and allowable stresses. BS 449 refers to the old steel grades where Grade 43 is S275, Grade 50 is S355 and Grade 55 is S460.

Notation for BS 449: Part 2

Symbols		Stress subscripts	
f	Applied stress	c or bc	Compression or bending compression
P	Permissible stress	t or bt	Tension or bending tension
l/r	Slenderness ratio	q	Shear
D	Overall section depth	b	Bearing
t	Flange thickness	e	Equivalent

Allowable stresses

The allowable stresses may be exceeded by 25% where the member has to resist an increase in stress which is solely due to wind forces – provided that the stresses in the section before considering wind are within the basic allowable limits.

Applied stresses are calculated using the gross elastic properties of the section, Z or A, where appropriate.

Allowable stress in axial tension P_t

Form	Steel grade	Thickness of steel mm	P_t N/mm^2
Sections, bars, plates, wide flats and hollow sections	43 (S275)	$t \leq 40$	170
		$40 < t \leq 100$	155

Source: BS 449: Part 2: 1969.

Maximum allowable bending stresses P_{bc} or P_{bt} to BS 449

Form	Steel grade	Thickness of steel mm	P_{bc} or P_{bt} N/mm²
Sections, bars, plates, wide flats and hollow sections	43 (S275)	$t \leq 40$	180
Compound beams – hot rolled sections with additional plates		$40 < t \leq 100$	165
Double channel sections acting as an I beam			
Plate girders	43 (S275)	$t \leq 40$	170
		$40 < t \leq 100$	155
Slab bases	All steels		185

Upstand webs or flanges in compression have a reduced capacity and need to be checked in accordance with clause 20, BS 449. These tabulated values of P_{bc} can be used only where full lateral restraint is provided, where bending is about the minor axis or for hollow sections in bending.

Source: BS 449: Part 2: Table 2: 1969.

Allowable compressive bending stresses to BS 449

The maximum allowable bending stress is reduced as the slenderness increases, to allow for the effects of buckling. The reduced allowable bending stress, P_{bc}, can be obtained from the following graph from the ratio of depth of section to thickness of flange (D/T) and the slenderness ($\lambda = L_e/r$).

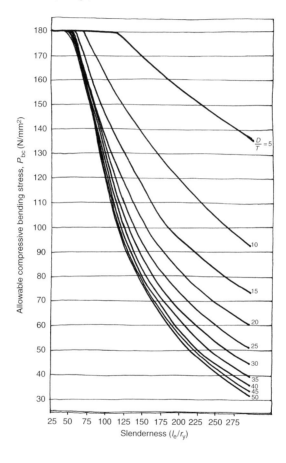

Allowable compressive stresses to BS 449

For uncased compression members, allowable compressive stresses must be reduced by 10% for thick steel sections: if $t > 40$ mm for Grade 43 (S275), $t > 63$ mm for Grade 50 (S355) and $t > 25$ mm for Grade 55 (S460). The allowable axial stress, P_c, reduces as the slenderness of the element increases as shown in the following chart:

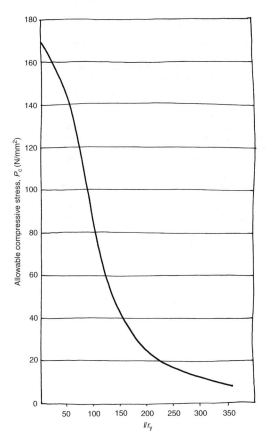

Allowable average shear stress P_v in unstiffened webs to BS 449

Form	Steel grade	Thickness mm	P_v* N/mm²
Sections, bars, plates, wide flats and hollow sections	43 (S275)	$d \leq 40$	110
		$40 < d \leq 100$	100
	50 (S355)	$d \leq 63$	140
		$63 < d \leq 100$	130
	55 (S460)	$d \leq 25$	170

*See Table 12 in BS 449: Part 2 for allowable average shear stress in stiffened webs.

Section capacity checks to BS 449

Combined bending and axial load

Compression: $\dfrac{f_c}{P_c} + \dfrac{f_{bc_x}}{P_{bc_x}} + \dfrac{f_{bc_y}}{P_{bc_y}} \leq 1.0$

Tension: $\dfrac{f_t}{P_t} + \dfrac{f_{bt}}{P_{bt}} \leq 1.0$ and $\dfrac{f_{bc_x}}{P_{bc_x}} + \dfrac{f_{bc_y}}{P_{bc_y}} \leq 1.0$

Combined bending and shear

$f_e = \sqrt{(f_{bt}^2 + 3f_q^2)}$ or $f_e = \sqrt{(f_{bc}^2 + 3f_q^2)}$ and $f_e < P_e$ and $(f_{bc}/P_o)^2 + (f_q'/P_q')^2 \leq 1.25$

Where f_e is the equivalent stress, f_q' is the average shear in the web, P_o is defined in BS 449 subclause 20 item 2b iii and P_q' is defined in clause 23. From BS 449: Table 1, the allowable equivalent stress $P_e = 250\,N/mm^2$ for Grade 43 (S275) steel $< 40\,mm$ thick.

Combined bending, shear and bearing

$f_e = \sqrt{(f_{bt}^2 + f_b^2 + f_{bt}f_b + 3f_q^2)}$ or $f_e = \sqrt{(f_{bc}^2 + f_b^2 + f_{bc}f_b + 3f_q^2)}$ and $f_e < P_e$ and $(f_{bc}/P_o)^2 + (f_q'/P_q')^2 + (f_{cw}/P_{cw}) \leq 1.25$

Source: BS 449: Part 2: 1969.

Connections to BS 449

Selected fillet weld working capacities for Grade 43 (S275) steel

Leg length s mm	Throat thickness a = 0.7s mm	Weld capacity* kN/mm
4	2.8	0.32
6	4.2	0.48
8	5.6	0.64
12	8.4	0.97

*When a weld is subject to a combination of stresses, the combined effect should be checked using the same checks as used for combined loads on sections to BS 449.

Selected full penetration butt weld working capacities for Grade 43 (S275) steel

Thickness mm	Shear capacity kN/mm	Tension or compression capacity* kN/mm
6	0.60	0.93
15	1.50	2.33
20	2.00	3.10
30	3.00	4.65

*When a weld is subject to a combination of stresses, the combined effect should be checked using the same checks as used for combined loads on sections to BS 449.

Source: BS 449: Part 2: 1969.

Allowable stresses in non-pre-loaded bolts to BS 449

Description	Bolt grade	Axial tension N/mm²	Shear N/mm²	Bearing N/mm²
Close tolerance and turned bolts	4.6	120	100	300
	8.8	280	230	350
Bolts in clearance holes	4.6	120	80	250
	8.8	280	187	350

Allowable stresses on connected parts of bolted connections to BS 449

Description	Allowable stresses on connected parts for different steel grades N/mm²		
	43 (S275)	50 (S355)	55 (S460)
Close tolerance and turned bolts	300	420	480
Bolts in clearance holes	250	350	400

Bolted connection capacity check for combined tension and shear to BS 449

$$\frac{f_t}{P_t} + \frac{f_s}{P_s} \leq 1.4$$

Source: BS 449: Part 2: 1969.

Selected working load bolt capacities for non-pre-loaded ordinary bolts in grade 43 (S275) steel to BS 449

Diameter of bolt, φ mm	Tensile stress area mm²	Tension capacity kN	Shear capacity		Bearing capacity for end distance = 2φ kN						
			Single kN	Double kN	Thickness of steel passed through (mm)						
					5	6	8	10	12	15	20
Grade 4.6											
6	20.1	1.9	1.6	3.2	7.5	9.0	12.0	15.0	18.0	22.5	30.0
8	36.6	3.5	2.9	5.9	10.0	12.0	16.0	20.0	24.0	30.0	40.0
10	58	5.6	4.6	9.3	12.5	15.0	20.0	25.0	30.0	37.5	50.0
12	84.3	8.1	6.7	13.5	15.0	18.0	24.0	30.0	36.0	45.0	60.0
16	157	15.1	12.6	25.1	*20.0*	*24.0*	*32.0*	40.0	48.0	60.0	80.0
20	245	23.5	19.6	39.2	*25.0*	*30.0*	40.0	50.0	60.0	75.0	100.0
24	353	33.9	28.2	56.5	*30.0*	*36.6*	*48.0*	60.0	72.0	90.0	120.0
30	561	53.9	44.9	89.8	*37.5*	*45.0*	*60.0*	*75.0*	90.0	112.5	150.0
Grade 8.8											
6	20.1	4.5	3.8	7.5	7.5	9.0	12.0	15.0	18.0	22.5	30.0
8	36.6	8.2	6.8	13.7	*10.0*	*12.0*	16.0	20.0	24.0	30.0	40.0
10	58	13.0	10.8	21.7	*12.5*	*15.0*	20.0	25.0	30.0	37.5	50.0
12	84.3	18.9	15.8	31.5	**15.0**	*18.0*	*24.0*	*30.0*	36.0	45.0	60.0
16	157	35.2	29.4	58.7	**20.0**	**24.0**	*32.0*	*40.0*	*48.0*	60.0	80.0
20	245	54.9	45.8	91.6	**25.0**	**30.0**	**40.0**	*50.0*	*60.0*	*75.0*	100.0
24	353	79.1	66.0	132.0	**30.0**	**36.0**	**48.0**	**60.0**	*72.0*	*90.0*	*120.0*
30	561	125.7	104.9	209.8	**37.5**	**45.0**	**60.0**	**75.0**	**90.0**	*112.5*	*150.0*

NOTES:

- 2 mm clearance holes for φ < 24 or 3 mm clearance holes for φ ≥ 24.
- Bearing values shown in **bold** are less than the single shear capacity of the bolt.
- Bearing values shown in *italic* are less than the double shear capacity of the bolt.
- Multiply tabulated bearing values by 0.7 if oversized or short slotted holes are used.
- Multiply tabulated bearing values by 0.5 if kidney shaped or long slotted holes are used.
- Shear capacity should be reduced for large packing, grip lengths or long joints.
- Tabulated tension capacities are nominal tension capacity = $0.8A_t p_t$ which accounts for prying forces.

Selected working load bolt capacities for non-pre-loaded countersunk ordinary bolts in grade 43 (S275)

Diameter of bolt, φ mm	Tensile stress area mm²	Tension capacity kN	Shear capacity		Bearing capacity for end distance = 2φ kN						
			Single kN	Double kN	Thickness of steel passed through (mm)						
					5	6	8	10	12	15	20
Grade 4.6											
6	20.1	1.9	1.6	3.2	4.7	6.2	9.2	12.2	15.2	19.7	27.2
8	36.6	3.5	2.9	5.9	–	7.0	11.0	15.0	19.0	25.0	35.0
10	58	5.6	4.6	9.3	–	–	11.9	16.9	21.9	29.4	41.9
12	84.3	8.1	6.7	13.5	–	–	–	18.8	24.8	33.8	48.8
16	157	15.1	12.6	25.1	–	–	–	–	30.0	42.0	62.0
20	245	23.5	19.6	39.2	–	–	–	–	33.8	48.8	73.8
24	353	33.9	28.2	56.5	–	–	–	–	–	46.5	76.5
Grade 8.8											
6	20.1	4.5	3.8	7.5	4.7	6.2	9.2	12.2	15.2	19.7	27.2
8	36.6	8.2	6.8	13.7	–	**7.0**	*11.0*	15.0	19.0	25.0	35.0
10	58	13.0	10.8	21.7	–	–	*11.9*	16.9	21.9	29.4	41.9
12	84.3	18.9	15.8	31.5	–	–	–	*18.8*	24.8	33.8	48.8
16	157	35.2	29.4	58.7	–	–	–	–	**30.0**	*42.0*	62.0
20	245	54.9	45.8	91.6	–	–	–	–	**33.8**	**48.8**	*73.8*
24	353	79.1	66.0	132.0	–	–	–	–	–	**46.5**	*76.5*

NOTES:
- Values are omitted from the table where the bolt head is too deep to be countersunk into the thickness of the plate.
- 2 mm clearance holes for φ < 24 or 3 mm clearance holes for φ ≥ 24.
- Tabulated tension capacities are nominal tension capacity = $0.8A_t p_t$ which accounts for prying forces.
- Bearing values shown in **bold** are less than the single shear capacity of the bolt.
- Bearing values shown in *italic* are less than the double shear capacity of the bolt.
- Multiply tabulated bearing values by 0.7 if oversized or short slotted holes are used.
- Multiply tabulated bearing values by 0.5 if kidney shaped or long slotted holes are used.
- Shear capacity should be reduced for large packing, grip lengths or long joints.
- Grade 4.6 $p_s = 160 \text{N/mm}^2$, $p_t = 240 \text{N/mm}^2$.
- Grade 8.8 $p_s = 375 \text{N/mm}^2$, $p_t = 560 \text{N/mm}^2$.
- Total packing at a shear plane should not exceed $\dfrac{4\phi}{3}$.
- Table based on Unbrako machine screw dimensions.

Stainless steel to BS 5950

Stainless steels are a family of corrosion and heat resistant steels containing a minimum of 10.5% chromium which results in the formation of a very thin self-healing transparent skin of chromium oxide – which is described as a passive layer. Alloy proportions can be varied to produce different grades of material with differing strength and corrosion properties. The stability of the passive layer depends on the alloy composition. There are five basic groups: austenitic, ferritic, duplex, martensitic and precipitation hardened. Of these, only austenitic and Duplex are really suitable for structural use.

Austenitic

Austenitic is the most widely used for structural applications and contains 17–18% chromium, 8–11% nickel and sometimes molybdenum. Austenitic stainless steel has good corrosion resistance, high ductility and can be readily cold formed or welded. Commonly used alloys are 304L (European grade 1.4301) and 316L (European grade 1.4401).

Duplex

Duplex stainless steels are so named because they share the strength and corrosion resistance properties of both the austenitic and ferritic grades. They typically contain 21–26% chromium, 4–8% nickel and 0.1–4.5% molybdenum. These steels are readily weldable but are not so easily cold rolled. Duplex stainless steel is normally used where an element is under high stress in a severely corrosive environment. A commonly used alloy is Duplex 2205 (European grade 1.44062).

Material properties

The material properties vary between cast, hot rolled and cold rolled elements.

Density	$78–80\,kN/m^3$
Tensile strength	$200–450\,N/mm^2$ 0.2% proof stress depending on grade.
Poisson's ratio	0.3
Modulus of elasticity	E varies with the stress in the section and the direction of the stresses. As the stress increases, the stiffness decreases and therefore deflection calculations must be done on the basis of the secant modulus.
Shear modulus	$76.9\,kN/mm^2$
Linear coefficient of thermal expansion	$17 \times 10^{-6}/°C$ for 304L (1.4301) $16.5 \times 10^{-6}/°C$ for 316L (1.4401) $13 \times 10^{-6}/°C$ for Duplex 2205 (1.4462)
Ductility	Stainless steel is much tougher than mild steel and so BS 5950 does not apply any limit on the thickness of stainless steel sections as it does for mild steel.

Elastic properties of stainless steel alloys for design

The secant modulus, $E_s = \dfrac{(E_{s1} + E_{s2})}{2}$, where

$$E_{si} = \cfrac{E}{\left[1 + k\left(\dfrac{f_{1 \text{ or } 2}}{P_y}\right)^m\right]}$$

where i = 1 or 2, $k = 0.002E/P_y$ and m is a constant.

Values of the secant modulus are calculated below for different stress ratios (f_i/P_y)

Values of secant modulus for selected stainless steel alloys for structural design

Stress ratio* $\dfrac{f_i}{P_y}$	Secant modulus kN/mm^2					
	304L		316L		Duplex 2205	
	Longitudinal	Transverse	Longitudinal	Transverse	Longitudinal	Transverse
0.0	200	200	190	195	200	205
0.2	200	200	190	195	200	205
0.3	199	200	190	195	199	204
0.4	197	200	188	195	196	200
0.5	191	198	184	193	189	194
0.6	176	191	174	189	179	183
0.7	152	173	154	174	165	168

*Where i = 1 or 2 for the applied stress in the tension and compression flanges respectively.

Typical stock stainless steel sections

There is no UK-based manufacturer of stainless steel and so all stainless steel sections are imported. Two importers who will send out information on the sections they produce are Valbruna and IMS Group. The sections available are limited. IMS has a larger range including hot rolled equal angles (from 20 × 20 × 3 up to 100 × 100 × 10), unequal angles (20 × 10 × 3 up to 200 × 100 × 13), I beams (80 × 46 up to 400 × 180), H beams (50 × 50 up to 300 × 300), channels (20 × 10 up to 400 × 110) and tees (20 × 20 × 3 up to 120 × 120 × 13) in 1.4301 and 1.4571. Valbruna has a smaller selection of plate, bars and angles in 1.4301 and 1.4404. Perchcourt are stainless steel section stockists based in the Midlands who can supply fabricators. Check their website for availability.

Source: Nickel Development Institute (1994).

Durability and fire resistance

Suggested grades of stainless steel for different atmospheric conditions

Stainless steel grade	Location											
	Rural			Urban			Industrial			Marine		
	Low	Med	High	Low	Med	High	Low	Med	High	Low	Med	High
304L (1.4301)	✓	✓	✓	✓	✓	(✓)	(✓)	(✓)	X	✓	(✓)	X
316L (1.4401)	○	○	○	○	✓	✓	✓	✓	(✓)	✓	✓	(✓)
Duplex 2205 (1.4462)	○	○	○	○	○	○	○	○	✓	○	○	✓

Where: ✓ = optimum specification, (✓) = may require additional protection, X = unsuitable, ○ = overspecified.

Note that this table does not apply to chlorinated environments which are very corrosive to stainless steel. Grade 304L (1.4301) can tarnish and is generally only used where aesthetics are not important; however, marine Grade 316L (1.4401) will maintain a shiny surface finish.

Corrosion mechanisms

Durability can be reduced by heat treatment and welding. The surface of the steel forms a self-healing invisible oxide layer which prevents ongoing corrosion and so the surface must be kept clean and exposed to provide the oxygen required to maintain the corrosion resistance.

Pitting   Mostly results in the staining of architectural components and is not normally a structural problem. However, chloride attack can cause pitting which can cause cracking and eventual failure. Alloys rich in molybdenum should be used to resist chloride attack.

Crevice corrosion   Chloride attack and lack of oxygen in small crevices, e.g. between nuts and washers.

Bi-metallic effects   The larger the cathode, the greater the rate of attack. Mild steel bolts in a stainless steel assembly would be subject to very aggressive attack. Austenitic grades typically only react with copper to produce an unsightly white powder, with little structural effect. Prevent bi-metallic contact by using paint or tape to exclude water as well as using isolation gaskets, nylon/Teflon bushes and washers.

Fire resistance

Stainless steels retain more of their strength and stiffness than mild steels in fire conditions, but typically as stainless steel structure is normally exposed, its fire resistance generally needs to be calculated as part of a fire engineered scheme.

Source: Nickel Development Institute (1994).

Preliminary sizing

Assume a reduced Young's modulus depending on how heavily stressed the section will be and assume an approximate value of maximum bending stress for working loads of 130 N/mm^2. A section size can then be selected for checking to BS 5950.

Stainless steel design to BS 5950: Part 1

The design is based on ultimate loads calculated on the same partial safety factors as for mild steel.

Ultimate mechanical properties for stainless steel design to BS 5950

Alloy type	Steel desig-nation	European grade (UK grade)	Minimum 0.2% proof stress N/mm^2	Ultimate tensile strength N/mm^2	Minimum elongation after fracture %
Basic austenitic[1]	X5CrNi 18-9	304L (1.4301)	210	520–720	45
Molybdenum austenitic[2]	X2CrNiMo 17-12-2	316L (1.4401)	220	520–670	40
Duplex	X2CrNi MoN 22-5-3	Duplex 2205 (1.4462)	460	640–840	20

Notes:
1. Most commonly used for structural purposes.
2. Widely used in more corrosive situations.

The alloys listed in the table above are low carbon alloys which provide good corrosion resistance after welding and fabrication.

As for mild steel, the element cross section must be classified to BS 5950: Part 1 in order to establish the appropriate design method. Generally this method is as given for mild steels; however, as there are few standard section shapes, the classification and design methods can be laborious.

Source: Nickel Development Institute (1994).

Connections

Bolted and welded connections can be used. Design data for fillet and butt welds requires detailed information about which particular welding method is to be used. The information about bolted connections is more general.

Bolted connections

Requirements for stainless steel fasteners are set out in BS EN ISO 3506 which split fixings into three groups: A = Austenitic, F = Ferritic and C = Martensitic. Grade A fasteners are normally used for structural applications. Grade A2 is equivalent to Grade 304L (1.4301) with a 0.2% proof stress of $210\,N/mm^2$ and Grade A4 is equivalent to Grade 316L (1.4401) with a 0.2% proof stress of $450\,N/mm^2$. There are three further property classes within Grade A: 50, 70 and 80 to BS EN ISO 3506. An approximate ultimate bearing strength for connected parts can be taken as $460\,N/mm^2$ for preliminary sizing.

Ultimate stress values for bolted connection design

Grade A property class	Shear strength* N/mm^2	Bearing strength* N/mm^2	Tensile strength* N/mm^2
50	140	510	210
70 (most common)	310	820	450
80	380	1000	560

*These values are appropriate with bolt diameters less than M24 and bolts less than 8 diameters long.

Sources: Nickel Development Institute (1994).

10
Composite Steel and Concrete

Composite steel and concrete flooring, as used today, was developed in the 1960s to economically increase the spans of steel framed floors while minimizing the required structural depths.

Composite flooring elements

Concrete slab   There are various types of slab: solid in situ, in situ on profiled metal deck and in situ on precast concrete units. Solid slabs are typically 125–150 mm thick and require formwork. The precast and metal deck systems both act as permanent form-work, which may need propping to control deflections. The profiled metal deck sheets have a 50–60 mm depth to create a 115–175 mm slab, which can span 2.5 to 3.6 m. Precast concrete units 75–100 mm thick with 50–200 mm topping can span 3–8 m.

Steelwork   Generally the steel section is sized to support the wet concrete and con-struction loads with limited deflection, followed by the full design loading on the com-posite member. Secondary beams carry the deck and are in turn supported on primary beams which are supported on the columns. The steel beams can be designed as simply supported or continuous. Long span beams can be adversely affected by vibration, and should be used with caution in dynamic loading situations.

Shear studs   Typically 19 mm diameter and 95 mm or 120 mm tall. Other heights for deep profiled decks are available with longer lead times. Larger diameter studs are avail-able but not many subcontractors have the automatic welding guns to fix them. Welded studs will carry about twice the load of proprietary 'shot fired' studs.

Economic arrangement   Secondary beam spacing is limited to about arrangement 2.5–3 m in order to keep the slab thickness down and its fire resistance up. The most economic geometrical arrangement is for the primary span to be about 3/4 of the secondary span.

Summary of material properties

The basic properties of steel and concrete are as set out in their separate sections.

Concrete grade	Normal weight concrete RC 30–50 and lightweight concrete RC 25–40.	
Density	Normal weight concrete 24 kN/m^3 and lightweight concrete 17 kN/m^3.	
Modular ratio ($\alpha_E = E_S/E_C$)	Normal concrete	$\alpha_E = 6$ short term and $\alpha_E = 18$ long term.
	Lightweight concrete	$\alpha_E = 10$ short term $\alpha_E = 25$ long term.
Steel grade	S275 is used where it is required in small quantities (less than 40 tonnes) or where deflection, not strength, limits the design. Otherwise S355 is more economical, but will increase the minimum number of shear studs which are required by the code.	

Durability and fire resistance

- The basic durability requirements of steel and concrete are as set out in their separate sections.
- Concrete slabs have an inherent fire resistance. The slab thickness may be controlled by the minimum thickness required for fire separation between floors, rather than by deflection or strength.
- Reinforcing mesh is generally added to the top face of the slab to control surface cracking. The minimum required is 0.1 per cent of the concrete area, but more may be required for continuous spans or in some fire conditions.
- Additional bars are often suspended in the troughs of profiled metal decks to ensure adequate stability under fire conditions. Deck manufacturers provide guidance on bar areas and spacing for different slab spans, loading and thickness for different periods of fire resistance.
- Precast concrete composite planks have a maximum fire resistance of about 2 hours.
- The steel frame has to be fire and corrosion protected as set out in the section on structural steelwork.

Preliminary sizing of composite elements

Typical span/depth ratios

Element	Typical spans m		Total structural depth (including slab and beams) for simply supported beams	
	Primary	Secondary	Primary	Secondary
Universal beam sections	6–10	8–18	$L/19$	$L/23$
Universal column sections	6–10	8–18	$L/22$	$L/29$
Fabricated sections	>12	>12	$L/15$	$L/25$
Fixed end/haunched beams (Haunch length $L/10$ with maximum depth $2D$)	>12	>12	$L/25$ (midspan)	$L/32$ (midspan)
Castellated beams (Circular holes $\phi = 2D/3$ at about 1.5ϕ c/c, D is the beam depth)	n/a	6–16	$L/17$	$L/20$
Proprietary composite trusses	>12	>12	$L/12$	$L/16$

Preliminary sizing

Estimate the unfactored moment which will be applied to the beam in its final (rather than construction) condition. Use an allowable working stress of 160N/mm^2 for S275, or 210N/mm^2 for S355, to estimate the required section modulus (Z) for a non-composite beam. A preliminary estimate of a composite beam size can be made by selecting a steel beam with 60–70% of the non-composite Z. Commercial office buildings normally have about 1.8 to 2.2 shear studs (19 mm diameter) per square metre of floor area. Deflections, response to vibration and service holes should be checked for each case.

Approximate limits on holes in rolled steel beams

Reduced section capacity due to holes through the webs of steel beams must be considered for both initial and detailed calculations.

Where D is the depth of the steel beam, limit the size of openings to $0.6D$ depth and $1.5D$ length in unstiffened webs, and to $0.7D$ and $2D$ respectively where stiffeners are provided above and below the opening. Holes should be a minimum of $1.5D$ apart and be positioned centrally in the depth of the web, in the middle third of the span for uniformly loaded beams. Holes should be a minimum of D from any concentrated loads and $2D$ from a support position. Should the position of the holes be moved off centre of depth of the beam, the remaining portions of web above and below the hole should not differ by a factor of 1.5 to 2.

Preliminary composite beam sizing tables for S275 and normal weight concrete

4 kN/m² live loading + 1 kN/m² for partitions

Primary span m	Secondary span m	No. of secondary beams per grid	Secondary beam spacing m	Beam sizes for minimum steel weight			Beam sizes for minimum floor depth		
				Primary beam	Secondary beam	Steel weight kN/m³	Primary beam	Secondary beam	Steel weight kN/m³
6	8	2	3.00	457 × 152 UB 67	406 × 178 UB 54	0.26	254 × 254 UC 132	254 × 254 UC 132	0.61
	12	2	3.00	533 × 210 UB 92	610 × 229 UB 113	0.45	305 × 305 UC 158	356 × 406 UC 287	1.09
	15	2	3.00	610 × 229 UB 101	762 × 267 UB 173	0.64	305 × 305 UC 198	356 × 406 UC 551	1.97
8	8	3	2.67	533 × 210 UB 92	356 × 171 UB 57	0.33	305 × 305 UC 198	254 × 254 UC 107	0.65
	12	3	2.67	610 × 229 UB 125	610 × 229 UB 101	0.48	305 × 305 UC 283	305 × 305 UC 283	1.30
	15	3	2.67	762 × 267 UB 147	762 × 267 UB 173	0.75	356 × 406 UC 287	356 × 406 UC 467	1.94
9	8	3	3.00	610 × 229 UB 101	406 × 178 UB 54	0.31	305 × 305 UC 240	254 × 254 UC 132	0.74
	12	3	3.00	686 × 254 UB 140	610 × 229 UB 113	0.49	356 × 406 UC 287	356 × 406 UC 287	1.20
	15	3	3.00	762 × 267 UB 173	762 × 267 UB 173	0.69	356 × 406 UC 393	356 × 406 UC 551	2.10
10	8	4	2.50	686 × 254 UB 140	356 × 171 UB 57	0.40	356 × 406 UC 287	254 × 254 UC 107	0.79
	12	4	2.50	838 × 292 UB 176	610 × 229 UB 101	0.55	356 × 406 UC 393	305 × 305 UC 283	1.46
	15	4	2.50	914 × 305 UB 201	762 × 267 UB 173	0.83	356 × 406 UC 467	356 × 406 UC 467	2.18
12	8	4	3.00	762 × 267 UB 173	406 × 178 UB 54	0.40	356 × 406 UC 467	254 × 254 UC 132	0.93
	12	4	3.00	914 × 305 UB 224	610 × 229 UB 113	0.56	356 × 406 UC 634	356 × 406 UC 287	1.49
	15	4	3.00	914 × 305 UB 289	762 × 267 UB 173	0.77	914 × 419 UB 289	356 × 406 UC 551	2.01

Check floor natural frequency <4.5 Hz.
Construction deflections limited to span/360 or 25 mm.

Preliminary composite beam sizing tables for S275 and lightweight concrete

4 kN/m² live loading + 1 kN/m² for partitions

Primary span m	Secondary span m	No. of secondary beams per grid	Secondary beam spacing m	Minimum steel weight			Minimum floor depth		
				Primary beam	Secondary beam	Steel weight kN/m³	Primary beam	Secondary beam	Steel weight kN/m³
6	8	2	3.00	457 × 152 UB 60	356 × 171 UB 57	0.27	254 × 254 UC 89	254 × 254 UC 89	0.41
	12	2	3.00	533 × 210 UB 82	533 × 210 UB 109	0.43	254 × 254 UC 132	305 × 305 UC 240	0.91
	15	2	3.00	533 × 210 UB 92	762 × 267 UB 147	0.55	305 × 305 UC 158	356 × 406 UC 467	1.66
8	8	3	2.67	457 × 152 UB 82	356 × 171 UB 51	0.29	305 × 305 UC 158	254 × 254 UC 89	0.53
	12	3	2.67	610 × 229 UB 101	533 × 210 UB 101	0.46	305 × 305 UC 240	305 × 305 UC 240	1.10
	15	3	2.67	610 × 229 UB 125	762 × 267 UB 134	0.59	305 × 305 UC 283	356 × 406 UC 393	1.66
9	8	3	3.00	610 × 229 UB 101	356 × 171 UB 57	0.32	305 × 305 UC 198	254 × 254 UC 89	0.54
	12	3	3.00	610 × 229 UB 125	533 × 210 UB 109	0.47	305 × 305 UC 283	305 × 305 UC 240	1.04
	15	3	3.00	762 × 267 UB 147	762 × 267 UB 147	0.59	356 × 406 UC 287	356 × 406 UC 467	1.75
10	8	4	2.50	610 × 229 UB 125	356 × 171 UB 51	0.36	305 × 305 UC 240	254 × 254 UC 89	0.66
	12	4	2.50	762 × 267 UB 147	533 × 210 UB 101	0.53	356 × 406 UC 287	305 × 305 UC 240	1.20
	15	4	2.50	838 × 292 UB 176	762 × 267 UB 134	0.65	356 × 406 UC 340	356 × 406 UC 393	1.80
12	8	4	3.00	762 × 267 UB 147	356 × 171 UB 57	0.37	356 × 406 UC 393	254 × 254 UC 89	0.79
	12	4	3.00	838 × 292 UB 194	533 × 210 UB 109	0.53	356 × 406 UC 551	305 × 305 UC 240	1.26
	15	4	3.00	914 × 305 UB 224	762 × 267 UB 147	0.64	356 × 406 UC 634	356 × 406 UC 467	1.98

Check floor natural frequency <4.5 Hz.

Composite design to BS 5950

BS 5950: Part 3 is based on ultimate loads and plastic design of sections, so the partial safety factors and design strengths of steel as BS 5950: Part 1 and BS 8110: Part 1 apply as appropriate. Steel sections should be classified for local buckling to determine whether their design should be plastic or elastic.

The composite moment capacity depends on the position of the neutral axis – whether in the concrete slab, the steel flange or the steel web. This depends on the relative strength of the concrete and steel sections. The steel beam should be designed for the non-composite temporary construction situation as well as for composite action in the permanent condition.

The code design methods are very summarized for this book, which only deals with basic moment capacity in relation to uniform loads. The code also provides guidance on how concentrated loads and holes in beams should be designed for in detail. Serviceability and vibration checks are also required.

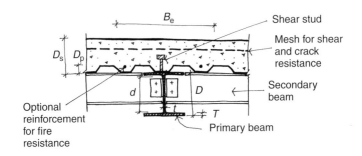

Effective slab breadth

Internal secondary beams: B_e = the lesser of secondary beam spacing or L/4.

Internal primary beams: B_e = the lesser of 0.8 × secondary beam spacing or L/4.

Edge beams: B_e = half of the appropriate primary or secondary value.

Composite plastic moment capacity for simply supported beams

Assuming that the steel section is compact and uniformly loaded, check that the applied moment is less than the plastic moment of the composite section. These equations are for a profiled metal deck slab where the shear is low ($F_v < 0.5P_v$):

Compression capacity of concrete slab, $R_c = 0.45\, F_{cu}B_e\,(D_s - D_p)$

Tensile capacity of steel element, $R_s = p_y\, A_{\text{area of steel section}}$

Tensile capacity of one steel flange (thickness T), $R_f = A_{\text{area of flange}}p_y$

Tensile capacity of steel web (thickness t and clear depth d), $R_w = p_y A_{\text{area of web}}$

Moment capacity of the fully restrained steel section, M_s

Plastic moment capacity of the composite section, M_c

Plastic moment capacity of the composite section after deduction of the shear area, M_f

Applied shear stress, F_v

Shear capacity, P_v

Plastic moment capacity for low shear – neutral axis in concrete slab: $R_c \geq R_s$

$$M_c = R_s\left(\frac{D}{2} + D_s - \frac{R_s(D_s - D_p)}{2R_c}\right)$$

Plastic moment capacity for low shear – neutral axis in steel flange: $R_s > R_c \geq R_w$

$$M_c = \frac{R_s D}{2} + \frac{R_c(D_s + D_p)}{2} - \frac{T(R_s + R_c)^2}{4R_f}$$

Plastic moment capacity for low shear – neutral axis in steel web: $R_c < R_w$

$$M_c = M_s + \frac{R_c(D + D_s + D_p)}{2} - \frac{dR_c^2}{4R_w}$$

Reduced plastic moment capacity for combined high shear and moment

$$M_{cv} = M_c - (M_c - M_f)\left(\frac{2F_v}{P_v} - 1\right)^2$$

Shear capacity

The shear capacity of the beam

$$P_v = 0.6\,p_y A_v.$$

Shear is considered low if the applied vertical shear load $F_v < 0.5P_v$. However, in simply supported beams, high shear and moment forces normally only coexist at the positions of heavy point loads. Therefore generally where there are no point loads and if $M_{applied} < M_c$ and $F_v < P_v$, no further checks are required.

For high shear, $F_v > 0.5P_v$, the web of the steel beam must be neglected from calculations for the reduced moment capacity, M_{cv}.

Longitudinal shear

Shear stud strengths for normal weight concrete are given in Table 5 of BS 5950: Part 3. Studs of 19 mm diameter have characteristic resistances of 80 kN to 100 kN for normal weight concrete, depending on the height and the concrete strength. The strength of the studs in lightweight concrete can be taken as 90% of the normal concrete weight values. Allowances and reductions must be made for groups of studs as well as the deck shape and its contact area (due to the profiled soffit) with the steel beam. The horizontal shear force on the interface between the steel and concrete should be estimated and an arrangement of shear studs selected to resist that force. The minimum spacing of studs is 6ϕ longitudinally and 4ϕ transversely, where ϕ is the stud diameter and the maximum longitudinal spacing of the studs is 600 mm.

Mesh reinforcement is generally required in the top face of the slab to spread the stud shear forces across the effective breadth of the slab (therefore increasing the longitudinal shear resistance) and to minimize cracking in the top of the slab.

Serviceability

For simply supported beams the second moment of area can be calculated on the basis of the midspan breadth of the concrete flange, which is assumed to be uncracked.

The natural frequency of the structure can be estimated by $f = 18/\sqrt{\delta}$ where δ is the elastic deflection (in mm) under dead and permanent loads. In most cases problems due to vibrations can be avoided as the natural frequency of the floor is kept greater than 4 Hz–4.5 Hz.

11
Structural Glass

Structural glass assemblies are those in which the self-weight of the glass, wind and other imposed loads are carried by the glass itself rather than by glazing bars, mullions or transoms, and the glass elements are connected together by mechanical connections or by adhesives.

Despite the increasing use of glass as a structural material over the last 25 years, there is no single code of practice which covers all of the issues relating to structural glass assemblies. Therefore values for structural design must be based on first principles, research, experience and load testing. The design values given in this chapter should be used very carefully with this in mind.

The following issues should be considered:

- Glass is classed as a rigid liquid as its intermolecular bonds are randomly arranged, rather than the crystalline arrangement normally associated with solids. Glass will behave elastically until the applied stresses reach the level where the interatomic bonds break. If sufficient stress is applied, these cracks will propagate and catastrophic failure will occur. The random arrangement of the interatomic bonds means that glass is not ductile and therefore failure is sudden.
- Cracks in glass propagate faster as temperature increases.
- Without the ability to yield, or behave plastically, glass can fail due to local over-stressing. Steel can redistribute high local stresses by local yielding and small plastic deformations, but glass cannot behave like this and high local stresses will result in brittle failure.
- Modern glass is not thought to deform or creep under long-term stresses. It behaves perfectly elastically and will return to its original shape when applied loads are removed. However, some old glass has been found to creep.
- Glass will generally fail as a result of the build-up of tensile stresses. Generally it is the outer surfaces of the glass which are subject to these stresses. Small flaws on glass surfaces encourage crack propagation which can lead to failure. Structural glass should be carefully checked for flaws.
- Annealed glass can also fail as a result of 'static fatigue'. There are various theories on why this occurs, but in simple terms, microcracks form and propagate under sustained loads resulting in failure of the glass. This means that the strength of glass is time dependent; in the short term glass can carry about twice the load that it can carry in the long term. Long-term stresses are kept low in design to prevent propagation of cracks. There is a finite time for static fatigue failure to occur for each type of glass and although this is beyond the scope of this book, this period can be calculated. It is about 15 days for borosilicate glass (better known as Pyrex), but is generally much longer for annealed glass.
- Thermal shock must also be considered for annealed glass. Temperature differences across a single sheet of glass can result in internal stresses. Glass elements which are partly in direct sunlight and partly shaded are at most risk of failure. Thermal shock cracks tend to start at the edge of the glass, travelling inwards at about 90°, but this type of failure can depend on many things including edge restraint, and manufacturers should be consulted for each situation. If thermal shock is expected to be an issue, toughened or tempered glass should be specified in place of annealed.
- Glass must come from a known and reliable source to provide reliable strength and minimal impurities.

Types of glass products

Annealed/float glass

Glass typically consists of frit: sand (silica 72%), soda ash (sodium carbonate 13%), lime-stone (calcium carbonate 10%) and dolomite (calcium magnesium carbonate 4%). This mixture is combined with broken glass (called cullet) at about 80% frit to 20% cullet, and is heated to 1500°C to melt it. It leaves the first furnace at about 1050°C and goes on to the forming process.

There are a number of forming processes, but structural glass is generally produced by the float glass method, which was developed in 1959 by Pilkington. The molten glass flows out of the furnace on to a bed of molten tin in a controlled atmosphere of nitro-gen and oxygen and is kept at high temperature. This means that defects and distor-tions are melted out of the upper and lower surfaces without grinding and polishing. The glass is progressively cooled as it is moved along the bath by rollers until it reaches about 600°C when the glass sheet becomes viscous enough to come into contact with the rollers without causing damage to the bottom surface. The speed at which the ribbon of glass moves along the tin bath determines the thickness of the glass sheet. Finally the glass is cooled in a gradual and controlled manner to 200°C in the annealing bay. The term 'annealed' means that the glass has been cooled carefully to prevent the build-up of residual stresses. The surfaces of float glass can be described by using the descriptions 'tin side' and 'air side' depending on which way the glass was lying in the float bath. For use as structural glass, the material should be free of impurities and discoloration. At failure, annealed glass typically breaks into large pieces with sharp edges. Annealed glass must therefore be specified carefully so that on failure it will not cause injury.

Toughened/fully tempered glass

Sheets of annealed glass are reheated to their softening point at about 700°C before being rapidly cooled. This can be done by hanging the glass vertically gripped by tongs with cooling applied by air jets, or by rolling the glass through the furnace and cooling areas. The rapid cooling of the glass causes the outer surfaces to contract quicker than the inner core. This means that a permanent precompression is applied to the glass, which can make its capacity for tensile stress three to four times better than annealed glass. The strength of toughened glass can depend on its country of origin. The values quoted in this book relate to typical UK strengths. The 'tin side' of toughened glass can be examined using polarizing filters to determine the residual stresses and hence the strength of the glass. Toughened glass cannot be cut or drilled after toughening, therefore glass is generally toughened for specific projects rather than being kept as a stock item. Toughened glass is more resistant to temperature dif-ferentials within elements than annealed glass and therefore it tends to be used externally in elements such as floors where annealed glass would normally be used in internal situations. Toughened glass can fail as a result of nickel sulphide impurities as described in the section on 'Heat soaked glass'. If specified to BS 6206, toughened glass is regarded as a safety glass because it fractures into small cubes (40 particles per 50 mm square) without sharp edges when broken. However, these cubes have the same density as crushed concrete and the design should prevent broken glass falling out of place to avoid injury to the public.

Partly toughened/heat tempered glass

Sheets of annealed glass are heated and then cooled in the same way as the toughen-ing process; however, the cooling is not as rapid. This means that slightly less permanent precompression is applied to the glass, which will make its capacity for tensile stress 1.5 to 2 times better than annealed glass. The residual strength can be specified depending on the proposed use. Heat tempered glass will not fail as a result of nickel sulphide impurities as described for toughened glass in the section on 'Heat soaked glass'. The fracture pattern is very like that of annealed glass, as the residual stresses are not quite enough to shatter the glass into the small cubes normally associated with toughened glass. Tempered glass can be laminated to the top of toughened glass to produce units resistant to thermal shock, which can remain in place with a curved shape even when both sheets of glass have been broken.

Heat soaked glass

The float glass process leaves invisible nickel sulphide (NS) impurities in the glass called inclusions. When the glass is toughened, the heat causes these inclusions to become smaller and unstable. After toughening, and often after installation, thermal movements and humidity changes can cause the NS inclusions in the glass to revert to their original form by expanding. This expansion causes the glass to shatter and failure can be quite explosive. Heat soaking can be specified to reduce the risk of NS failure of toughened glass by accelerating the natural phenomenon. This accelerated fatigue will tend to break flawed glass during a period of prolonged heating at about 300°C. Heat soaking periods are the subject of international debate and range between 2 and 8 hours. The German DIN standard is considered the most reliable code of practice. Glass manufacturers indicate that one incidence of NS failure is expected in every 4 tonnes of toughened glass, but after heat soaking this is thought to reduce to about 1 in 400 tonnes.

Laminated glass

Two or more sheets of glass are bonded, or laminated, together using plastic sheet or liquid resin interlayers. The interlayer is normally polyvinyl butyral (PVB) built up in sheets of 0.38 mm and can be clear or tinted. It is normal to use four of these layers (about 12 mm) in order to allow for the glass surface ripple which is produced by the rollers used in the float glass manufacturing process. Plastic sheets are used for larger numbers and sizes of panels. Liquid resins are more suited to curved glass and to small-scale manufacturing, as the glass sheets have to be kept spaced apart in order to obtain a uniform thickness of resin between the sheets. The bonding is achieved by applying heat and pressure in an autoclave. If the glass breaks in service, the interlayer tends to hold the fragmented glass to the remaining sheet until the panel can be replaced. Laminates can be used for safety, bullet proofing, solar control and acoustic control glazing. Toughened, tempered, heat soaked and annealed glass sheets can be incorporated and combined in laminated panels. Ideally the glass should be specified so that toughened or tempered glass is laminated with the 'tin side' of the glass outermost, so that the glass strength can be inspected if necessary. Laminated panels tend to behave monolithically for short time loading at temperatures below 70°C, but interlayer creep means that the layers act separately under long-term loads.

Summary of material properties

Density   25–26 kN/m^3
Compressive strength $F_{cu} = 1000$ N/mm^2

Tensile strength   Strength depends on many factors including: duration of loading, rate of loading, country of manufacture, residual stresses, temperature, size of cross section, surface finish and micro cracks. Fine glass fibres have tensile strengths of up to 1500 N/mm^2 but for the sections used in structural glazing, typical characteristic tensile strengths are: 45 N/mm^2 for annealed, 80 N/mm^2 for tempered glass and 120 N/mm^2 for toughened. Patterned or wired glass can carry less load.

Modulus of elasticity   70–74 kN/mm^2
Poisson's ratio   0.22–0.25
Linear coefficient of
thermal expansion 8×10^{-6}/°C

Typical glass section sizes and thicknesses

The range of available glass section sizes changes as regularly as the plant and facilities in the glass factories are updated or renewed. There are no standard sizes, only maximum sizes. Manufacturers should be contacted for up to date information about the sheet sizes available. The contact details for Pilkington, Solaglass Saint Gobain, Firman, Hansen Glass, QTG, European, Bishoff Glastechnik and Eckelt are listed in the chapter on useful addresses. Always check that the required sheet size can be obtained and installed economically.

Annealed/float glass

The typical maximum size is 3210 mm × 6000 mm although sheets up to 3210 mm × 8000 mm can be obtained on special order or from continental glass manufacturers.

Typical float glass thicknesses

Thickness mm	3*	4	5*	6	8*	10	12	15	19	25
Weight kg/m²	7.5	10.0	12.5	15.0	20.0	25.0	30.0	37.5	47.5	62.5

*Generally used in structural glazing laminated units.

Toughened/fully tempered glass

The sizes of toughened sheets are generally smaller than the sizes of float glass available. Toughened glass in 25 mm is currently still only experimental and is generally not available.

Thickness mm	4	5	6	8	10	12	15	19
Sheet size* mm × mm	1500 × 2200	2000 × 4200	2000 × 4200	2000 × 4200	2000 × 4200	2000 × 4200	1700 × 4200	1500 × 4200

*Larger sizes are available from certain UK and European suppliers.

Heat tempered/partly toughened glass

Normally only produced in 8 mm thick sheets for laminated units. Manufacturers should be consulted about the availability of 10 mm and 12 mm sheets. Sheet sizes are the same as those for fully toughened glass.

Heat soaked glass

Sheet sizes are limited to the size of the heat soaking oven, typically about 2000 mm × 6000 mm.

Laminated glass

Limited only by the size of sheets available for the different types of glass and the size of autoclave used to cure the interlayers.

Curved glass

Curved glass can be obtained in the UK from Pilkington with a minimum radius of 750 mm for 12 mm glass; a minimum radius of 1000 mm for 15 mm glass and a minimum radius of 1500 mm for 19 mm glass. However, Sunglass in Italy and Cricursa in Spain are specialist providers who can provide a minimum radius of 300 mm for 10 mm glass down to 100 mm for 4 mm to 6 mm glass.

Durability and fire resistance

Durability

Glass and stainless steel components are inherently durable if they are properly specified and kept clean. Glass is corrosion resistant to most substances apart from strong alkalis. The torque of fixing bolts and the adhesives used to secure them should be checked approximately every 5 years and silicone joints may have to be replaced after 25–30 years depending on the exposure conditions. Deflection limits might need to be increased to prevent water ingress caused by rotations at the framing and sealing to the glass.

Fire resistance

Fire resistant glasses are capable of achieving 60 minutes of stability and integrity when specially framed using intumescent seals etc. There are several types of fire resisting glass which all have differing amounts of fire resistance. The wire interlayer in Georgian wired glass maintains the integrity of the pane by holding the glass in place as it is softened by the heat of a fire. Intumescent interlayers in laminated glass expand to form an opaque rigid barrier to contain heat and smoke. Prestressed borosilicate glass (better known as Pyrex) can resist heat without cracking but must be specially made to order and is limited to 1.2 m by 2 m panels.

Typical glass sizes for common applications

The following are typical sizes from Pilkington for standard glass applications. The normal design principles of determinacy and redundancy should also be considered when using these typical sizes. These designs are for internal use only. External use requires more careful consideration of thermal effects, where it may be more appropriate to specify toughened glass instead of annealed glass.

Toughened glass barriers

Horizontal line load kN/m	Toughened glass thickness* mm
0.36	12
0.74	15
1.50	19
3.00	25

*For 1.1 m high barrier, clamped at foot.

Toughened glass infill to barriers bolted between uprights

Loading		Limiting glass span for glass thickness m			
UDL kN/m^2	Point load kN	6 mm*	8 mm*	10 mm	12 mm
0.5	0.25	1.40	1.75	2.10	2.40
1.0	0.50	0.90	1.45	1.75	2.05
1.5	1.50	–	–	1.20	1.60

*Not suitable if free path beside barrier is >1.5 m as it will not contain impact loads as Class A to BS 6206.

Laminated glass floors and stair treads

UDL kN/m^2	Point load kN	Glass thickness (top + bottom annealed)* mm + mm	Typical use
1.5	1.4	19 + 10	Domestic floor or stair
5.0	3.6	25 + 15	Dance floor
4.0	4.5	25 + 25	Corridors
4.0	4.0	25 + 10	Stair tread

*Based on a floor sheet size of 1 m^2 or a stair tread of 0.3 × 1.5 m supported on four edges with a minimum bearing length equal to the thickness of the glass unit. The 1 m^2 is normally considered to be the maximum size/weight which can be practically handled on site.

Glass mullions or fins in toughened safety glass

Mullion height m	Mullion thickness/depth for wind loading mm*			
	1.00 kN/m²	1.25 kN/m²	1.50 kN/m²	1.75 kN/m²
<2.0	19/120	19/130	25/120	25/130
2.0–2.5	19/160	25/160	25/170	–
2.5–3.0	25/180	25/200	–	–
3.0–3.5	25/230	–	–	–
3.5–4.0	25/280	–	–	–

*Assuming restraint at head and foot plus sealant to main panels. Normal maximum spacing is approximately 2 m.

Glass walls and planar glazing

Suspended structural glass walls can typically be up to 23 m high and of unlimited length, while ground supported walls are usually limited to a maximum height of 9 m.

Planar glazing is limited to a height of 18 m with glass sheet sizes of less than 2 m² so that the weights do not exceed the shear capacity of the planar bolts and fixings.

In a sheltered urban area, 2 × 2 m square panels will typically need a bolt at each of the four corners; 2 × 3.5 m panels will need six bolts and panels taller than 3.5 m will need eight bolts.

Source: Pilkington (2002).

Structural glass design

Summary of design principles

- Provide alternative routes within a building so that users can choose to avoid crossing glass structures.
- Glass is perfectly elastic, but failure is sudden.
- Deflection and buckling normally govern the design. Deflections of vertical panes are thought to be acceptable to span/150, while deflections of horizontal elements should be limited to span/360.
- Glass works best in compression, although bearing often determines the thickness of beams and fins.
- For designs in pure tension, the supports should be designed to distribute the stresses uniformly across the whole glass area.
- Glass can carry bending both in and out of its own plane.
- Use glass in combination with steel or other metals to carry tensile and bending stresses.
- The sizes of glass elements in external walls can be dictated by energy efficiency regulations as much as the required strength.
- Keep the arrangement of supports simple, ensuring that the glass only carries predictable loads to avoid failure as a result of stress concentrations. Isolate glass from shock and fluctuating loads with spring and damped connections.
- Sudden failure of the glass elements must be allowed for in the design by provision of redundancy, alternative load paths, and by using the higher short-term load capacity of glass.
- Glass failure should not result in a disproportionate collapse of the structure.
- Generally long-term stresses in annealed glass are kept low to prevent failure as a result of static fatigue, i.e. time dependent failure. For complex structures with simple loading conditions, it is possible to stress glass elements for a calculated failure period in order to promote failure by static fatigue.
- The effects of failure and the method of repair or replacement must be considered in the design, as well as maintenance and access issues.
- The impact resistance of each element should be considered to establish an appropriate behaviour as a result of damage by accident or vandalism.
- It is good practice to laminate glass sheets used overhead. Sand blasting, etching and fritting can be used to provide slip resistance and modesty for glass to be used underfoot.
- Toughened glass elements should generally be heat soaked to avoid nickel sulphide failure if they are to be used to carry load as single or unlaminated sheets.
- Consider proof testing elements/components if the design is new or unusual, or where critical elements rely on the additional strength of single ply toughened glass.
- Glass sheet sizes are limited to the standard sizes produced by the manufacturers and the size of sheets which cutting equipment can handle.
- When considering large sheet sizes, thought must be given to the practicalities of weight, method of delivery and installation and possible future replacement.
- Inspect glass delivered to site for damage or flaws which might cause failure.
- Check that the glass can be obtained economically, in the time available.

Codes of practice and design standards

There is no single code of practice to cover structural glass, although the draft Eurocode pr EN 13474 'Glass in Building' is the nearest to an appropriate code of practice for glass design. It is thought to be slightly conservative to account for the varying quality of glass manufacture coming from different European countries. Other useful references are BS 6262, Building Regulations Part N, Glass and Glazing Federation Data Sheets, Pilkington Design Guidance Sheets; the IStructE Guide to Structural Use of Glass in Buildings and the Australian standard AS 1288.

Glass can carry load in compression, tension, bending, torsion and shear, but the engineer must decide how the stresses in the glass are to be calculated, what levels of stress are acceptable, what factor of safety is appropriate and how can unexpected or changeable loads be avoided. Overdesign will not guarantee safety.

Although some design methods use fracture mechanics or Weibull probabilities, the simplest and most commonly used design approach is elastic analysis.

Guideline allowable stresses

The following values are for preliminary design using elastic analysis with unfactored loads and are based on the values available in pr EN 13474.

Glass type	Characteristic bending strength N/mm^2	Loading condition	Typical factor of safety	Typical allowable bending stress N/mm^2
Annealed	45	Long term	6.5	7
		Short term	2.5	20
Heat tempered	70	Long term	3.5	20
		Short term	2.4	30
Toughened	120	Long term	3.0	40
		Short term	2.4	50

Connections

Connections must transfer the load in and out of glass elements in a predictable way avoiding any stress concentrations. Clamped and friction grip connections are the most commonly used for single sheets. Glass surfaces are never perfectly smooth and connections should be designed to account for differences of up to 1 mm in the glass thickness. Cut edges can have tolerances of 0.2–0.3 mm if cut with a CNC laser, otherwise dimension tolerances can be 1.0–1.5 mm.

Simple supports

The sheets of glass should sit perfectly on to the supports, either in the plane, or perpendicular to the plane, of the glass. Gaskets can cause stress concentrations and should not be used to compensate for excessive deviation between the glass and the supports. The allowable bearing stress is generally limited to about 0.42–1.5 N/mm^2 depending on the glass and setting blocks used.

Friction grip connections

Friction connections use patch plates to clamp the glass in place and are commonly used for single ply sheets of toughened glass. More complex clamped connections can use galvanized fibre gaskets and holes lined with nylon bushes to prevent stress concentrations. Friction grip bolt torques should be designed to generate a frictional clamping force of $N = F/\mu$, where the coefficient of friction is generally $\mu = 0.2$.

Holes

Annealed glass can be drilled. Toughened or tempered glass must be machined before toughening. The Glass and Glazing Federation suggest that the minimum clear edge distance should be the greater of 30 mm or 1.5 times the glass thickness (t). The minimum clear corner distance and minimum clear bolt spacing should be $4t$. Holes should be positioned in low stress areas, should be accurately drilled and the hole diameter must not be less than the glass thickness.

Bolted connections

Bolted connections can be designed to resist loads, both in and out of the plane of the glass. Pure bolted connections need to be designed for strength, tolerance, deflection, thermal and blast effects. They can be affected by minor details (such as drilling accuracy or the hole lining/bush) and this is why proprietary bolted systems are most commonly used. Extensive testing should be carried out where bolted connections are to be specially developed for a project.

Non-silicon adhesives

The use of adhesives (other than silicon) is still fairly experimental and as yet is generally limited to small glass elements. Epoxies and UV cure adhesives are among those which have been tried. It is thought that failure strengths might be about ten times those of silicon, but suitable factors of safety have not been made widely available. Loctite and 3M have some adhesive products which might be worth investigating/testing.

Structural silicones

Sealant manufacturers should be contacted for assistance with specifying their silicon products. This assistance can include information on product selection, adhesion, compatibility, thermal/creep effects and calculation of joint sizes. Data from one project cannot automatically be used for other applications.

Structural silicon sealant joints should normally be a minimum of 6 mm $\times$ 6 mm, with a maximum width to depth ratio of 3:1. If this maximum width to depth ratio is exceeded, the glass sheets will be able to rotate causing additional stresses in the silicon. A simplified design approach to joint rotation can be used (if the glass deflection is less than $L/100$) where reduced design stresses are used to allow additional capacity in the joint to cover any rotational stresses. If joint rotation is specifically considered in the joint design calculations, higher values of allowable design stresses can be used.

Dow Corning manufacture two silicon adhesives for structural applications. Dow Corning 895 is one part adhesive, site applied silicon used for small-scale remedial applications or where a two-sided structurally bonded system has to be bonded on site. Dow Corning 993 is a two part adhesive, normally factory applied. The range of colours is limited and availability should be checked for each product and application.

Technical data on the Dow Corning silicones is set out below:

Dow Corning silicon	Young's modulus kN/mm²	Type of stress	Failure stress N/mm²	Loading condition	Typical allowable design stress N/mm²
933 (2 part)	0.0014	Tension/ compression	0.95	Short-term/live loads. Design stress for comparison with simplified calculations not allowing for stresses due to joint rotation	0.140
				Short-term/live loads. Design stress where the stresses due to joint rotation for a particular case have been specifically calculated	0.210
				Long-term/dead loads	n/a
		Shear	0.68	Short-term/live loads	0.105
				Long-term/dead loads	0.011
895 (1 part)	0.0009	Tension/ compression	1.40	Short-term/live loads. Design stress for comparison with simplified calculations not allowing for stresses due to joint rotation	0.140
				Short-term/live loads. Design stress where the stresses due to joint rotation for a particular case have been specifically calculated	0.210
				Long-term/dead loads	n/a
		Shear	1.07	Short-term/live loads	0.140
				Long-term/dead loads	0.007

Source: Dow Corning (2002).

12
Building Elements, Materials, Fixings and Fastenings

Waterproofing

Although normally detailed and specified by an architect, the waterproofing must coordinate with the structure and the engineer must understand the implications of the waterproofing on the structural design.

Damp proof course

A damp proof course (DPC) is normally installed at the top and bottom of external walls to prevent the vertical passage of moisture through the wall. Cavity trays and weep holes are required above the position of elements which bridge the cavity, such as windows or doors, in order to direct any moisture in the cavity to the outside. The inclusion of a DPC will normally reduce the flexural strength of the wall.

DPCs should:

- Be bedded both sides in mortar to prevent damage.
- Be lapped with damp proof membranes in the floor or roof.
- Be lapped in order to ensure that moisture will flow over and not into the laps.
- Not project into cavities where they might collect mortar and bridge the cavity.

Different materials are available to suit different situations:

- Flexible plastic sheets or bitumen impregnated fabric can be used for most DPC locations but can be torn if not well protected and the bituminous types can sometimes be extruded under high loads or temperatures.
- Semi-rigid sheets of copper or lead are expensive but are most effective for intricate junctions.
- Rigid DPCs are layers of slate or engineering brick in Class I mortar and are only used in the base of retaining walls or freestanding walls. These combat rising damp and (unlike the other DPC materials) can transfer tension through the DPC position.

Damp proof membrane

Damp proof membranes (DPMs) are sheet or liquid membranes which are typically installed at roof and ground floor levels. In roofs they are intended to prevent the ingress of rain and at ground floor level they are intended to prevent the passage of moisture from below by capillary action. Sheet membranes can be polyethylene, bituminous or rubber sheets, while liquid systems can be hot or cold bitumen or epoxy resin.

Basement waterproofing

Basement waterproofing is problematic as leaks are only normally discovered once the structure has been occupied. The opportunity for remedial work is normally limited, quite apart from the difficulty of reaching externally applied tanking systems. Although an architect details the waterproofing for the rest of the building, sometimes the engineer is asked to specify the waterproofing for the basement. In this case very careful co-ordination of the lapping of the waterproofing above and below ground must be achieved to ensure that there are no weak points. Basement waterproofing should always be considered as a three-dimensional problem.

It is important to establish whether the system will be required to provide basic resistance to water pressure, or whether special additional controls on water vapour will also be required.

Basement waterproofing to BS 8102

BS 8102 sets out guidance for the waterproofing of basement structures according to their use. The following table has been adapted from Table 1 in BS 8102: 1990 to include some of the increased requirements suggested in CIRIA Report 139.

Methods of basement waterproofing

The following types of basement waterproofing systems can be used individually or together depending on the building requirements:

Tanked   This can be used internally or externally using painted or sheet membranes. Externally it is difficult to apply and protect under building site conditions, while internally water pressures can blow the waterproofing off the wall; however, it is often selected as it is relatively cheap and takes up very little space.

Integral   Concrete retaining walls can resist the ingress of water in differing amounts depending on the thickness of the section, the applied stresses, the amount of reinforcement and the density of the concrete. The density of the concrete is directly related to how well the concrete is compacted during construction. Integral structural waterproofing systems require a highly skilled workforce and strict site control. However, moisture and water vapour can still pass through a plain wall and additional protection should be added if this moisture will not be acceptable for the proposed basement use. BS 8007 provides guidance on the design of concrete to resist the passage of water, but this still does not stop water vapour. Alternatively, Caltite or Pudlo additives can be used with a BS 8110 structure to create 'waterproof concrete. This is more expensive than standard concrete but this can be offset against any saving on the labour and installation costs of traditional forms of waterproofing.

Drained   Drained cavity and floor systems allow moisture to penetrate the retaining wall. The moisture is collected in a sump to be pumped away. Drained cavity systems tend to be expensive to install and can take up quite a lot of basement floor area, but they are thought to be much more reliable than other waterproofing systems. Draining ground water to the public sewers may require a special licence from the local water authority. Access hatches for the inspection and maintenance of internal gulleys should be provided where possible.

Basement waterproofing to BS 8102

Grade of basement to BS 8102	Basement use	Performance of water proofing	Form of construction	Comments
1	Car parking, plant rooms (excluding electrical equipment) and workshops	Some water seepage and damp patches tolerable (typical relative humidity >65%)	**Type B** – RC to BS 8110 (with crack widths limited to 0.3 mm)	Provides integral protection and needs waterstops at construction joints. Medium risk. Consider ground chemicals for durability and effect on finishes. The BS 8102 description of a workshop is not as good as the workshop environment described in the Building Regulations
2	Workshops and plantrooms requiring drier environment. Retail storage areas	No water penetration but moisture tolerable (typical relative humidity = 35–50%)	**Type A**	Requires drainage to external basement perimeter below the level of the wall/floor membrane lap. Medium risk with multiple membrane layers and strict site control
			Type B – RC to BS 8007	Provides integral protection and needs waterstops at construction joints. Medium risk. Consider ground chemicals for durability and effect on finishes. Additional tanking is likely to be needed to meet retail storage requirements
3	Ventilated residential and working areas, offices, restaurants and leisure centres	Dry environment, but no specific control on moisture vapour (typical relative humidity 40–60%)	**Type A**	Not recommended unless drainage is provided above the wall/floor membrane lap position and the site is relatively free draining. High risk
			Type B – RC to BS 8007	Provides integral protection and needs waterstops at construction joints. Medium risk. Consider ground chemicals for durability and effect on finishes. Additional tanking is recommended
			Type C – wall and floor cavity system	A drained cavity allows the wall to leak and it is therefore foolproof. Sumps may need back-up pumps. High safety factor
4	Archives and computer stores	Totally dry environment with strict control of moisture vapour (typical relative humidity = 35% for books – 50% for art storage)	**Type A**	Unlikely to be able to provide the controlled conditions required. Very high risk
			Type B – RC to BS 8007 plus vapour barrier	High risk. Medium risk with addition of a drainage cavity to reduce water penetration
			Type C – wall and floor cavity system with vapour barrier to inner skin and floor cavity with DPM	Medium risk. Addition of a water resistant concrete wall would provide the maximum possible safety for sensitive environments

NOTES:

1. **Type A** = tanked construction, **Type B** = integral structural waterproofing and **Type C** = drained protection.
2. Relative humidity indicates the amount of water vapour in the air as a percentage of the maximum amount of water vapour which would be possible for air at a given temperature and pressure. Typical values of relative humidity for the UK are about 40–50% for heated indoor conditions and 85% for unheated external conditions.

Source: BS 8102:1990.

Remedial work

Failed basement systems require remedial work. Application of internal tanking in this situation is not normally successful. The junction of the wall and floor is normally the position where water leaks are most noticeable.

An economical remedial method is to turn the existing floor construction into a drained floor by chasing channels in the existing floor finishes around the perimeter. Additional channels may cross the floor where there are large areas of open space. Proprietary plastic trays with perforated sides and bases can be set into the chases, connected up and drained to a sump and pump. New floor finishes can then be applied over the original floor and its new drainage channels, to provide ground water protection with only a small thickness of additional floor construction.

The best way to avoid disrupting, distressing and expensive remedial work is to design and detail a good drained cavity system in the first place!

Screeds

Screeds are generally specified by an architect as a finish to structural floors in order to provide a level surface, to conceal service routes and/or as a preparation for application of floor finishes. Historically screeds fail due to inadequate soundness, cracking and curling and therefore, like waterproofing, it is useful for the engineer to have some background knowledge. Structural toppings generally act as part of a precast structural floor to resist vertical load or to enhance diaphragm action. The structural issues affecting the choice of screed are: type of floor construction, deflection, thermal or moisture movements, surface accuracy and moisture condition.

Deflection

Directly bonded screeds can be successfully applied to solid reinforced concrete slabs as they are generally sufficiently rigid, while floating screeds are more suitable for flexible floors (such as precast planks or composite metal decking) to avoid reflective cracking of the screed. Floating screeds must be thicker than bonded screeds to withstand the applied floor loadings and are laid on a slip membrane to ensure free movement and avoid reflective cracking.

Thermal/moisture effects

Drying shrinkage and temperature changes will result in movement in the structure, which could lead to the cracking of an overlying bonded screed. It is general practice to leave concrete slabs to cure for 6 weeks before laying screed or applying rigid finishes such as tiles, stone or terrazzo. For other finishes the required floor slab drying times vary. If movement is likely to be problematic, joints should be made in the screed at predetermined points to allow expansion/contraction/stress relief.

Sand:cement screeds must be cured by close covering with polythene sheet for 7 days while foot traffic is prevented and the screed is protected from frost. After this the remaining free moisture in the screed needs time to escape before application of finishes. This is especially true if the substructure and finish are both vapour proof as this can result in moisture being trapped in the screed. Accurate prediction of screed drying times is difficult, but a rough rule is 4 weeks per 25 mm of screed thickness (to reach about 75% relative humidity). Accelerated heating to speed the drying process can cause the screed to crack or curl, but dehumidifiers can be useful.

Surface accuracy

The accuracy of surface level and flatness of a laying surface is related to the type of base, accuracy of the setting out and the quality of workmanship. These issues should be considered when selecting the overall thickness of the floor finishes to avoid problems with the finish and/or costly remedial measures.

Precast concrete hollowcore slabs

The values for the hollowcore slabs set out below are for precast prestressed concrete slabs by Bison Concrete Products. The prestressing wires are stretched across long shutter beds before the concrete is extruded or slip formed along beds up to 130 m long. The prestress in the units induces a precamber. The overall camber of associated units should not normally exceed L/300. Some planks may need a concrete topping (not screed) to develop their full bending capacity or to contribute to diaphragm action. Minimum bearing lengths of 100 mm are required for masonry supports, while 75 mm is acceptable for supports on steelwork or concrete. Planks are normally 1200 mm wide at their underside and are butted up tight together on site. The units are only 1180 to 1190 mm wide at the top surface and the joints between the planks are grouted up on site. Narrower planks are normally available on special order in a few specific widths. Special details, notches, holes and fixings should be discussed with the plank manufacturer early in the design.

Typical hollowcore working load capacities

Nominal hollowcore plank depth mm	Fire resistance hours	Clear span for imposed loads[1] m				
		1.5 kN/m^2	3.0 kN/m^2	5.0 kN/m^2	10.0 kN/m^2	15.0 kN/m^2
100[2]	up to 2	5.0	4.8	4.2	3.3	2.8
150[2]	up to 2	7.5	7.3	6.4	5.0	3.2
200	2	10.0	9.1	8.0	6.4	5.5
250	2	11.5	10.4	9.2	7.4	6.3
300	2	14.9	13.5	12.1	10.0	8.7
350	2	16.5	15.0	13.6	11.2	9.8
400	2	17.6	16.1	14.6	12.1	10.6
450	2	19.0	17.5	15.9	13.4	11.7

NOTES:
1. 1.5 kN/m^2 for finishes included in addition to self-weight of plank.
2. 35 mm screed required for 2 hour fire resistance.
3. The reinforcement pattern within a Bison section will vary according to the design loading specified.
4. Ask for 'Sound Slab' where floor mass must be greater than 300 kg/m^3.

Source: Bison Concrete Products (2007). Note that this information is subject to change at any time. Consult the latest Bison literature for up to date information.

Bi-metallic corrosion

When two dissimilar metals are put together with an 'electrolyte' (normally water) an electrical current passes between them. The further apart the metals are on the galvanic series, the more pronounced this effect becomes.

The current consists of a flow of electrons from the anode (the metal higher in the galvanic series) to the cathode, resulting in the 'wearing away' of the anode. This effect is used to advantage in galvanizing where the zinc coating slowly erodes, sacrificially protecting the steelwork. Alloys of combined metals can produce mixed effects and should be chosen with care for wet or corrosive situations in combination with other metals.

The amount of corrosion is dictated by the relative contact surface (or areas) and the nature of the electrolyte. The effect is more pronounced in immersed and buried objects. The larger the cathode, the more aggressive the attack on the anode. Where the presence of electrolyte is limited, the effect on mild steel sections is minimal and for most practical building applications where moisture is controlled, no special precautions are needed. For greater risk areas where moisture will be present, gaskets, bushes, sleeves or paint systems can be used to separate the metal surfaces.

The galvanic series

Anode

Magnesium

Zinc

Aluminium

Carbon and low alloy steels (structural steel)

Cast iron

Lead

Tin

Copper, brass, bronze

Nickel (passive)

Titanium

Stainless steels (passive)

Cathode

Structural adhesives

There is little definite guidance on the use of adhesives in structural applications which can be considered if factory controlled conditions are available. Construction sites rarely have the quality control which is required. Adhesive manufacturers should be consulted to ensure that a suitable adhesive is selected and that it will have appropriate strength, durability, fire resistance, effect on speed of fabrication, creep, surface preparation, maintenance requirements, design life and cost. Data for specific products should be obtained from manufacturers.

Adhesive families

Epoxy resins	Good gap filling properties for wide joints, with good strength and durability; low cure shrinkage and creep tendency and good operating temperature range. The resins can be cold or hot cure, in liquid or in paste form but generally available as two part formulations. Relatively high cost limits their use to special applications.
Polyurethanes	Very versatile, but slightly weaker than epoxies. Good durability properties (resistance to water, oils and chemicals but generally not alkalis) with operating temperatures of up to 60°C. Moisture is generally required as a catalyst to curing, but moisture in the parent material can adversely affect the adhesive. Applications include timber and stone, but concrete should generally be avoided due to its alkalinity.
Acrylics	Toughened acrylics are typically used for structural applications which generally need little surface preparation of the parent material to enhance bond. They can exhibit significant creep, especially at higher operating temperatures and are best suited to tight fitting (thin) joints for metals and plastics.
Polyesters	Polyesters exhibit rapid strength gain (even in extremely low temperatures) and are often used for resin anchor fixings etc. However they can exhibit high cure shrinkage and creep, and have poor resistance to moisture.
Resorcinol-formaldehydes (RF) and phenol-resorcinol-formaldehydes (PRF)	Intended for use primarily with timber. Curing can be achieved at room temperature and above. These adhesives are expensive but strong, durable, water and boil proof and will withstand exposure to salt water. They can be used for internal and external applications, and are generally used in thin layers, e.g. finger joints in glulam beams.
Phenol-formaldehydes (PF)	Typically used in factory 'hot press' fabrication of structural plywood. Cold curing types use strong acids as catalysts which can cause staining of the wood. The adhesives have similar properties to RF and PRF adhesives.
Melamine-urea-formaldehydes (MUF) and urea-formaldehydes (UF)	Another adhesive typically used for timber, but these need protection from moisture. These are best used in thin joints (of less than 0.1 mm) and cure above 10°C.
Caesins	Derived from milk proteins, these adhesives are less water resistant than MUF and UF adhesives and are susceptible to fungal attack.
Polyvinyl acetates and elastomerics	Limited to non-loadbearing applications indoors as they have limited moisture resistance.
Adhesive tapes	Double sided adhesive tapes are typically contact adhesives and are suitable for bonding smooth surfaces where rapid assembly is required. The tapes have a good operating temperature range and can accommodate a significant amount of strain. Adhesive tapes are typically used for metals and/or glass in structural applications.

Surface preparation of selected materials in glued joints

Surface preparation is essential for the long-term performance of a glued joint and the following table describes the typical steps for different materials. Specific requirements should normally be obtained from the manufacturer of the adhesive.

Material	Surface preparation	Typical adhesive
Concrete	1. Test parent material for integrity 2. Grit blast or water jet to remove the cement rich surface, curing agents and shutter oil, etc. 3. Vacuum dust and clean surface with solvent approved by the glue manufacturer 4. Apply a levelling layer to the roughened concrete surface before priming for the adhesive	Epoxies are commonly used with concrete, while polyesters are used in resin fixings and anchors. Polyurethanes are not suitable for general use due to the alkalinity of the concrete
Steel and cast iron	1. Degrease the surface 2. Mechanically wire brush, grit blast or water jet to remove mill scale and surface coatings 3. Vacuum dust then prime surface before application of the adhesive	Epoxies are the most common for use with structural iron/steel. Where high strength is not required acrylic or polyurethane may be appropriate, but only where humidity can be controlled or creep effects will not be problematic
Zinc coated steel	1. Test the steel/zinc interface for integrity 2. Degrease the surface 3. Lightly abrade the surface and avoid rupturing the zinc surface 4. Vacuum dust and then apply an etch primer 5. Thoroughly clean off the etch primer and prime the surface for the adhesive	Epoxies are suitable for structural applications. Acrylics are not generally compatible with the zinc surface
Stainless steel	Factory method: 1. Acid etch the surface and clean thoroughly 2. Apply primer Site method: 1. Degrease the surface with solvent 2. Grit blast 3. Apply chemical bonding agent, e.g. silane	Toughened epoxies are normally used for structural applications
Aluminium	Factory method: 1. Degrease with solvent 2. Use alkaline cleaning solution 3. Acid etch, then neutralize 4. Prime surface before application of the adhesive Site method (as 1 and 2): 3. Grit blast 4. Apply a silane primer/bonding agent	Epoxies and acrylics are most commonly used. Anodized components are very difficult to bond
Timber	1. Remove damaged parent material 2. Dry off contact surfaces and ensure both surfaces have a similar moisture content (which is also less than 20) 3. Plane to create a clean flat surface or lightly abrade for sheet materials 4. Vacuum dust then apply adhesive promptly	Epoxies are normally limited to special repairs. RF and PRF adhesives have long been used with timber. Durability of the adhesive must be carefully considered. They are classified: WBP – Weather Proof and Boil Proof BR – Boil Resistant MR – Moisture Resistant INT – Interior
Plastic and fibre composites	1. Dust and degrease surface 2. Abrade surface to remove loose fibres and resin rich outer layers 3. Remove traces of solvent and dust	Epoxies usual for normal applications. In dry conditions polyurethanes can be used, and acrylics if creep effects are not critical
Glass	1. Degreasing should be the only surface treatment. Abrading or etching the surface will weaken the parent material 2. Silane primer is occasionally used	Structural bonding tape or modified epoxies. The use of silicon sealant adhesives if curing times are not critical

Fixings and fastenings

Although there are a great number of fixings available, the engineer will generally specify nails, screws or bolts. Within these categories there are variations depending on the materials to be fixed. The fixings included here are standard gauges generally available in the UK.

Selected round wire nails to BS 1202

Length mm	Diameter (standard wire gauge (swg) and mm)						
	11 swg 3.0 mm	10 3.35	9 3.65	8 4.0	7 4.5	6 5.0	5 5.6
50	•	•					
75		•	•	•			
100				•	•	•	
125						•	
150							•

Selected wood screws to BS 1210

Length mm	Diameter (standard gauge (sg) and mm)						
	6 sg 3.48 mm	7 3.50	8 4.17	10 4.88	12 5.59	14 6.30	16 6.94
25	•	•	•	•	•		
50	•	•	•	•	•	•	•
75	•		•	•	•	•	•
100				•	•	•	•
125				•	•	•	•

Selected self-tapping screws to BS 4174

Self-tapping screws can be used in metal or plastics, while thread cutting screws are generally used in plastics or timber.

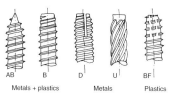

Selected ISO metric black bolts to BS 4190 and BS 3692

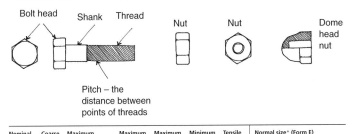

Pitch – the distance between points of threads

Nominal diameter mm	Coarse pitch mm	Maximum width of head and nut mm		Maximum height of head mm	Maximum thickness of nut (black) mm	Minimum distance between centres mm	Tensile stress area mm²	Normal size* (Form E) round washers to BS 4320		
		Across flats	Across corners					Inside diameter mm	Outside diameter mm	Nominal thickness mm
M6	1.00	10	11.5	4.375	5.375	15	20.1	6.6	12.5	1.6
M8	1.25	13	15.0	5.875	6.875	20	36.6	9.0	17.0	1.6
M10	1.50	17	19.6	7.450	8.450	25	58.0	11.0	21.0	2.0
M12	1.75	19	21.9	8.450	10.450	30	84.3	14.0	24.0	2.5
M16	2.00	24	27.7	10.450	13.550	40	157.0	18.0	30.0	3.0
M20	2.50	30	34.5	13.900	16.550	50	245.0	22.0	37.0	3.0
M24	3.00	36	41.6	15.900	19.650	60	353.0	26.0	44.0	4.0
M30	3.50	46	53.1	20.050	24.850	75	561.0	33.0	56.0	4.0

*Larger diameter washers as Form F and Form G are also available to BS 4320.

Length* mm	Bolt size						
	M6	M8	M10	M12	M16	M20	M24
30	•	•					
50	•	•	•	•	•		
70		•	•	•	•	•	•
100			•	•	•	•	•
120			•	•	•	•	•
140					•	•	•
150				•	•	•	
180				•			

*Intermediate lengths are available.

M6, M8, M10 and M12 threaded bar (called studding) is also available in long lengths.

Spanner and podger dimensions

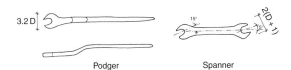

Podger   Spanner

Selected metric machine screws to BS 4183

Available in M3 to M20, machine screws have the same dimensions as black bolts but they are threaded full length and do not have a plain shank. Machine screws are often used in place of bolts and have a variety of screw heads:

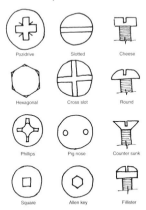

Selected metric countersunk allen key machine screws

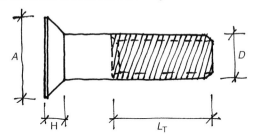

Nominal diameter mm	Coarse pitch mm	Maximum width of head mm	Maximum depth of tapered head mm	Minimum threaded length mm
M3	0.50	6.72	1.7	18
M4	0.70	8.96	2.3	20
M5	0.80	11.20	2.8	22
M6	1.00	13.44	3.3	24
M8	1.25	17.92	4.4	28
M10	1.50	22.40	5.5	32
M12	1.75	26.88	6.5	36
M16	2.00	33.60	7.5	44
M20	2.50	40.32	8.5	52
M24	3.00	40.42	14.0	60

Selected coach screws to BS 1210

Typically used in timber construction. The square head allows the screw to be tightened by a spanner.

Length* mm	Diameter			
	6.25	7.93	9.52	12.5
25	•	•		
37.5	•	•	•	
50	•	•	•	
75	•	•	•	•
87.5	•	•		
100	•	•	•	•
112	•			
125	•	•	•	•
150		•	•	•
200				•

Selected welding symbols to BS 449

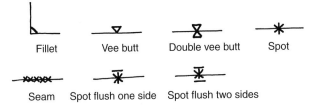

Fillet   Vee butt   Double vee butt   Spot

Seam   Spot flush one side   Spot flush two sides

Cold weather working

Cold weather and frosts can badly affect wet trades such as masonry and concrete; however, rain and snow may also have an effect on ground conditions, make access to the site and scaffolds difficult, and cause newly excavated trenches to collapse. Site staff should monitor weather forecasts to plan ahead for cold weather.

Concreting

Frost and rain can damage newly laid concrete which will not set or hydrate in temperatures below 1°C. At lower temperatures, the water in the mixture will freeze, expand and cause the concrete to break up. Heavy rain can dilute the top surface of a concrete slab and can also cause it to crumble and break up.

- Concrete should not be poured below an air temperature of 2°C or if the temperature is due to fall in the next few hours. Local conditions, frost hollows or wind chill may reduce temperatures further.
- If work cannot be delayed, concrete should be delivered at a minimum temperature of 5°C and preferably at least 10°C, so that the concrete can be kept above 5°C during the pour.
- Concrete should not be poured in more than the lightest of rain or snow showers and poured concrete should be protected if rain or snow is forecast. Formwork should be left in place longer to allow for the slower gain in strength. Concrete which has achieved 5 N/mm^2 is generally considered frost safe.
- Mixers, handling plant, subgrade/shuttering, aggregates and materials should be free from frost and be heated if necessary. If materials and plant are to be heated, the mixing water should be heated to 60°C. The concrete should be poured quickly and in extreme cases, the shuttering and concrete can be insulated or heated.

Bricklaying

Frost can easily attack brickwork as it is usually exposed on both sides and has little bulk to retain heat. Mortar will not achieve the required strength in temperatures below 2°C. Work exposed to temperatures below 2°C should be taken down and rebuilt. If work must continue and a reduced mortar strength is acceptable, a mortar mix of 1 part cement to 5 to 6 parts sand with an air entraining agent can be used. Accelerators are not recommended and additives containing calcium chloride can hold moisture in the masonry resulting in corrosion of any metalwork in the construction.

- Bricks should not be laid at air temperatures below 2°C or if the temperature is due to fall in the next few hours. Bricklaying should not be carried out in winds of force 6 or above, and walls without adequate returns to prevent instability in high winds should be propped.
- Packs, working stacks and tops of working sections should be covered to avoid soaking, which might lead to efflorescence and/or frost attack. An airspace between any polythene and the brickwork will help to prevent condensation. Hessian and bubble wrap can be used to insulate. The protection should remain in place for about 7 days after the frost has passed. In heavy rain, scaffold boards nearest the brickwork can be turned back to avoid splashing, which is difficult to clean off.
- If bricks have not been dipped, a little extra water in the mortar mix will allow the bricks to absorb excess moisture from the mortar and reduce the risk of expansion of the mortar due to freezing.

Effect of fire on construction materials

This section is a brief summary of the effect of fire on structural materials to permit a quick assessment of how a fire may affect the overall strength and stability of a structure.

It is necessary to get an accurate history of the fire and an indication of the temperatures achieved. If this is not available via the fire brigade, clues must be gathered from the site on the basis of the amount of damage to the structure and finishes. At 150°C paint will be burnt away, at 240°C wood will ignite, at 400–500°C PVC cable coverings will be charred, zinc will melt and run off and aluminium will soften. At 600–800°C aluminium will run off and glass will soften and melt. At 900–1000°C most metals will be melting and above this, temperatures will be near the point where a metal fire might start.

The effect of heat on structure generally depends on the temperature, the rate and duration of heating, and the rate of cooling. Rapid cooling by dousing with water normally results in the cracking of most structural materials.

Reinforced concrete

Concrete is likely to blacken and spall, leaving the reinforcement exposed. The heat will reduce the compressive strength and elastic modulus of the section, resulting in cracking and creep/permanent deflections. For preliminary assessment, reinforced concrete heated to 100–300°C will have about 85% of its original strength, by 300–500°C it will have about 40% of its original strength and above 500°C it will have little strength left. As it is a poor conductor of heat only the outer 30–50 mm will have been exposed to the highest temperatures and therefore there will be temperature contours within the section which may indicate that any loss of strength reduces towards the centre of the section. At about 300°C concrete will tend to turn pink and at about 450–500°C it will tend to become a dirty yellow colour. Bond strengths can normally be assumed to be about 70% of pre-fire values.

Prestressed concrete

The concrete will be affected by fire as listed for reinforced concrete. More critical is the behaviour of the steel tendons, as non-recoverable extension of the tendons will result in loss of prestressing forces. For fires with temperatures of 350–400°C the tendons may have about half of their original capacity.

Timber

Timber browns at 120–150°C, blackens at 200–250°C and will ignite and char at temperatures about 400°C. Charring may not affect the whole section and there may be sufficient section left intact which can be used in calculations of residual strength. Charring can be removed by sandblasting or planing. Large timber sections have often been found to perform better in fire than similarly sized steel or concrete sections.

Brickwork

Bricks are manufactured at temperatures above 1000°C, therefore they are only likely to be superficially or aesthetically damaged by fire. It is the mortar which can lose its strength as a result of high temperatures. Cementitious mortar will react very similarly to reinforced concrete, except without the reinforcement and section mass, it is more likely to be badly affected. Hollow blocks tend to suffer from internal cracking and separation of internal webs from the main block faces.

Steelwork

The yield strength of steel at 20°C is reduced by about 50% at 550°C and at 1000°C it is 10% or less of its original value. Being a good conductor of heat, the steel will reach the same temperature as the fire surrounding it and transfer the heat away from the area to affect other remote areas of the structure. Steelwork heated up to about 600°C can generally be reused if its hardness is checked. Cold worked steel members are more affected by increased temperature. Connections should be checked for thread stripping and general soundness. An approximate guide is that connections heated to 450°C will retain full strength, to 600°C will retain about 80% of their strength and to 800°C will retain only about 60% of their strength.

Aluminium

Aluminium is extracted from ore and has little engineering use in its pure form. Aluminium is normally alloyed with copper, magnesium, silicon, manganese, zinc, nickel and chromium to dramatically improve its strength and work hardening properties.

Aluminium has a stiffness of about one third of that for steel and therefore it is much more likely to buckle in compression than steel. The main advantage of aluminium is its high strength:weight ratio, particularly in long span roof structures. The strength of cold worked aluminium is reduced by the application of heat, and therefore jointing by bolts and rivets is preferable to welding.

For structural purposes wrought aluminium alloy sections are commonly used. These are shaped by mechanical working such as rolling, forging, drawing and extrusion. Heat treatments are also used to improve the mechanical properties of the material. This involves the heating of the alloy followed by rapid cooling, which begins a process of ageing resulting in hardening of the material over a period of a few days following the treatment. The hardening results in increased strength without significant loss of ductility. Wrought alloys can be split into non-heat treatable and heat treatable according to the amount of heat treatment and working received. The temper condition is a further classification, which indicates the processes which the alloy has undergone to improve its properties. Castings are formed from a slightly different family of aluminium alloys.

Summary of material properties

Density	$27.1 \, kN/m^3$
Poisson's ratio	0.32
Modulus of elasticity, E	$70 \, kN/mm^2$
Modulus of rigidity, G	$23 \, kN/mm^2$
Linear coefficient of thermal expansion	$24 \times 10^{-6}/°C$

Notation for the classification of structural alloys

Heat treatable alloys	T4	Heat treated – naturally aged
	T6	Heat treated – artificially aged
Non-heat treatable alloys	F	Fabricated
	O	Annealed
	H	Strain hardened

Summary of main structural aluminium alloys to BS 8118

Values of limiting stresses depend on whether the products are extrusions, sheet, plate or drawn tubes.

Alloy		Temper	Types of product*	Typical thicknesses mm	Durability	Approx. loss of strength due to welding (%)	Limiting stresses		
							P_y N/mm²	P_a/P_t N/mm²	P_v N/mm²
Heat treatable	6063	T4 T6	Thin walled extruded sections and tubes as used in curtain walling and window frames	1–150 1–150	B B	0 50	65 160	85 175	40
	6082	T4 T6	Solid and hollow extrusions	1–150 1–20 20–150		0 50 50	115 255 270	145 275 290	70 155 160
Non-heat treatable	5083	O F H22	Sheet and plate. Readily welded. Often used for plating and tanks	0.2–80 3–25 0.2–6	A	0 0 45	105 130 235	150 170 270	65 75 140
	LM 5	F	Mainly sand castings in simple shapes with high surface polish	–	A	–	Strengths of castings determined in consultation with castings manufacturer. Approx. values:		
	LM 6	F	Good for complex shaped castings / Sand castings / Chill castings	– –	B	–	40–120	70–140	25–75

*British Aluminium Extrusions do a range of sections in heat treatable aluminium alloys.

Source: BS 8118: Part 1: 1991.

Durability

Corrosion protection guidelines are set out in BS 8118: Part 2. Each type of alloy is graded as A or B. Corrosion protection is only required for A rated alloys in severe industrial, urban or marine areas. Protection is required for B rated alloys for all applications where the material thickness is less than 3 mm, otherwise protection is only required in severe industrial, urban or marine areas and where the material is immersed in fresh or salt water.

Substances corrosive to aluminium include: timber preservatives; copper naphthanate, copper-chrome-arsenic or borax-boric acid; oak, chestnut and western red cedar unless they are well seasoned; certain cleaning agents and building insulation. Barrier sealants (e.g. bituminous paint) are therefore often used.

Fire protection

Aluminium conducts heat four times as well as steel. Although this conductivity means that 'hot spots' are avoided, aluminium has a maximum working temperature of about 200 to 250°C (400°C for steel) and a melting temperature of about 600°C (1200°C for steel). In theory fire protection could be achieved by using thicker coatings than those provided for steel, aluminium is generally used in situations where fire protection is not required. Possible fire protection systems might use ceramic fibre, intumescent paints or sacrificial aluminium coatings.

Selected sizes of extruded aluminium sections to BS 1161

Section type	Range of sizes (mm)	
	Minimum	**Maximum**
Equal angles	30 × 30 × 2.5	120 × 120 × 10
Unequal angles	50 × 38 (web 3, flange 4)	140 × 105 (web 8.5, flange 11)
Channels	60 × 30 (web 5, flange 6)	240 × 100 (web 9, flange 13)
I sections	60 × 30 (web 4, flange 6)	160 × 80 (web 7, flange 11)
Tee sections	50 × 38 × 3	120 × 90 × 10

Rolled plates in thicknesses of 6.5–155 mm can be obtained in widths up to 3 m and lengths up to 15 m.

Structural design to BS 8118: Part 1

Partial safety factors for applied loads

BS 8118 operates a two tier partial safety factor system. Each load is first factored according to the type of load and when loads are combined, their total is factored according to the load combination. Dynamic effects are considered as imposed loads and must be assessed to control vibration and fatigue. This is not covered in detail in BS 8118 which suggests 'special' modelling.

Primary load factors

Load type	γ_{f1}
Dead	1.20 or 0.80
Imposed	1.33
Wind	1.20
Temperature effects	1.00

Secondary load factors for load combinations

Load combinations	γ_{f2}
Dead load	1.0
Imposed or wind load giving the most severe loading action on the component	1.0
Imposed or wind load giving the second most severe loading action on the component	0.8
Imposed or wind load giving the third most severe loading action on the component	0.6
Imposed or wind load giving the fourth most severe loading action on the component	0.4

Partial safety factors for materials depending on method of jointing

Type of construction	γ_m	
	Members	Joints
Riveted and bolted	1.2	1.2
Welded	1.2	1.3 or 1.6
Bonded/glued	1.2	3.0

Comment on aluminium design to BS 8118

As with BS 5950 for steel, the design of the structural elements depends on the classification of the cross section of the element. An initial estimate of bending strength would be $M_b = p_y S/\gamma_m$ but detailed reference must be given to the design method in the code. Strength is usually limited by local or overall buckling of the section and deflections often govern the design.

Source: BS 8118: Part 1: 1997.

13
Sustainability

Sustainability is steadily moving into mainstream building projects through the Building Regulations and other legislative controls. It is a particularly difficult area to cover as it is a relatively new topic, involving changes in public opinion, as well as traditional construction industry practices. It is therefore difficult for the engineer to find good practice guidance beyond the minimum standards found in legislation and there are often no 'right answers'.

Context

Sustainability was defined by the 1987 Brundtland Report as *meeting the needs of the present without compromising the ability of future generations to meet their own needs*. This is also often described as the *Triple Bottom Line*, which aims to balance environmental, social and economic factors. As this balance depends on each individual's moral framework, it is generally very difficult to get agreement about what constitutes sustainable design.

At present, the environmental sustainability debate is focussed on climate change and there is consensus within the scientific community that this is happening due to carbon emissions (although there is less agreement about whether these changes are due to human activity (anthropogenic) or natural cycles). While the debate about causes and action continues, *The Precautionary Principle* states that the effects of climate change are potentially so bad, that action should be taken now to reduce carbon emissions. The 2006 Stern Review supported this approach by concluding that it was likely to be more economic to pay for sustainable design now (in an attempt to mitigate climate change) rather than wait and potentially pay more to deal with its consequences. These are the principles which form the basis of the UK government's drive to reduce carbon emissions through policy on transport and energy, and legislative controls such as Planning and Building Regulations.

Environmental indicators

As the built environment is a huge consumer of resources and source of pollution, the construction industry has a significant role to play. It is estimated that building construction's use and demolition account for nearly half of the UK's carbon emissions, impacting on the environment as follows:

- Climate change (CO_2)
- Ozone depletion (CFC, HCFC)
- Ecological loss
- Fossil fuel depletion
- Land and materials depletion
- Water depletion
- Waste generation
- Acid rain (SO_2, NO_x)
- Toxicity and health (VOCs).

Although there are many indicators and targets in relation to environmental impact, carbon emissions are generally used as a simplified, or key indicator, of environmental sustainability. Other, more sophisticated, methods include the *Ecopoint* system developed by the UK's Building Research Establishment (BRE) or *Eco-Footprinting* as supported by the World Wildlife Fund. Although there is significant controversy surrounding the Eco-Footprinting methodology it does provide some simplified concepts to help to understand the scale of the climate change problem:

1. The earth's renewable resources are currently being consumed faster than they can be regenerated.
2. Three planets worth of resources would be required if all of the world's population had a westernized lifestyle.

Climate change predictions for the UK

The following scenarios are predicted for the UK:

- The climate will become warmer – by possibly 2 to 3.5°C by the 2080s.
- Hot summers will be more frequent and extreme cold winters less common.
- In 2004 one day per summer was expected to reach 31°C, but by the 2080s this could be nearer 10 days, with one day per summer reaching 38.5°C.
- Summers will become drier and winters wetter.
- There will be less snowfall.
- Heavy winter precipitation events will become more frequent.
- Sea levels will continue to rise by an estimated 26–86 cm by the 2080s.
- Extreme sea levels will return more frequently.

These changes to the climate are likely to have the following implications for building design:

- Increased flooding events (coastal, river and urban/flash).
- Buildings less weather-tight in face of more inclement weather.
- Increased foundation movement on clay soils due to drier summers.
- Increased summer overheating.
- Disruption of site activities due to inclement weather.
- Potential modifications to design loadings (e.g. wind) and durability predictions for building materials (in particular sealants, jointing materials, plastics, coatings and composites).

Source: UKCIP02 Climate Change Scenarios (funded by DEFRA, produced by the Tyndall & Hadley Centres for UKCIP). Copyright BRE, reproduced from Good Building Guide 63 with permission.

Sustainability scenarios and targets

In addition to dealing with the effects of climate change as it happens, sustainable design must also consider how buildings will help to achieve governmental carbon emissions reductions targets. In broad terms, environmental impact can be reduced by applying the following principles in order: **Reduce – Reuse – Recycle – Specify Green**. However this sentiment is meaningless without specific targets and this is where the individual's view affects what action (if any) is taken.

As climate change involves complex interactions between human and natural systems over the long term, scenario planning is a frequently used assessment tool. Scenarios can be qualitative, narrative or mathematical predictions of the future based on different actions and outcomes. In 2001, the Intergovernmental Panel on Climate Change (IPCC) identified more than 500 mathematical scenarios and over 120 narrative scenarios, which fall into four simplified categories:

1. *Pessimist* – climate change is happening; little can be done to prevent or mitigate this.
2. *Economy paramount* – business should continue as usual, unless the environment affects the economy.
3. *High-tech optimist* – technological developments and a shift to renewable energy will provide the required efficiencies.
4. *Sustainable development* – self-imposed restrictions and lifestyle change with energy efficiencies in developed countries, to leave capacity for improved quality of life in developing countries.

UK policy is broadly based on the findings of the IPCC study and at the time of writing, the draft UK Climate Change Bill 2008 proposes the following carbon reduction targets based on a *Sustainable Development* agenda (although there are many organizations campaigning for higher targets):

- 26–32% reduction by 2020 compared to 1990 emissions.
- 60% reduction by 2050 compared to 1990 emissions.

These targets are likely to have a significant impact on the average UK lifestyle and it is unlikely that the required carbon emissions reductions will be achieved by voluntary lifestyle changes:

> In 1990, carbon emissions were about 10.9 tonnes of CO_2 per person in the UK and we can assume that this represents the emissions generated by the average UK lifestyle. Government policy aims to reduce emissions by 60% by 2050 which equates to 4.4 tonnes of CO_2 per person per year. However a UK citizen who holidays in the UK, only travels on foot or by bike, likes a cool house, conserves energy and has lower than average fuel bills, while buying energy from renewable sources, eating locally grown fresh food and producing less than the average amount of household waste (most of which would be recycled), might generate about 5.6 tonnes of CO_2 per year.

As carbon emissions targets are non-negotiable – if less is done in one area, more will need to be achieved in another area. For some time, UK government policy has concentrated on making alterations to electricity generation to reduce carbon emissions (hence the public consultation on nuclear power in 2007). This has not been entirely successful as the proliferation of computers, personal electronics and air conditioning have offset most of the savings made by altering electricity supply.

Therefore as the built environment is thought to be the second largest consumer of raw materials after food production, the construction industry is considered to be a sector where disproportionate reductions can be made to offset against other areas. Increasing controls are being applied via Planning, Building Regulations, Energy Performance of Buildings Directive and other initiatives such as the Code for Sustainable Homes. This legislation aims to achieve 'zero carbon' housing by 2016 and it is likely that similar controls are to be applied to commercial buildings by 2019. It is highly likely that, until the nuclear debate is settled, construction professionals will be expected to deliver increasingly more efficient buildings to minimize the impact on other areas of the average UK lifestyle.

It is worth noting that buildings can vary their energy in use by up to a factor of 3 depending on how the people inside choose to operate them, thus carbon emission reductions from energy efficient buildings are not guaranteed. Carbon reductions from the built environment will only be delivered if we all choose to use and operate our buildings more efficiently.

For climate change sceptics, increasing energy costs and security of energy supply are two alternative reasons why energy efficiency measures are being pursued.

Sustainable building design priorities

In addition to the wider issues regarding targets, sustainable building design is difficult to grasp because of the many conflicting design constraints, as well as the need to involve the client, design team, contractors and building users in decision-making. If substantial carbon emission savings are to be achieved, it is likely that similar substantial changes will also be required to broader industry practices, such as changes to contract structures, standard specifications and professional appointments being extended to include post occupation reviews. However, while the industry infrastructure develops to support sustainable design in mainstream practice, there are still many issues for designers to tackle.

Order of priorities

To ensure that design efforts to improve environmental sustainability are efficiently targeted, a broad hierarchy should be applied by the design team:

- **Building location** – Transportation of building users to and from a rural location can produce more carbon emissions over the building's life than those produced by the building services. Encourage clients to choose a location which encourages the use of public transport.
- **Energy in use** – The energy used by building services can account for 60–80% of the total carbon emissions produced in the life of a building. Heavily serviced buildings can produce almost double the lifetime emissions of naturally ventilated buildings. Selection of natural ventilation and simple services can substantially reduce carbon emissions – as well as protecting clients against future energy costs or shortages.
- **Embodied energy** – The relative importance of embodied energy increases as energy in use is reduced. If an efficient location and services strategy has been selected, embodied energy can account for about 40% of a building's lifetime emissions. This should leave plenty of scope for some emissions reductions, although higher embodied energy materials can be justified if they can contribute to the thermal performance of the building and therefore reduce the energy in use emissions.
- **Renewables** – After efficiencies have been made in all other areas, renewable energy sources can be harnessed to reduce carbon emissions – assuming that the lifetime energy savings justify the embodied energy used in the technology!

Design team actions

In the early stages of design, design teams should aim to:

- Choose an efficient building shape to minimize energy in use and details to minimize air leakage and heat loss.
- Design simple buildings with reduced interfaces.
- Assess the implications of design life – lightweight short-lived construction, versus heavier, flexible, durable buildings.
- Anticipate change and make those changes easy to achieve. Consider the effect of future adaptions based on fashion and the expected service life of different building layers: 5–15 years for the fixtures and fittings, 5–20 years for the space plan, 5–30 years for building services, 30–60 years for the façade and 60–200 years for the structure.
- Establish whether the structure should contribute to the thermal performance of the building.
- Agree reduction targets for pollution, waste and embodied and operational energy.
- Consider a specification catchment area (or radius around the site) from which all the materials for the project might come, to minimize transport emissions and benefit the local economy and society.

Actions for the structural engineer

Structural engineers are most likely to have direct control of, or influence over:

- Assessment of structures and foundations for reuse.
- On-site reuse of materials from demolitions or excavations.
- Minimum soil movements around and off site.
- Selection of a simple structural grid and efficient structural forms.
- Detailing structures with thermal mass to meet visual/aesthetic requirements.
- Balancing selection of design loadings to minimize material use, versus provision of future flexibility/adaptability/deconstruction.
- Use of reclaimed, recycled, 'A-rated' or 'green' building materials.
- Use of specifications to ensure material suppliers use environmental management systems (e.g. ISO 14001 or EMAS).
- Avoidance of synthetic chemicals, polyvinyl chloride (PVC), etc.
- Limiting numbers of building materials to reduce waste.
- Design to material dimensions to limit off-cuts and waste.
- Assessment of embodied energy and potential reductions.
- Assessment of prefabrication to minimize waste, if the carbon emissions resulting from transport do not outweigh the benefits.
- Specifications to reduce construction and packaging waste.
- Drainage systems to minimize run-off.
- Use of flood protection measures and flood-resistant materials.
- Keep good records to help enable future reuse or refurbishment.

Exposed slabs and thermal mass

Thermal mass is the name given to materials which (when exposed to air flows) regulate temperature by slowly absorbing, retaining and releasing heat; preventing rapid temperature fluctuations. This property is quantified by the specific heat capacity (kJ/kgK). There are three situations where a services engineer might use exposed thermal mass to minimize or eliminate the need for mechanical heating and cooling, and therefore make considerable energy and carbon emissions savings:

- **Night-time cooling** – air is let into buildings overnight to pre-cool walls and slabs, to increase their ability to absorb heat during the day. In the UK, this system is particularly suited to offices, but can be difficult to achieve in urban areas where heat tends to be retained, reducing the temperature differentials required to make passive cooling systems work.
- **Passive solar heating** – heat from the sun is collected during daylight hours, stored in the building fabric and then slowly released overnight. In the UK, this system is particularly suited to domestic houses which require heating throughout most of the year.
- **Temperature stabilization** – where day and night-time temperatures vary significantly above and below the average temperature. Not typically required in the UK.

Although the decision to use thermal mass is generally driven by the desire for an energy-efficient building services strategy, the need for exposed thermal mass means that the architect and building users will be more concerned about how the structural elements look. Extra care has to be taken when selecting and specifying exposed elements in terms of finish, details, erection and protection from damage and weather. Thermal mass is likely to become increasingly important in the light of increasingly hot summers and UK government policy to reduce carbon emissions.

Typical specific heat capacity of different building materials

Material	Typical specific heat capacity (kJ/m³K)
Water	4184
Granite	2419
Concrete	2016
Sandstone	1806
Clay tiles	1428
Compressed earth block	1740
Rammed earth	1675
Brick	1612
Earth wall (adobe)	1300
Wood	806
Rockwool insulation	25
Fibreglass insulation	10

Source: Adapted from data in Guide A, CIBSE (2006).

Embodied energy

The embodied energy of a material is the energy used to extract, process, refine and transport it for use. Typically the more processing steps, or distance travelled, the higher the embodied energy – which is often reflected in its price. The higher the embodied energy, the higher the carbon emissions generated by production.

In many cases it is possible to justify higher embodied energy if there is some other benefit, for example increased design loadings resulting in a more flexible building or concrete slabs providing thermal mass to regulate temperature. Although energy in use is more significant, it is still worth reducing embodied energy when this can be achieved without compromising performance standards or incurring other adverse environmental impacts.

Typical embodied energy contribution of building elements

Volume and service life for different materials affect each building element's environmental impact.

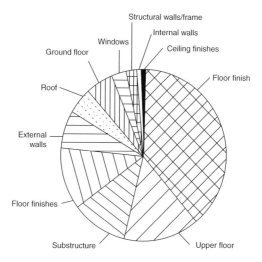

Although all specification choices are important, designers might concentrate on the building elements which have the greatest environmental impact:

● Floor construction, floor surfacing and floor finishes.
● External walls and windows.
● Roofs.

General strategies for reduction of embodied energy

It is a complex area and each case should be studied independently using the best method available at the time. Therefore as the available figures for the embodied energy of typical building materials can vary by up to a factor of 10, it is better to follow general guidelines if no site specific data is available for a particular project. The general themes are:

- Don't build more than you need – optimize rather than maximize space.
- Design long life, durable, simple and adaptable buildings.
- Modify or refurbish, rather than demolish or extend.
- 'High-tech' normally means higher levels of embodied energy.
- Consider higher design loadings to maximize the building life.
- Reuse material found on, or excavated from, site.
- Source materials locally.
- Use salvaged materials in preference to recycled materials.
- Use recycled materials in preference to new materials.
- Use high grade salvaged or recycled materials, not just as bulk fill, etc.
- Select low embodied energy materials.
- Give preference to materials produced using renewable energy.
- Specify standard sizes and avoid energy-intensive fillers.
- Avoid wasteful material use and recycle off-cuts and leftovers.
- Design for demountability for reuse or recycling.

Typical embodied energy values for UK building materials

Materials		Typical embodied energy* MJ/kg
Concrete	1:2:4 Mass concrete	0.99
	1% Reinforced section	1.81
	2% Reinforced section	2.36
	3% Reinforced section	2.88
	Precast	2.00
Steel	General	24.00
	Bar and rod	19.7
	Stainless	51.50
Facing bricks		8.20
Lightweight concrete blocks		3.50
Aggregate (general)		0.15
Timber	Sawn softwood	7.40
	Sawn hardwood	7.80
	Plywood	15.00
	Glue-laminated	11.00
Polycarbonate		112.90
Glass	Annealed	13.50 ± 5.00
	Toughened	16.20–20.70
Insulation	Mineral wool	16.60
	Polyurethane	72.10
	Sheeps' wool	3.00
Wallpaper		36.40
Plaster	Gypsum	1.8
	Plasterboard	2.7
Aluminium		154.30
Stone	Granite	5.90
	Imported granite	13.90
	Limestone	0.24
	Marble	2.00
	Slate	0.1–1.0
Drainage	Clay pipe	6.19–7.86
	PVC pipe	67.50

*These figures are for illustrative purposes only. Embodied energy figures can vary up to a factor of 10 from site specific data. The Green Guide to Specification might be a more useful guide for lay persons.

Source: University of Bath (2006).

Construction waste

It is estimated that a minimum of 53 million tonnes of construction and demolition waste are produced annually in the UK. Of this, 24 million tonnes is recycled and 3 million tonnes is reclaimed – leaving 26 million tonnes being dumped in landfill, with associated air and water pollution. However with waste disposal costs rising and the UK government's plan to halve the amount of construction waste going to landfill by 2012, waste reduction strategies are likely to become more widespread. Some of the first steps towards this are the creation of WRAP (Waste and Resources Action Programme) and introduction of compulsory Site Waste Management Plans for all sites in England from April 2008.

The benefits of reducing construction waste are threefold:

- Reduction of carbon emissions associated with less transport and processing.
- Reduction of waste going to landfill.
- Reduction of raw material use.

Using the **Reduce – Reuse – Recycle – Specify Green** hierarchy, construction waste can be minimized by using the specification and contract documents to:

- Encourage use of reusable protection and packaging systems for deliveries.
- Reduce soil movements by reusing material on site.
- Increase recovery and segregation of waste for reuse on site or recycling elsewhere (preferably locally).
- Use of efficient installation and temporary works systems.
- Design simple, efficient buildings with minimum materials and interfaces to reduce waste and off-cuts.
- Design buildings for flexibility, adaption and demountability.

Main potential for waste recovery and reuse after demolition

Steel, masonry, concrete and timber comprise the vast bulk of construction materials and all offer possibilities for reuse where fixings have been designed to facilitate this. Timber tends to be susceptible to poor practice and is not reused as often as steel and masonry. Glass and plastics tend to have limited reuse potential, and are generally more suited to recycling.

Material	Value	Potential for reuse	Potential for recycling
Concrete	High	• Precast concrete elements. • Large concrete pieces to form thermal store for passive heating system.	• Crushed for use as aggregate in concrete mixes. • Crushed for use as unbound fill.
Masonry	Medium	• Bricks and blocks if used with soft mortar. • Stone and slate.	• Crushed for use as aggregate in low strength concrete mixes. • Crushed for use as unbound fill.
Metals	High	• Steel beams and columns if dismantled rather than cut with thermal lance.	• Aluminium and copper. • Steel beams and reinforcement.
Timber	Medium	• Generally reused in non-structural applications for indemnity reasons. • High value joinery.	• Chipped for use in landscaping, engineered timber products, etc., if not prevented by remains of old fixings and preservatives (treated timber is considered hazardous waste). • Composting. • Energy production.
Glass	High	• Re-glazing.	• Crushed for use as sand or fine aggregate in unbound or cement-bound applications. • Crushed for use in shot blasting, water filtration, etc.

Reclaimed materials

Reclaimed materials are considered to be any materials that have been used before either in buildings, temporary works or other uses and are reused as construction materials without reprocessing. Reclaimed materials may be adapted and cut to size, cleaned up and refinished, but they fundamentally are being reused in their original form.

This is the purest and most environmentally friendly form of recycling and therefore, where possible, should be investigated before the use of recycled or reprocessed materials in line with the **Reduce – Reuse – Recycle – Specify Green** hierarchy.

Although recycled content is generally resolved by manufacturers and their quality control processes, the use of reclaimed materials is generally more difficult as it must be resolved by the design team, within the limits of the site and project programme. The following should be considered if substantial amounts of reclaimed materials are proposed:

- Early discussions with reclaimed materials dealers will help to identify materials that are easily available at the right quality and quantity.
- Basic modern salvage direct from demolition is often cheap or free, whereas older antique or reclaimed materials from salvage yards and stockholders (particularly in large quantities) may be quite costly.
- Material specifications need to be flexible to allow for the normal variations in reclaimed materials. Specifications should outline the essential performance properties required of a material, avoiding specification of particular products.
- Early design information helps in the sourcing of reclaimed materials, which might have considerably longer lead times than for off-the-shelf materials.
- It can be helpful to use agreed samples as part of the specification process – indicating acceptable quality, colour and state of wear and tear, etc.
- Identify nearby demolition projects and negotiate for reclaimed materials which might be useful.
- The building contractor will often need to set up relationships with new suppliers in the salvage trade.
- Additional storage space on, or near, the site is essential. Reclaimed materials do not fall into the 'just in time' purchasing process normally used by contractors.
- Use of experience, provenance, visual inspections, testing and/or clear audit trails to confirm material standards and satisfy indemnity requirements. Reclaimed materials inspectors are available in some areas if certain aspects are outside the design team's expertise.

Recycled materials

Recycled materials are considered to be any materials that have been taken from the waste stream and reprocessed or remanufactured to form part of a new product. A further refinement of this is:

- *Recycling* is where materials can be reclaimed with broadly the equivalent value to their original, for example steel, paper, aluminium, glass and so on.
- *Downcycling* is where materials are reclaimed but can only be used in a lesser form than previously, for example crushed concrete frame used as hardcore.

Ideally downcycling should be limited, but different materials are particularly suited to specific reprocessing techniques; metals being easily recycled, concrete most easily downcycled and timber relatively easily reused.

Most building designs achieve reasonable amounts of recycled content without explicitly trying as many manufacturers have traditionally used high levels of recycling. For example, a typical steel framed, masonry clad building might achieve 15–20% and a timber framed building about 10% (by value). WRAP (the Waste and Resources Action Plan) predict that most building projects could achieve a further 5–10% recycled content by specifying similar products (with higher recycled contents) without affecting the cost or affecting the proposed building design. This means that typical buildings should easily be able to achieve 20–25% (by value) recycled content without any radical action.

Although this sort of target reduces the amount of waste going to landfill, it does very little in the context of carbon emissions reductions. On the basis of the 30–60% reductions stated in the UK's Climate Change Bill, recycled content targets might be more meaningful in the 25–40% (by value) range.

Recycled content is normally calculated as a percentage of the **material value** to:

- Maximize carbon emissions savings – high value items generally have higher embodied energy values.
- Encourage action across the whole specification – rather than allowing design teams to concentrate on small savings on high volume materials.

WRAP have an online tool for calculating the recycled content of projects for most construction materials.

Finally, after deciding on a target for the project, adequate time must be set aside to ensure that appropriate materials and/or recycled contents are specified, as well as monitoring on site whether the materials specified are actually those being used.

Design for demountability

Although the use of reclaimed and recycled materials deals with waste from past building operations, building designers should perhaps consider the role of demountability in limiting the amount of waste produced in the future, in line with the **Reduce – Reuse – Recycle – Specify Green** hierarchy.

Basic principles of demountability

Specific considerations for demountability are as follows:

- Anticipate change and design/detail the building to allow changes for fit-out, replanning, major refurbishment and demolition to be made easily.
- Pay particular attention to the differential weathering/wearing of surfaces and allow for those areas to be maintained or replaced separately from other areas.
- Detail the different layers of the building so that they can be easily separated.
- Try to use durable components which can be reused and avoid composite elements, wet/applied finishes, adhesives, resins and coatings as these tend to result in contamination of reclaimed materials on demolition.
- Try to specify elements which can be overhauled, renovated or redecorated as a 'second hand' appearance is unlikely to be acceptable.
- Selection of small elements (e.g. bricks) allows design flexibility and therefore allows more scope for reuse.
- Adopt a fixing regime which allows all components to be easily and safely removed, and replaced through the use of simple/removable fixings.
- Ensure the client, design team, contractors and subcontractors are briefed.
- Develop a detailed Deconstruction Plan with each design stage and ensure this forms part of the final CDM Building Manual submitted to the client.
- Carefully plan services to be easily identified, accessed and upgraded or maintained with minimum disruption. Design the services to allow long-term servicing rather than replacement.
- There is unlikely to be a direct cost benefit to most clients and design for complete demountability, however it should be possible to make some degree of provision on most projects without cost penalty.

Demountable structure

For structure, design for demountability should probably be the last design consideration after flexible design loadings and layout, as the latter aim to keep the structural materials out of the waste stream for as long as possible. However with rising landfill costs, demountability of structures may become more interesting in the future.

Connections are probably the single most important aspect of designing for deconstruction. The best fixings are durable and easily removable without destroying the structural integrity and finish of the joined construction elements. Dry fixing techniques are preferable and recessed or rebated connections which involve mixed materials should be avoided.

Implications of connections on deconstruction

Type of connection	Advantages	Disadvantages
Nail fixing	• Quick construction. • Cheap.	• Difficult to remove and seriously limits potential for timber reuse. • Removal usually destroys a key area of element.
Screw fixing	• Relatively easy to remove with minimum damage.	• Limited reuse. • Breakage during removal very problematic.
Rivets	• Quick installation.	• Difficult to remove without destroying a key area of element.
Bolt fixing	• Good strength. • Easily reused.	• Can seize up, making removal difficult. • Fairly expensive.
Clamped	• Easily reused. • Reduced fabrication.	• Limited choice of fixings. • Expensive.
Traditional/Tenon	• Quick installation. • Easily reused.	• Additional fabrication. • Expensive labour.
Mortar	• Strength can be varied. • Soft lime mortar allows easy material reclamation.	• Cement mortars difficult to reuse and prevent reclamation of individual units. • Lime mortar structures require more mass to resist tension compared to cement mortars.
Adhesives	• Strong and efficient. • Durable. • Strength can be varied. • Good for difficult geometry.	• Likely to prevent effective reuse of parent materials. • Relatively few solvents available for separation of bonded layers at end of life. • Adhesive not easily recycled or reused.

Green materials specification

Assessment and specification of environmentally friendly materials is incredibly complex as research is ongoing and good practice is under constant review. There are a number of companies who provide advice on this area, but one of the best and simplest sources of information is *The Green Guide to Specification*. The most important issue is to feed this into the design early to inform decision-making, rather than trying to justify a design once complete.

Environmental rating for selected suspended floors

Structural element	Summary rating
Beam and block floor with screed	A
Hollowcore slabs with screed	A
Hollowcore slabs with structural topping	B
In situ reinforced concrete slab	C
In situ reinforced concrete ribbed slab	B
In situ reinforced concrete waffle slab	B
Omni-deck-type precast lattice and structural topping	B
Omni-deck-type precast lattice with polystyrene void formers and structural topping	B
Holorib-type in situ concrete slab with mesh	B
Solid prestressed composite planks and structural topping	C

NOTES:

1. Suspended floor assessment based on 7.5 m grid, 2.5 kN/m^2 design load and 60 year building life.
2. Timber joisted floors are typically not viable for this sort of arrangement, but perform significantly better in environmental terms than the flooring systems listed.
3. Weight reductions in profiled slabs outweigh the environmental impact of increased amounts of shuttering.

Source: Green Guide to Specification (2002).

Toxicity, health and air quality

Issues regarding toxicity generate considerable debate. Industrial chemicals are very much part of our lives and are permitted by UK and European laws based on risk assessment analysis. Chemicals present in everyday objects, such as paints, flooring and plastics, leach out into air and water. However many chemicals have not undergone risk assessment and assessment techniques are still developing. Environmental organizations, such as Greenpeace and the World Health Organization, suggest that we should substitute less or non-hazardous materials wherever possible to protect human health and the environment on the basis of the *Precautionary Principle*. With such large quantities specified, small changes to building materials could make considerable environmental improvements.

Summary of toxins, associated problems and substitutes

Material or product	Typical uses	Associated problem	Possible substitutes
Any containing VOCs	Petrochemical-based paints, plastics, flooring, etc.	See note.	Water or vegetable oil based products (e.g. linoleum, ceramic tile, wood, etc.).
Polyvinyl chloride (PVC or 'vinyl')	About 50% of all PVC is used in construction: pipes, conduit, waterproofing, roof membranes, door and window frames; flooring and carpet backing, wall coverings, furniture and cable sheathing.	Lifetime VOC emissions; regular combination with heavy metals and release of hydrochloric acid if burnt.	Aim for biobased plastics such as polyethylene terephthalate (PET), polyolefins (PE, PP, etc.) or second choice polyethylene terephthalate (PET), polyolefins (PE, PP, etc.) but try to avoid polyurethane (PU), polystyrene (PS), acrylonitrile butadiene styrene (ABS), polycarbonate (PC).
Phthalates	Used to make PVC flexible. Typically in vinyl flooring, carpet backing and PVC wall or ceiling coverings.	Bronchial irritants; potential asthma triggers and have been linked to developmental problems.	Subject to ongoing research and risk assessment, but options being considered include: adipates, citrates and cyclohexyl-based plasticizers.
Polychloroprene (neoprene)	Geotextile, weather stripping, water seals, expansion joint filler, gaskets and adhesives.	Same as PVC.	
Composite wood products and insulation (using urea or phenol formaldehyde)	Panelling, furniture, plywood, chipboard, MDF, adhesives and glues.	Formaldehyde is a potent eye, upper respiratory and skin irritant and is a carcinogen.	
Preserved wood	Chromium copper arsenic (CCA), creosote and pentachlorophenol, PCP, lindane, tributyl tin oxide, dichlofluanid, permethrin.	Carcinogenic.	No treatment, boron-based compounds, Cu and Zn naphthanates or acypectas zinc.
Heavy metals	Flashings, roofing, solder, switches, thermostats, thermometers, fluorescent lamps, paints and PVC products as stabilizers.	Lead, mercury and organotins are particularly damaging to the brains of children. Cadmium can cause kidney and lung damage.	
Halogenated flame retardants (HFRs)	Flame retardants used on polyurethane (PR) and polyisocyanurate (PIR) insulation in buildings.	PBDEs and other brominated flame retardants (BFRs) disrupt thyroid and oestrogen hormones, causing problems with the brain and reproductive system.	Area of ongoing research looking at halogen-free phosphorous compounds for PIR products. Rarely available.

NOTES:
1. Volatile organic compounds (VOCs) are thousands of different chemicals (e.g. formaldehyde and benzene) which evaporate readily in air. VOCs are associated with dizziness, headaches, eye, nose and throat irritation or asthma, but some can also cause cancer, provoke longer term damage to the liver, kidney and nervous system.
2. Greenpeace publish a list of Chemicals for Priority Action (after OSPAR, 1998) which they are lobbying to have controlled.

Sustainable timber

Timber is generally considered a renewable resource as harvested trees can be replaced by new saplings. However this is not always the case, with deforestation and illegal logging devastating ancient forests around the world. Since 1996 a number of certification schemes have been set up to help specifiers select 'legal' and 'sustainable' timber. The *Chain of Custody Certification* standards address management planning, harvesting, conservation of biodiversity, pest and disease management and social impacts of the forestry operations. At present only 7% of the world's forests are certified; mainly located in the Northern Hemisphere and are relatively free from controversy. Sustainable forest management has great significance for the world climate and if all building specifications insist on certified timber, industry practices worldwide will be forced to improve.

The first action is to consider where timber is to be used on a project: structure, temporary works, shuttering, joinery, finishes, etc. and avoid selection of materials which are highly likely to come from illegal or unsustainable sources. At specification stage, options which should be included for all types of timber to be used are:

- Certified timber from an official scheme.
- Timber from independently certified and reliable suppliers, with documentary evidence that supplies are from legal and well-managed forests.
- Timber from suppliers that have adopted a formal Environmental Purchasing Policy (such as those freely available from Forests Forever, WWF 95+) for those products and that can provide evidence of commitment to that policy.
- Timber for illegal sources must not be used and any timber must be shown to come from legal sources. Environmental statements alone are not to be used as demonstration that materials are from a sustainable source.
- Where possible provide a list of FSC accredited suppliers and request evidence of certified timber purchase. This might include requests for custody certificate numbers, copies of invoices and delivery notes.

Despite some weaknesses in some certification schemes, they are a significant step towards sourcing of sustainable timber. The main certification schemes are as follows:

● **COC – Chain of Custody**
Independent audit trail to prevent timber substitution and ensure an unbroken chain from well-managed forest to user.

● **FSC – Forest Stewardship Council**
Independent, non-profit organization implementing a COC system. Greenpeace consider this to be the only scheme which is truly effective.

● **PEFC – Programme for the Endorsement of Forest Certification Schemes**
Global umbrella organization for forestry industry bodies covering about 35 certification schemes including FSC, although with weaker social and environmental criteria.

● **MTCC – Malaysian Timber Certification Council**
● **CSA – Canada Standards Authority**
● **SFI – Sustainable Forests Initiative**

Timber preservatives

Although many hardwoods can be left untreated, softwood (whether internal or external) is routinely treated to provide durability via resistance to rot and insect infestation. This practice has only really developed since about 1940 and may be because the fast grown timber used today does not have the same natural resistance to decay as the close-ringed, slow grown timber which used to be standard (and can still be sourced at a premium today). Alternatively it may be because clients want the warranties and guarantees which preservative companies provide. However the reasons for considering avoiding timber preservatives are as follows:

- Many preservatives release toxins to air, surface water and soil, to which workers and consumers are exposed.
- Impregnating wood hampers the sustainable reuse of wood.
- Treated timber is classed as 'Hazardous Waste' and should not be burned or sent to landfill to avoid air and water pollution.

Specifying timber preservatives

Timber will generally only deteriorate when its moisture content is higher than 20%. Therefore with careful detailing, preservative treatment might be reduced or avoided, in accordance with BS 5589 and BS 5268: Part 5.

If preservatives are required, try to use non-toxic boron-based compounds such as borate oxide. However the treatment can only be carried out on green timber with a moisture content of over 50%. As this is well above the desired moisture content at installation, sufficient time will need to be allowed in the programme for suppliers to be sourced and the timber to be seasoned (preferably by air, rather than kiln, drying).

If time is an issue, copper or zinc naphthanates, acypectas zinc, ammoniacal copper quaternary (ACQ), copper azole and copper citrate can be considered – as a last resort. However creosote, arsenic, chromium salts, dieldrin (banned in the UK), PCP, lindane, tributyl tin oxide, dichlofluanid, permethrin and copper chrome arsenate should be completely avoided.

Cement substitutes

Old Portland cement production (firing limestone and clay in kilns at high temperature) produces about 1 tonne of CO_2 for every tonne of cement produced. Cement production accounts for 5% of all European carbon emissions. In addition to improving energy efficiency at cement production plants, carbon emissions savings can be made by using cement substitutes, which also improve concrete durability. One drawback is that construction programmes need to allow for the slower curing times of the substitutes which work in two ways:

1. Hydration and curing like portland cement, although slightly slower.
2. 'Pozzolans' providing silica that reacts with the hydrated lime which is an unwanted by-product of concrete curing. While stronger and more durable in the end, pozzolans take longer to set, although this can be mitigated slightly by reducing water content.

Ground granulated blast furnace slag aggregate cement

Ground granulated blast furnace slag aggregate cement (GGBS) is a by-product of iron and steel production. Molten slag is removed from the blast furnaces, rapidly quenched in water and then ground into a fine cementitious powder. GGBS tends to act more like Portland cement than a pozzolan and can replace Portland cement at rates of 30–70%, up to a possible maximum of 90%. As the recovery and production of 1 tonne of GGBS produces about 0.1 tonne of CO_2 considerable carbon emissions savings can be made. It is common practice in the UK for ready mixed concrete companies to produce concrete with a cementitious component of 50% GGBS and 50% Portland cement. Concrete using GGBS tends to be lighter in colour than those with Portland cements and can be considered as an alternative to white cement (which results in higher carbon emissions than Portland cement) for aesthetics or integration with daylighting strategies.

Pulverized fuel ash

Pulverized fuel ash (PFA) is a by-product of burning coal in power stations and is also known as 'Fly ash'. The ash is removed from flue gases using electrostatic precipitators and is routinely divided into two classes: 'Type C' and 'Type F' according to the lime (calcium) content. Type F has a higher calcium content and acts more like a pozzolan than Type C, which has pozzolanic and Portland cement qualities. Both types can be used in concrete production, replacing Portland cement at rates of 10–30%, though there have been examples of over 50% replacement. Available from ready mix suppliers, concrete mixed with PFA cement substitute tends to be darker in colour than Portland cement mixes.

Non-hydraulic and naturally hydraulic lime

Fired at lower temperatures than Portland cement and with the ability to reabsorb CO_2 while curing (as long as the volume of lime material itself does not prevent this), has led some lime manufacturers to claim that lime products are responsible for 50% less carbon emissions than similar cement products. Lime products are not commonly used in new-build projects, but are generating increasing interest. Traditionally lime is used for masonry bedding and lime-ash floors. Being softer than cement, lime allows more movement and reduces the need for masonry movement joints (as long as the structure has sufficient mass to resist tensile stresses), as well as allowing easier recycling of both the masonry units and the lime itself.

Magnesite

The idea of replacement of the calcium carbonate in Portland cement with magnesium carbonate (magnesite or dolomite) dates back to the nineteenth century. Less alkaline than Portland cement mixes, they were not pursued due to durability problems. However MgO cement uses 'reactive' magnesia that is manufactured at much lower temperatures than Portland cement (reducing emissions by about 50%), is more recyclable than PC, is expected to provide improved durability and to have a high propensity for binding with waste materials. It is claimed that magnesite can be used in conjunction with other cement replacements without such problems as slow curing times. The main barrier to use seems to be that although magnesite is an abundant mineral, it is expensive to mine. Research is currently being carried out and further information is available from the University of Cambridge, TecEco and the BRE.

Sustainable aggregates

Sustainable aggregates fall into two categories:

1. **Recycled aggregates** – derived from reprocessing materials previously used in construction.
2. **Secondary aggregates** – usually by-products of other industrial processes not previously used in construction. Secondary aggregates can be further sub-divided into manufactured and natural, depending on their source.

Although the UK is a leading user of sustainable aggregates, with about 25% of the total UK aggregate demand being met with sustainable products, there is scope for this to be expanded further. Recycled aggregates (RAs) can be used in unbound, cement bound and resin bound applications subject to various controls.

RAs can be purchased directly from demolition sites or from suitably equipped processing centres, and the quality of the product depends on the selection, separation and processing techniques used. RA can be produced on site, at source or off site in a central processing plant, with economic and environmental benefits maximized with on-site processing.

Many materials have a strong regional character, with china clay sand from South West England, slate waste from North Wales and metallurgical slag from South Wales, Yorkshire and Humberside. Clearly the biggest economic and environmental benefits will be gained when materials are used locally.

Typical uses for UK recycled and secondary aggregates

Aggregate type	Potential for reuse				Notes
	Unbound aggregate	Concrete aggregate	Lightweight aggregate	Building components aggregate	
Recycled aggregate					
Crushed concrete (RCA)	High	High	None	Some	Alkali silica reaction (ASR), frost resistance and weathering should be considered.
Crushed masonry (RA)	High	High	High	Some	
Ceramic waste	Some	Some	None	Some	
Recycled glass	Some	High	Some	Some	
Spent rail ballast	High	Low	None	High	
Mixed plastic	None	Low	Some	High	
Scrap tyres	None	None	Low	Some	
Secondary aggregate – manufactured					
Blast furnace slag (Lytag)	High	High	High	High	Regular use. BS EN and BRE IP18/01:2001 guidance.
Steel slag	None	None	None	None	See BRE Reports.
Non-ferrous slags	Low	Some	None	Low	Unbound use to comply with BRE Digest SD1:2001.
Pulverized fuel ash (PFA)	Some	Some	High	Some	ASR, sulphates, frost resistance and weathering should be considered.
Incinerator bottom ash (IBA) from municipal waste incinerators	Some	Some	Some	Some	
Furnace bottom ash (FBA)	Some	Some	High	High	Fully utilized in concrete blocks.
Used foundry sand	Some	High	High	Some	
Sewage sludge as synthetic aggregate	None	None	High	None	Use as a substitute for natural gypsum.
FGD gypsum (desulphogypsum results from desulphurization of coal fired power station flue gases)	None	None	None	High	
Secondary aggregate – natural					
Slate waste	High	Some	High	High	
China clay sand	Some	High	None	High	
Burnt colliery spoil	Some	Low	None	Low	Sulphate content can be high. Unbound use to comply with BRE Digest SD1:2001.
Unburnt colliery spoil	High	None	Some	None	
Clay waste	None	None	High	None	

Detailed descriptions of these materials can be found within the Recycled Content Specifier Tool available on the WRAP website.

Unbound use

RAs are highly suitable for use under floor slabs and for pipe bedding as well as for general fill materials in building construction. Contaminants such as metals, plastic and wood should normally be kept below 2% and the grading should be suitable for full compaction where this is required.

The only other consideration when using them is that the fine fractions of both RCA and RA could be contaminated with sulphate salts (e.g. from some types of gypsum plaster) to a degree sufficient to cause sulphate attack on concrete in contact with it. Therefore the soluble sulphate content of material, containing fine RA should be tested and appropriate precautions taken.

RAs in concrete mixes

Based on the **Reduce – Reuse – Recycle – Specify Green** hierarchy, and the desire to avoid *Downcycling*, engineers should be aiming to use sustainable aggregates in concrete mixes, in addition to lower grade unbound applications such as fill and pipe bedding and so on.

Most ready mix concrete suppliers can offer concrete containing RAs, but mixes are of limited availability and may not be available at the right time, in the right place or in the required quantities. It is therefore not practical to insist on their use on every project at present, but expressing a preference in specifications should encourage their use where possible.

BS 8500 gives limits on the permitted composition of recycled coarse aggregates as well as guidance on where and how their use in concrete is permitted, but the use of fine aggregates is not covered at present as their increased water demand generally leads to low strength mixes. The use of concrete containing recycled coarse aggregates is restricted to the least severe exposure classes and is not yet practical for use in site batching. Designated concrete mixes, which require strength tests to be carried out, are the easiest way to specify recycled content while maintaining quality control.

BS 8500: Part 2 Clause 4.3 defines two categories of coarse recycled aggregate, that is, Recycled Concrete Aggregate (RCA) consisting primarily of crushed concrete (i.e. where less than 5% is crushed masonry) and Recycled Aggregate (RA) which may include a higher proportion of masonry and must meet a default value for aggregate drying shrinkage of 0.075%.

RA is limited to use in concrete with a maximum strength class of C16/20 and in only the mildest exposure conditions, whereas RCA can be used up to strength class C40/50 and in a wider range of exposure conditions, but is generally restricted to use in non-aggressive soils (DC-1 conditions).

Although it is generally accepted that the use of coarse RCA to replace up to 30% of the natural coarse aggregate will have an insignificant effect on the properties of concrete, for BS 8500 designated concretes RC25–RC50, the amount of RCA or RA is restricted to 20% by weight of the total coarse aggregate fraction unless the specifier gives permission to relax this requirement.

Therefore where appropriate for exposure conditions, specification clauses should include the following in order to promote the use of RAs:

● The use of recycled materials (RCA or RA), if available, as coarse aggregate is the preferred option.
● The proportion of RA or RCA (as a mass fraction of the total coarse aggregate) is permitted to exceed 20%.

14
Useful Mathematics

Trigonometric relationships

Addition formulae

$\sin(A \pm B) = \sin A \cos B \pm \cos A \sin B$
$\cos(A \pm B) = \cos A \cos B \mp \sin A \sin B$
$\tan(A \pm B) = \dfrac{\tan A \pm \tan B}{1 \mp \tan A \tan B}$

Sum and difference formulae

$\sin A + \sin B = 2\sin\frac{1}{2}(A + B)\,\cos\frac{1}{2}(A + B)$
$\sin A - \sin B = 2\cos\frac{1}{2}(A + B)\,\sin\frac{1}{2}(A - B)$
$\cos A + \cos B = 2\cos\frac{1}{2}(A + B)\,\cos\frac{1}{2}(A - B)$
$\cos A - \cos B = -2\sin\frac{1}{2}(A + B)\,\sin\frac{1}{2}(A - B)$

$\tan A + \tan B = \dfrac{\sin(A + B)}{\cos A \cos B}$
$\tan A - \tan B = \dfrac{\sin(A - B)}{\cos A \cos B}$

Product formulae

$2\sin A \cos B = \sin(A - B) + \sin(A + B)$
$2\sin A \sin B = \cos(A - B) - \cos(A - B)$
$2\cos A \cos B = \cos(A - B) + \cos(A + B)$

Multiple angle and powers formulae

$\sin 2A = 2\sin A \cos A$
$\cos 2A = \cos^2 A - \sin^2 A$
$\cos 2A = 2\cos^2 A - 1$
$\cos 2A = 1 - 2\sin^2 A$
$\tan 2A = \dfrac{2\tan A}{1 - \tan^2 A}$
$\sin^2 A + \cos^2 A = 1$
$\sec^2 A = \tan^2 A + 1$

Relationships for plane triangles

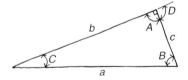

Pythagoras for right angled triangles:

$$b^2 + c^2 = a^2$$

Sin rule:

$$\frac{a}{\sin A} = \frac{b}{\sin B} = \frac{c}{\sin C}$$

$$\sin A = \frac{2}{bc}\sqrt{s(s-a)(s-b)(s-c)},$$

where $s = (a + b + c)/2$

Cosine rule:

$$a^2 = b^2 + c^2 - 2bc \cos A$$
$$a^2 = b^2 + c^2 + 2bc \cos D$$

$$\cos A = \frac{b^2 + c^2 - a^2}{2bc}$$

Special triangles

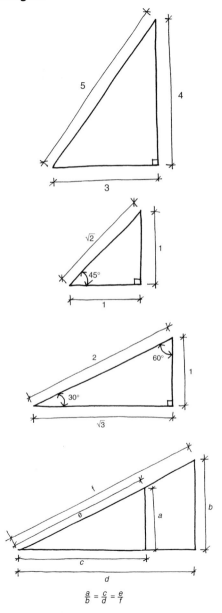

$$\frac{a}{b} = \frac{c}{d} = \frac{e}{f}$$

Algebraic relationships

Quadratics

$ax^2 + bx + c = 0$ $x = \dfrac{-b \pm \sqrt{b^2 - 4ac}}{2a}$

$x^2 + 2xy + y^2 = (x + y)^2$
$x^2 - y^2 = (x + y)(x - y)$
$x^3 - y^3 = (x - y)(x^2 + xy + y^2)$

Powers

$a^x a^y = a^{x+y}$ $\dfrac{a^x}{a^y} = a^{x-y}$ $(a^x)^y = a^{xy}$

Logarithms

$x \equiv e^{\log_e x} \equiv e^{\ln x}$
$x \equiv \log_{10}(10^x) \equiv \log_{10}(\text{antilog}_{10}\, x) \equiv 10^{\log_{10} x}$

$e = 2.71828$
$\ln x = \dfrac{\log_{10} x}{\log_{10} e} = 2.30259\, \log_{10} x$

Equations of curves

Circle

$x^2 + y^2 = a^2$

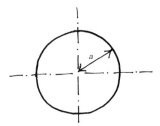

Ellipse

$\dfrac{x^2}{a^2} + \dfrac{y^2}{b^2} = 1$

Hyperbola

$$\frac{x^2}{a^2} - \frac{y^2}{b^2} = 1$$

Parabola

$$y^2 = ax$$

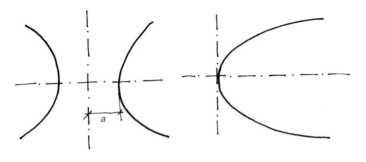

Circular arc

$$R = \left(d^2 + \frac{L^2}{4}\right)\frac{1}{2d}$$

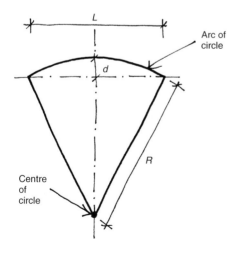

L

Arc of circle

d

R

Centre
of
circle

Rules for differentiation and integration

$$\frac{d}{dx}(uv) = u\frac{dv}{dx} + v\frac{du}{dx}$$

$$\frac{d}{dx}\left(\frac{u}{v}\right) = \frac{1}{v^2}\left(v\frac{du}{dx} - u\frac{dv}{dx}\right)$$

$$\frac{d}{dx}(uvw) = uv\frac{dw}{dx} + uw\frac{dv}{dx} = vw\frac{du}{dx}$$

$$\int (uv)dx = u\int (v)dx - \int \frac{du}{dx}\int (v)dx$$

Standard differentials and integrals

$$\frac{d}{dx}x^n = nx^{n-1}$$

$$\frac{d}{dx}\ln x = \frac{1}{x}$$

$$\frac{d}{dx}e^{ax} = ae^{ax}$$

$$\frac{d}{dx}a^x = a^x \ln a$$

$$\frac{d}{dx}x^x = x^x(1 + \ln x)$$

$$\frac{d}{dx}\sin x = \cos x$$

$$\frac{d}{dx}\cos x = -\sin x$$

$$\frac{d}{dx}\tan x = \sec^2 x$$

$$\frac{d}{dx}\cot x = -\operatorname{cosec}^2 x$$

$$\frac{d}{dx}\sin^{-1} x = \frac{1}{\sqrt{1 - x^2}}$$

$$\frac{d}{dx}\cos^{-1} x = \frac{-1}{\sqrt{1 - x^2}}$$

$$\frac{d}{dx}\tan^{-1} x = \frac{1}{1 + x^2}$$

$$\frac{d}{dx}\cot^{-1} x = \frac{-1}{1 + x^2}$$

$$\int x^n dx = \frac{x^{n+1}}{n+1} \quad n \neq 1$$

$$\int \frac{1}{x}dx = \ln x$$

$$\int e^{ax}dx = \frac{e^{ax}}{a} \quad a \neq 0$$

$$\int a^x dx = \frac{a^x}{\ln a} \quad a > 0, a \neq 0$$

$$\int \ln x \, dx = x(\ln x - 1)$$

$$\int \sin x \, dx = -\cos x$$

$$\int \cos x \, dx = \sin x$$

$$\int \tan x \, dx = -\ln(\cos x)$$

$$\int \cot x \, dx = \ln(\sin x)$$

$$\int \sec^2 x \, dx = \tan x$$

$$\int \operatorname{cosec}^2 x \, dx = -\cot x$$

$$\int \frac{1}{\sqrt{1 - x^2}}dx = \sin^{-1} x \quad |x| < 1$$

$$\int \frac{1}{\sqrt{1 + x^2}}dx = \tan^{-1} x$$

Useful Addresses

Advisory organizations

Aluminium Federation Ltd
National Metal Forming Centre, 47 Birmingham Road,
W Bromwich B70 6PY
www.alfed.org.uk
tel: 0121 601 6363
fax: 0870 138 9714

Ancient Monuments Society
St Anne's Vestry Hall, 2 Church Entry, London
EC4V 5HB
www.ancientmonumentssociety.org.uk
tel: 020 7236 3934
fax: 020 7329 3677

Arboricultural Advisory & Information Service
Alice Holt Lodge, Wrecclesham, Farnham GU10 4LH
www.treehelp.info
tel: 09065 161147
fax: 01420 22000

Arboricultural Association
Ampfield House, Ampfield, Romsey, Hampshire
S051 9PA
www.trees.org.uk
tel: 01794 368717
fax: 01794 368978

Architects Registration Board (ARB)
8 Weymouth Street, London W1W 5BU
www.arb.org.uk
tel: 020 7580 5861
fax: 020 7436 5269

Asbestos Removal Contractors Association (ARCA)
Arca House, 237 Branston Road, Burton-upon-
Trent, Staffordshire DE14 3BT
www.arca.org.uk
tel: 01283 531126
fax: 01283 568228

Association for Project Safety
12 Stanhope Place, Edinburgh EH12 5HH
www.aps.org.uk
tel: 0131 346 9020
fax: 0131 346 9029

Association for the Conservation of Energy
Westgate House, 2a Prebend Street, London N1 8PT
www.ukace.org
tel: 020 7359 8000
fax: 020 7359 0863

Association of Consultancy and Engineering
Alliance House, 12 Caxton Street, London SW1 0QL
www.acenet.co.uk
tel: 020 7222 6557
fax: 020 7222 0750

Brick Development Association Ltd (BDA)
Woodwide House, Winkfield, Windsor, Berkshire
SL4 2DX
www.brick.org.uk
tel: 01344 885651
fax: 01344 890129

British Adhesives & Sealants Association (BASA)
5 Alderson Road, Worksop, Notts, S80 1UZ
www.basa.uk.com

tel: 01909 480888
fax: 01909 473834

British Architectural Library
RIBA, 66 Portland Place, London W1N 4AD
www.riba-library.com

tel: 020 7307 3708
fax: 020 7589 3175

British Board of Agrément (BBA)
Bucknalls Lane, Garston, Watford,
Herts WD25 9BA
www.bbacerts.co.uk

tel: 01923 665300
fax: 01923 665301

British Cement Association (BCA)
Riverside House, 4 Meadows Business Park,
Camberley GU17 9AB
www.cementindustry.co.uk

tel: 01276 608700
fax: 01276 608701

**British Constructional Steelwork Association
Ltd (BCSA)**
4 Whitehall Court, London SW1A 2ES
www.steelconstruction.org

tel: 020 7839 8566
fax: 020 7976 1634

British Library
96 Euston Road, London NW1 2DB
www.bl.uk

tel: 020 7412 7676
fax: 020 7412 7954

British Precast Concrete Federation (BPCF)
60 Charles Street, Leicester LE1 1FB
www.britishprecast.org

tel: 0116 253 6161
fax: 0116 251 4568

**British Rubber and Polyurethane Products
Association Ltd (BRPPA)**
6 Bath Place, Rivington Street, London EC2A 3JE
www.brppa.co.uk

tel: 020 7457 5040
fax: 020 7972 9008

British Safety Council (BSC)
70 Chancellor's Road, London W6 9RS
www.britishsafetycouncil.org

tel: 020 8741 1231
fax: 020 8741 4555

British Stainless Steel Association
Broomgrove, 59 Clarkehouse Road, Sheffield
S10 2LE
www.bssa.org.uk

tel: 0114 267 1260
fax: 0114 266 1252

British Standards Institution (BSI)
389 Chiswick High Road, London W4 4AL
www.bsi-global.com

tel: 020 8996 9001
fax: 020 8996 7001

British Stone
Kent House, 77 Compton Road, Wolverhampton
WV3 9QH
www.stonesofbritain.co.uk

tel: 01902 717789
fax: 01902 717789

British Waterways Board
64 Clarendon Road, Watford WD17 1DA
www.britishwaterways.com

tel: 01923 201120
fax: 01923 201400

British Wood Preserving & Damp Proofing Association (BWPDA)
1 Gleneagles House, Vernon Gate, Derby DE1 1UP
www.bwpda.co.uk

tel: 01332 225100
fax: 01332 225101

Building Centre
26 Store Street, London WC1E 7BT
www.buildingcentre.co.uk

tel: 020 7692 4000
fax: 020 7580 9641

Building Research Advisory Service
Bucknalls Lane, Garston, Watford WD25 9XX
www.bre.co.uk

tel: 01923 664664
fax: 01923 664098

Building Research Establishment (BRE)
Bucknalls Lane, Garston, Watford WD25 9XX
www.bre.co.uk

tel: 01923 664000
fax: 01923 664787

Building Services Research and Information Association (BSRIA)
Old Bracknell Lane West, Bracknell, Berks RG12 7AH
www.bsria.co.uk

tel: 01344 465600
fax: 01344 465626

CADW – Welsh Historic Monuments
Plas Carew, Unit 5–7, Cefncoed, Parc Nantgarw,
Cardiff CF15 7QQ
www.cadw.wales.gov.uk

tel: 01443 336 000
fax: 01443 336 001

Cares (UK Certification Authority for Reinforcing Steels)
Pembroke House, 21 Pembroke Road, Sevenoaks,
Kent TN13 1XR
www.ukcares.com

tel: 01732 450000
fax: 01732 455917

Cast Metal Federation
47 Birmingham Road, West Bromwich Road,
West Bromwich B70 6PY
www.castmetalsfederation.com

tel: 0121 601 6390
fax: 0121 601 6391

Castings Development Centre
Castings Technology International, Advanced
Manufacturing Park,
Brunel Way, Rotherham S60 5WG

tel: 0114 254 1144
fax: 0114 254 1155

Commission for Architecture and the Built Environment (CABE)
1 Kemble Street, London WC2B 4AN
www.cabe.org.uk

tel: 020 7070 6700
fax: 020 7070 6777

Concrete Repair Association (CRA)
Tournai Hall, Evelyn Woods Road,
Hampshire GU11 2LL
www.cra.org.uk

tel: 01252 357835

Concrete Society
Riverside House, 4 Meadows Business Park, Station
Approach, Blackwater, Camberley GU17 9AB
www.concrete.org.uk

tel: 01276 607140
fax: 01276 607141

Construction Fixings Association
65 Dean Street, Oakham LE15 6AF
www.fixingscfa.co.uk

tel: 01664 474755
fax: 01664 474755

Construction Industry Research & Information Association (CIRIA)
Classic House, 174–180 Old Street, London
EC1V 9BP
www.ciria.org.uk

tel: 020 7549 3300
fax: 020 7253 0523

Copper Development Association (CDA)
Unit 5, Grovelands Business Centre, Boundary
Way, Hemel Hempstead HP2 7TE
www.cda.org.uk

tel: 01442 275705
fax: 01442 275716

CORUS Construction Centre
PO Box 1, Brigg Road, Scunthorpe DN16 1BP
www.corusconstruction.com

tel: 01724 405060
fax: 01724 405600

Council for Aluminium in Building
Bank House, Bond's Mill, Stonehouse, Glos GL10 3RF
www.c-a-b.org.uk

tel: 01453 828 851
fax: 01453 828 861

Design Council
34 Bow Street, London WC2E 7DL
www.designcouncil.org.uk

tel: 020 7420 5200
fax: 020 7420 5300

English Heritage
PO Box 569, Swindon SN2 2YP
www.english-heritage.org.uk

tel: 0870 333 1181
fax: 01793 414926

Environment Agency
PO Box 544, Rotherham S60 1BY
www.environment-agency.gov.uk

tel: 01709 389201

Environment & Heritage Service (EHS) – Northern Ireland
Waterman House, 5–33 Hill Street, Belfast BT1 2LA
www.ehsni.gov.uk

tel: 028 9054 3095
fax: 028 9054 3111

European Glaziers Association (UEMV)
Gothersgade 160, DK-1123 Copenhagen, Denmark
www.uemv.com

tel: +45 33 13 65 10
fax: +45 33 13 65 60

European Stainless Steel Advisory Body (Euro-Inox)
241 Route d'Alon, L-1150 Luxembourg
www.euro-inox.org

tel: +352 26 10 30 50
fax: +352 26 10 30 51

Federation of Manufacturers of Construction Equipment & Cranes
Orbital House, 85 Croydon Road,
Caterham, Surrey CR3 6PD
www.coneq.org.uk

tel: 01883 334499
fax: 01883 334490

Federation of Master Builders
FMB Headquarters, 14/15 Great James Street,
London WC1N 3DP
www.fmb.org.uk

tel: 020 7242 7583
fax: 020 7404 0296

Federation of Piling Specialists
Forum Court, 83 Copers Cope Road, Beckenham,
Kent BR3 1NR
www.fps.org.uk

tel: 020 8663 0947
fax: 020 8663 0949

Fire Protection Association (FPA)
London Road, Moreton in Marsh, Glos GL56 0RH
www.thefpa.co.uk

tel: 01608 812500
fax: 01608 812501

Forest Stewardship Council (FSC)
11–13 Great Oak Street, Llanidloes, Powys SY18 6BU
www.fsc-uk.org.uk

tel: 01686 413916
fax: 01686 412176

Friends of the Earth
26–28 Underwood Street, London N1 7JQ
www.foe.co.uk

tel: 020 7490 1555
fax: 020 7490 0881

Galvanizers' Association
Wren's Court, 56 Victoria Road, Sutton Coldfield,
W. Midlands B72 1SY
www.hdg.org.uk

tel: 0121 355 8838
fax: 0121 355 8727

Georgian Group
6 Fitzroy Square, London W1P 6DX
www.georgiangroup.org.uk

tel: 020 7387 1720
fax: 020 7387 1721

Glass and Glazing Federation
44–48 Borough High Street, London SE1 1XB
www.ggf.org.uk

tel: 0870 042 4255
fax: 0870 042 4266

Glue Laminated Timber Association
Chiltern House, Stocking Lane, High Wycombe
HP14 4ND
www.glulam.co.uk

tel: 01494 565180
fax: 01494 565487

Health and Safety Executive (HSE)
Rose Court, 2 Southwark Bridge, London SE1 9HS
www.hse.gov.uk

tel: 020 7556 2102
fax: 020 7556 2109

Historic Scotland
Longmore House, Salisbury Place, Edinburgh EH9 1SH
www.historic-scotland.gov.uk

tel: 0131 668 8600
fax: 0131 668 8669

HM Land Registry
Lincoln's Inn Fields, London WC2A 3PH
www.landreg.gov.uk

tel: 020 7917 8888
fax: 020 7955 0110

Institution of Civil Engineers (ICE)
1–7 Great George Street, London SW1P 3AA
www.ice.org.uk

tel: 020 7222 7722
fax: 020 7222 7500

Institution of Structural Engineers (IStructE)
11 Upper Belgrave Street, London SW1X 8BH
www.istructe.org

tel: 020 7235 4535
fax: 020 7235 4294

London Metropolitan Archives
40 Northampton Road, London EC1 0HB
www.cityoflondon.gov.uk

tel: 020 7332 3820
fax: 020 7833 9136

Meteorological Office
Fitzroy Road, Exeter, Devon EX1 3PB
www.meto.gov.uk

tel: 01392 885680
fax: 01392 885681

National Building Specification Ltd (NBS)
The Old Post Office, St Nicholas Street, Newcastle upon Tyne NE1 1RH
www.thenbs.co.uk

tel: 0191 232 9594
fax: 0191 232 5714

National House-Building Council (NHBC)
Buildmark House, Chiltern Avenue, Amersham, Bucks HP6 5AP
www.nhbc.co.uk

tel: 01494 735363
fax: 01494 735201

Network Rail
40 Melton Street, London NW1 2EE
www.networkrail.co.uk

tel: 020 7557 8000
fax: 020 7557 9000

Nickel Development Institute (NIDI)
The Holloway, Alvechurch, Birmingham B48 7QA
www.nickelinstitute.org

tel: 01527 584 777
fax: 01527 585 562

Ordnance Survey
Romsey Road, Southampton SO16 4GU
www.ordnancesurvey.co.uk

tel: 08456 050505
fax: 023 8079 2615

Paint Research Association (PRA)
14 Castle Mews, High Street, Hampton, Middx TW12 2NP
www.pra-world.com

tel: 020 8487 0800
fax: 020 8487 0801

Plastics and Rubber Advisory Service, British Plastics Federation (BPF)
6 Bath Place, Rivington Street, London EC2A 3JE
www.bpf.co.uk

tel: 020 7457 5000
fax: 020 7457 5045

Pyramus and Thisbe Club
Administration Office, Rathdale House, 30 Back Road, Rathfriland, Belfast BT34 5QF
www.partywalls.org.uk

tel: 028 4063 2082
fax: 028 4063 2083

Quarry Products Association
38–44 Gillingham Street, London SW1V 1HU
www.qpa.org

tel: 020 7963 8000
fax: 020 7963 8001

Royal Incorporation of Architects in Scotland (RIAS)
15 Rutland Square, Edinburgh EH1 2BE
www.rias.org.uk

tel: 0131 229 7545
fax: 0131 228 2188

Royal Institute of British Architects (RIBA)
66 Portland Place, London W1B 1AD
www.architecture.com

tel: 020 7580 5533
fax: 020 7255 1541

Royal Institution of Chartered Surveyors (RICS)
12 Great George Street, London SW1P 3AD
www.rics.org

tel: 0870 333 1600
fax: 020 7334 3811

Royal Society of Architects in Wales
Bute Building, King Edward VII Avenue, Cathays Park,
Cardiff CF1 3NB
www.architecture.com

tel: 029 2087 4753
fax: 029 2087 4926

Royal Society of Ulster Architects (RSUA)
2 Mount Charles, Belfast BT7 1NZ
www.rsua.org.uk

tel: 028 9032 3760
fax: 028 9023 7313

Scottish Building Standards Agency
Scottish Government, Denholm House, Almondvale
Business Park, Livingston EH54 6GA
www.sbsa.gov.uk

tel: 01506 600400
fax: 01506 600401

Society for the Protection of Ancient Buildings
37 Spital Square, London E1 6DY
www.spab.org.uk

tel: 020 7377 1644
fax: 020 7247 5296

Stainless Steel Advisory Service
Broomgrove, 59 Clarkehouse Street, Sheffield
S10 2LE
www.bssa.org.uk

tel: 0114 267 1260
fax: 0114 266 1252

Stationery Office (previously HMSO)
PO Box 29, Norwich NR3 1GN
www.tso.co.uk

tel: 0870 600 5522
fax: 0870 600 5533

Steel Construction Institute (SCI)
Silwood Park, Buckhurst Road, Ascot, Berks SL5 7QN
www.steel-sci.org.uk

tel: 01344 636 525
fax: 01344 636 570

Stone Federation Great Britain (SFGB)
Channel Business Centre, Ingles Manor, Castle Hill Ave,
Folkestone CT20 2RD
www.stone-federationgb.org.uk

tel: 01303 856123
fax: 01303 856117

Surface Engineering Association
Federation House, 10 Vyse Street, Birmingham B18 6LT
www.sea.org.uk

tel: 0121 237 1123
fax: 0121 237 1124

**Thermal Spraying & Surface Engineering
Association (TSSEA)**
38 Lawford Lane, Bilton, Rugby, Warwickshire CV22 7JP
www.tssea.co.uk

tel: 0870 760 5203
fax: 0870 760 5206

Timber Trade Federation
Building Centre, 26 Store Street, London WC1E 7BT
www.ttf.co.uk

tel: 020 3205 0067

TRADA Technology Ltd
Stocking Lane, Hughenden Valley, High
Wycombe HP14 4ND
www.tradatechnology.co.uk

tel: 01494 569600
fax: 01494 565487

UK Cast Stone Association
15 Stonehill Court, The Arbours, Northampton NN3 3RA
www.ukcsa.co.uk

tel: 01604 405666
fax: 01604 405666

UK Climate Impact Programme (UKCIP)
Oxford University Centre for the Environment,
Dyson Perrins Building, South Parks Road,
Oxford OX1 3QY
www.ukcip.org.uk

tel: 01865 285717
fax: 01865 285710

Victorian Society
1 Priory Gardens, Bedford Park, London W4 1TT
www.victorian-society.org.uk

tel: 020 8994 1019
fax: 020 8747 5899

Waste + Resources Action Plan (WRAP)
The Old Academy, 21 Horse Fair, Banbury
OX16 0AH
www.wrap.org.uk

tel: 01295 819 900
fax: 01295 819 911

Water Authorities Association
1 Queen Anne's Gate, London, SW1H 9BT
www.water.org.uk

tel: 020 7344 1844
fax: 020 7344 1853

Water Jetting Association
17 Judiths Lane, Sawtrey, Huntingdon,
Cambridgeshire PE28 5XE
www.waterjetting.org.uk

tel: 01487 834034
fax: 01487 832232

Wood Panel Industries Federation
28 Market Place, Grantham, Lincolnshire
NG31 6LR
www.wpif.org.uk

tel: 01476 563707
fax: 01476 579314

Manufacturers

3M Tapes & Adhesives UK Ltd
3M Centre, Cain Road, Bracknell, RG12 8HT
www.3m.co.uk

tel: 01344 858000
fax: 01344 858278

Angle Ring Company Ltd
Bloomfield Road, Tipton, West Midlands DY4 9EH
www.anglering.co.uk

tel: 0121 557 7241
fax: 0121 522 4555

Aplant Acrow/Ashtead Group plc
102 Dalton Ave, Birchwood Park, Warrington
WA3 6YE
www.aplant.com

tel: 01925 281000
fax: 01925 281001

BGT Bischoff Glastechnik
Alexanderstraße 2, 705015 Bretten, Germany
www.bgt-bretten.de

tel: +49 7252 5030
fax: +49 7252 503283

BRC Building Products
Carver Road, Astonfields Industrial Estate, Stafford
ST16 3BP
www.brc-uk.co.uk

tel: 01785 222288
fax: 01785 240029

Caltite/Cementaid (UK) Ltd
1 Baird Close, Crawley, West Sussex RH10 9SY
www.cementaid.com

tel: 01293 447878
fax: 01293 447880

Catnic
Corus UK Ltd, Pontypandy Industrial Estate,
Caerphilly CF83 3GL
www.catnic.com

tel: 029 2033 7900
fax: 0870 0241809

Civil Marine
London Road, West Thurrock, Grays, Essex RM20 3NL
www.civilmarine.co.uk

tel: 01708 864813
fax: 01708 865907

CORUS Group
30 Millbank, London SW1P 4WY
www.corusgroup.com/
www.corusconstruction.com

tel: 020 7717 4444
fax: 020 7717 4455

Cricursa
Cami de Can Ferran s/n, Pol. Industrial Coll de la Manya,
08403 Granollers, Spain
www.cricursa.com

tel: +34 93 840 4470

Dow Corning Ltd
Meriden Business Park, Copse Drive, Allesley, Coventry
CV5 9RG
www.dowcorning.com

tel: 01676 528000
fax: 01676 528001

Eckelt Glass
Zentrale/Produktion, Resthofstaße 18, 4400 Steyr,
Austria
www.eckelt.at

tel: +43 72528940
fax: +43 725289424

European Glass Ltd
European House, Abbey Point, Abbey Road, London
NW10 7DD
www.europeanglass.co.uk

tel: 020 8961 6066
fax: 020 8961 1411

F. A. Firman (Harold Wood) Ltd
19 Bates Road, Harold Wood, Romford, Essex
RM3 0JH
www.firmanglass.com

tel: 01708 374534
fax: 01708 340511

Finnforest
46 Berth, Tilbury, Freeport, Tilbury, Essex RM18 7HS
www.finnforest.co.uk

tel: 01375 856 855
fax: 01375 856 264

Hansen Brick Ltd
Stewartby, Bedfordshire MK43 9LZ
www.hansen.co.uk

tel: 0870 5258258
fax: 01234 762040

Hansen Glass
Hornhouse Lane, Kirkby L33 7YQ
www.hansenglass.co.uk

tel: 0151 545 3000
fax: 0151 545 3003

IG Lintels Ltd
Avondale Road, Cwmbran, Gwent NP44 1XY
www.igltd.co.uk

tel: 01633 486486
fax: 01633 486465

IMS Group (Special Steels)
Arley Road, Saltley, Birmingham, West Midlands
B8 1BB
www.ims-uk.com

tel: 0121 326 3100
fax: 0121 326 3105

James Latham plc
Unit 3, Swallow Park, Finway Road, Hemel Hempstead
HP2 7QU
www.lathamtimber.co.uk

tel: 01442 849100
fax: 01442 239287

Loctite UK (Henkel Technologies)
Technologies House, Wood Lane End, Hemel Hempstead
HP2 4RQ
www.loctite.co.uk

tel: 01442 278 000
fax: 01442 278 293

Perchcourt Ltd
Unit 6B, Heath Street Industrial Estate, Smethwick,
Warley, B66 2QZ
www.perchcourt.co.uk

tel: 0121 555 6272
fax: 0121 555 6176

Permasteelisa
26 Mastmaker Road, London E14 9UB
www.permasteelisa.com

tel: 020 7531 4600
fax: 020 7531 4610

Pilkington UK Ltd
Prescot Road, St Helens WA10 3TT
www.pilkington.com

tel: 01744 28882
fax: 01744 692660

Pudlo/David Bell Group plc
Huntingdon Road, Bar Hill, Cambridge CB3 8HN
www.pudloconcrete.co.uk

tel: 01954 780687
fax: 01954 782912

Quality Tempered Glass (QTG)
Concorde Way, Millennium Business Park, Mansfield,
Notts NG19 7JZ

tel: 01623 416300
fax: 01623 416303

Richard Lees Steel Decking Ltd
Moor Farm Road West, The Airfield, Ashbourne,
Derbyshire DE6 1HN
www.rlsd.com

tel: 01335 300999
fax: 01335 300888

RMD Kwikform
Brickyard Lane, Aldridge, Walsall WS9 8BW
www.rmdformwork.co.uk

tel: 01922 743743
fax: 01922 743400

Solaglass Saint Gobain
Binley One, Herald Way, Binley, Coventry CV3 2ND
www.saint-gobain.co.uk

tel: 024 76 547400
fax: 024 76 547799

SPS Unbrako Machine Screws
SPS Technologies Ltd, 4444 Lee Road, Cleveland,
Ohio 441282902
www.unbrako.com

tel: +1 216 581 3000
fax: +1 800 225 5777

Staytite Self Tapping Fixings
Coronation Road, Cressex Business Park,
High Wycombe HP12 3RP
www.staytite.com

tel: 01494 462322
fax: 01494 464747

Sunglass
via Piazzola 13E, 35010 Villafranca, Padova, Italy
www.sunglass.it

tel: +39 049 90500100
fax: +39 049 9050964

Supreme Concrete Ltd
Coppingford Road, Sawtry, Huntingdon PE28 5GP
www.supremeconcrete.co.uk

tel: 01487 833312
fax: 01487 833348

Tarmac Topfloor Ltd
Weston Underwood, Ashbourne, Derbyshire DE6 4PH
www.tarmac.co.uk/topfloor

tel: 01332 868 400

TecEco Pty. Ltd
497 Main Road, Glenorchy, Tasmania 7010, Australia
www.tececo.com

tel: +61 3 6249 7868
fax: +61 3 6273 0010

Valbruna UK Ltd
Oldbury Road, West Bromwich, West Midlands B70 9BT
www.valbruna.co.uk

tel: 0121 553 5384
fax: 0121 500 5095

W. J. Leigh
Tower Works, Kestor Road, Bolton BL2 2AL
www.leighspaints.co.uk

tel: 01204 521771
fax: 01204 382115

Zero Environment Ltd
PO Box 1659, Warwick CV35 8ZD
www.zeroenvironment.co.uk

tel: 01926 624966
fax: 01926 624926

Further Reading

Suggested further reading

1 General Information

ACE (1998). *Standard Conditions of Service Agreement B1*, 2nd Edition. Association of Consulting Engineers.

Blake, L. S. (1989). *Civil Engineer's Reference Book*, 4th Edition. Butterworth-Heinemann.

CPIC (1998). *Selected CAWS Headings from Common Arrangement of Work Sections*, 2nd Edition. CPIC.

DD ENV 1991: Eurocode 1. *Basis of Design and Actions on Structures*. BSI.

Hunt, T. (1999). *Tony Hunt's Sketchbook*. Architectural Press.

2 Statutory Authorities and Permissions

DETR (1997). *The Party Wall etc. Act: explanatory booklet*. HMSO.

HSE (2001). *Managing Health & Safety in Construction. CDM Regulations 1994. Approved Code of Practice*. HSE.

HSE (2001). *Health & Safety in Construction*. HSE.

Information on UK regional policies:
www.defra.gov.uk/www.dft.gov.uk/www.odpm.gov.uk/www.wales.gov.uk
www.scotland.gov.uk/www.nics.gov.uk

PTC (1996). *Party Wall Act Explained. A Commentary on the Party Wall Act 1996*. Pyramus & Thisbe Club.

3 Design Data

BS 648: 1970. *Schedule of Weights of Building Materials*. BSI.

BS 5606: 1990. *Guide to Accuracy in Building*.

BS 6180: 1995. *Code of Practice for Protective Barriers In and About Buildings*. BSI.

BS 6399 *Loading for Buildings*. Part 1: 1996. *Code of Practice for Dead and Imposed Loads*. Part 2: 1997. *Code of Practice for Wind Loads*. Part 3: 1988. *Code of Practice for Imposed Roof Loads*. BSI.

CIRIA Report 111 (1986). *Structural Renovation of Traditional Buildings*. CIRIA.

Hunt, T. (1997). *Tony Hunt's Structures Notebook*. Architectural Press.

Information on the transportation of abnormal indivisible loads: www.dft.gov.uk

Lisborg, N. (1967). *Principles of Structural Design*. Batsford.

Lyons, A. R. (1997). *Materials for Architects and Builders – An Introduction*. Arnold.

Morgan, W. (1964). *The Elements of Structure*. Pitman.

Richardson, C. (2000). The Dating Game. *Architect's Journal*, **23/3/00**, 56–59, **30/3/00**, 36–39. **6/4/00**, 30–31.

4   Basic Shortcut Tools for Structural Analysis

Bolton, A. (1978). Natural Frequencies of Structures for Designers. *The Structural Engineer*. **Vol. 9/No. 56A**, 245–253.

Brohn, D. M. (2005). *Understanding Structural Analysis*. New Paradigm Solutions.

Calvert, J. R. & Farrer, R. A. (1999). *An Engineering Data Book*. Macmillan Press.

Carvill, J. (1993). *Mechanical Engineer's Data Book*. Butterworth-Heinemann.

Gere, J. M. & Timoshenko, S. P. (1990). *Mechanics of Materials*, 3rd SI Edition. Chapman Hall.

Hambly, E. (1994). *Structural Analysis by Example*. Archimedes.

Heyman, J. (2005). Theoretical analysis and real-world design. *The Structural Engineer*. **Vol. 83 No. 8**. 19 Apr 2005 pp 14–17.

Johansen, K. W. (1972). *Yield Line Formulae for Slabs*. Cement and Concrete Association.

Megson, T. H. G. (1996). *Structural and Stress Analysis*. Butterworth-Heinemann.

Mosley, W. H. & Bungey, J. H. (1987). *Reinforced Concrete Design*, 3rd Edition. Macmillan.

Sharpe, C. (1995). *Kempe's Engineering Yearbook*, 100th Edition. M-G Information Services Ltd.

Wood, R. H. (1961). *Plastic and Elastic Design of Slabs and Plates*. Thames & Hudson.

5   Geotechnics

Berezantsev, V. G. (1961). Load Bearing Capacity and Deformation of Piled Foundations. *Proc. of the 5th International Conference on Soil Mechanics. Paris.* **Vol. 2**, 11–12.

BS 5930: 1981. *Code of Practice for Site Investigations*. BSI.

BS 8004: 1986. *Code of Practice for Foundations*. BSI.

BS 8002: 1994. *Code of Practice for Earth Retaining Structures*. BSI.

Craig, R. F. (1993). *Soil Mechanics*, 5th Edition. Chapman Hall.

DD ENV 1997: Eurocode 8. *Geotechnical Design.* BSI.

Environment Agency (1997). *Interim Guidance on the Disposal of Contaminated Soils*, 2nd Edition. HMSO.

Environment Agency (2002). *Contaminants in Soil: Collation of Toxological Data and Intake Values for Humans.* CLR Report 9, HMSO.

Hansen, J. Brinch (1961). A General Formula for Bearing Capacity. *Danish Geotechnical Institute Bulletin.* **No. 11**. Also Hansen, J. Brinch (1968). A Revised Extended Formula for

Bearing Capacity. *Danish Geotechnical Institute Bulletin.* **No. 28**. Also Code of Practice for Foundation Engineering (1978), *Danish Geotechnical Institute Bulletin.* **No. 32**.

ICRCL (1987). *Guidance on the Redevelopment of Contaminated Land*, 2nd Edition. Guidance Note 59/83, DoE.

Kelly, R. T. (1980). Site Investigation and Material Problems. *Proc. of the Conference on the Reclamation of Contaminated Land*. Society of Chemical Industry. **B2**, 1–14.

NHBC. *National House-Building Council Standards.*

Terzaghi, K. & Peck, R. B. (1996). *Soli Mechanics in Engineering Practice*, 3rd Edition. Wiley.

Tomlinson, M. J. (2001). *Foundation Design and Construction*, 7th edition. Pearson.

6   Timber and Plywood

BS 5268: Part 2: 2002. *Structural Use of Timber.* BSI.

DD ENV 1995: Eurocode 6. *Design of Timber Structures.* BSI.

Ozelton, E. C. & Baird, J. A. (2002). *Timber Designers' Manual*, 3rd Edition. Blackwell.

7   Masonry

BS 5977: Part 1: 1981. *Lintels. Method for assessment of load*. BSI.

BS 5628 *Code of practice for masonry*. Part 1: 1992. *Structural use of unreinforced masonry*. Part 2: 2001. *Materials & components, design & workmanship*. BSI.

CP111: 1970. *Code of Practice for the Design of Masonry in Building Structures*. BSI.

Curtin, W. G., Shaw, G. & Beck, J. K. (1987). *Structural Masonry Designers' Manual*, 2nd Edition. BSP Professional Books.

DD ENV. 1996. Eurocode 7. *Design of Masonry Structures*. BSI.

Heyman, J. (1995). *The Stone Skeleton*. Cambridge University Press.

Howe, J. A. (1910). *The Geology of Building Stones*. Edward Arnold. Reprinted Donhead Publishing (2000).

Manual for the Design of Plain Masonry in Building Structures (1997). ICE/IStructE.

8   Reinforced Concrete

Bennett, D. (2007). *Architectural In-situ Concrete*. RIBA Publishing.

BS 5328: Parts 1 to 4: 1997. *Concrete Specification and Testing*. BSI.

BS 4483: 1985. BS 4483. 1998. *Steel Fabric for the Reinforcement of Concrete*. BSI.

BS 8110: Part 1: 1997. *Structural Use of Concrete. Code of Practice for Design and Construction*. BSI.

BS 8666: 2000. *Specification for scheduling, dimensioning, bending and cutting of steel reinforcement for concrete*. BSI.

DD ENV 1992: Eurocode 2. *Design of Concrete Structures*. BSI.

DD ENV 1992–4: Eurocode 3. *Liquid Retaining and Containing Structures*. BSI.

Goodchild, C. H. (1997). *Economic Concrete Frame Elements*. BCA.

Manual for the Design of Reinforced Concrete Building Structures (2002). ICE/IStructE.

Mosley, W. H. & Bungey, J. H. (1987). *Reinforced Concrete Design*, 3rd Edition. Macmillan.

Neville, A. M. (1977). *Properties of Concrete*. Pitman.

9   Steel

Baddoo, N. R. & Burgan, B. A. (2001). *Structural Design of Stainless Steel*. Steel Construction Institute, P291.

BS 449: Part 2: 1969. (as amended) *The Use of Structural Steel in Building*. BSI.

BS 5950 *Structural Use of Steelwork in Buildings*. Part 1: 2000. *Code of Practice for Design – Rolled and Welded Sections*. Part 5: 1998. *Code of Practice for Design – Cold Formed Thin Gauge Sections*. BSI.

DD ENV 1993: Eurocode 4. *Design of Steel Structures*. BSI.

MacGinley, T. J. & Ang, T. C. (1992). *Structural Steelwork Design to Limit State Theory*, 2nd Edition. Butterworth-Heinemann.

Manual for the Design of Steelwork Building Structures (2002). ICE/IStructE.

Nickel Development Institute (1994). *Design Manual for Structural Stainless Steel*. NIDI, 12011.

Owens, G. W. & Knowles, P. R. (1994). *SCI Steel Designer's Manual*, 5th Edition. Blackwell Science.

SCI (2001). *Steelwork Design Guide to BS 5950: Part 1: 2000 Volume 1 Section Properties and Member Capacities*, 6th Edition. Steel Construction Institute, P202.

SCI (1995). *Joints in Steel Construction: Moment Connections Volumes 1 + 2*. BCSA 207/95.

SCI (2005). *Joints in Steel Construction: Simple Connections Volumes 1 + 2*. BCSA P212.

10   Composite Steel and Concrete

BS 5950 *Structural Use of Steelwork in Buildings*. Part 3: 1990. *Code of Practice for Design – Composite Construction*. Part 4: 1994. *Code of Practice for Design – Composite Slabs with Profiled Metal Sheeting*. BSI.

DD ENV 1994: Eurocode 5. *Design of Composite Steel and Concrete Structures*. BSI.

Noble, P. W. & Leech, L. V. (1986). *Design Tables for Composite Steel and Concrete Beams*. Constrado.

SCI (1990). *Commentary on BS 5950: Part 3: Section 3.1 Composite Beams*. Steel Construction Institute, P78.

11   Glass

Glass and Mechanical Strength Technical Bulletin (2000). Pilkington.

pr EN 13474. Glass in Building – Design of Glass Panes. Part 1, *Basis for Design*. Part 2, *Design for Uniformly Distributed Loads*. BSI.

Structural Use of Glass in Buildings (1999). IStructE.

12   Building Elements

BRE (2002). *Thermal Insulation Avoiding Risks. A Good Practice Guide Supporting Building Regulation Requirements*, 3rd Edition. BR 262. CRC Ltd.

BS 8007: 1987. *Code of Practice for Design of Concrete Structures for Retaining Aqueous Liquids.* BSI.

BS 8102: 1990. *Protection of structures against water from the ground.* BSI.

BS 8118: Part 1: 1991. *Structural Use of Aluminium. Code of Practice for Design.* BSI.

CIRIA (1995). *Water-resisting basement construction – a guide. Safeguarding new and existing basements against water and dampness.* CIRIA, Report 139.

CIRIA (1998). *Screeds, Flooring and Finishes – Selection, Construction & Maintenance.* CIRIA, Report 184.

DD ENV 1999: Eurocode 9. *Design of Aluminium Structures.* BSI.

General BRE Publications: *BRE Digests, Good Building Guides & Good Repair Guides.*

Guide to the Structural Use of Adhesives (1999). IStructE.

Russell, J. R. & Ferry, R. L. (2002). *Aluminium Structures.* Wiley.

13   Sustainability

Addis, W. & Schouten, J. (2004). *Design for Deconstruction.* CIRIA.

Anderson, J. & Shiers, D. (2002). *The Green Guide to Specification.* Blackwell Science.

Berge, B. (2001). *The Ecology of Building Materials.* Architectural Press.

Bioregional (2002). *Toolkit for Carbon Neutral Developments.* BedZed Construction Materials Report.

CIBSE (2007). *Sustainability.* Guide L.

Friends of the Earth (1996). *The Good Wood Guide.* Friends of the Earth.

FSC (1996). Forest Stewardship Council and the Construction Sector. Factsheet.

IStructE (1999). *Building for a Sustainable Future: Construction without Depletion.*

RCEP (2000). *Energy – the Changing Climate.* Royal Commission on Environmental Pollution 22nd Report.

WCED (1987). Our Common Future (the Brundtland Report). Oxford.

Woolley, T. & Kimmins, S. (2000). The Green Building Handbook Volume 2. E&F Spon.

WRAP (2007). Reclaimed Building Products Guide. Waste + Resources Action Programme.

WRAP (2007). *Recycled Content Toolkit*. Online at www.wrap.org.uk.

WRAP (2008). *Choosing Construction Products: Guide to Recycled Content of Mainstream Construction Products*. Waste + Resources Action Programme.

WWF (2004). *The Living Planet Report.*

Sources

ACE (2004). *Standard Conditions of Service Agreement B1*, 2nd Edition. Association of Consulting Engineers. General summary of normal conditions.

Anderson, J. & Shiers, D. (2002). *Green Guide to Specification*, 3rd Edition.

Angle Ring Company Limited (2002). *Typical bend radii for selected steel sections*.

Bison Concrete Products (2008). *Loading data for hollowcore precast planks*.

Berezantsev, V. G. (1961). Load Bearing Capacity and Deformation of Piled Foundations. *Proc. of the 5th International Conference on Soil Mechanics. Paris*, **Vol. 2**, 11–12.

Bolton, A. (1978). Natural Frequencies of Structures for Designers. *The Structural Engineer* 56A(9):245–253. Table 1.

BRE Digests 299, 307, 345. Extracts on durability of timber. Reproduced by permission of Building Research Establishment.

BRE Good Building Guide 63, Climate Change Prediction for the UK.

BS 449: Part 2: 1969 (as amended) *The Use of Structural Steel in Building*. BSI. Tables 2, 11 and 19.

BS 4483: 1985. BS 4483: 1998. *Steel fabric for the reinforcement of concrete*. BSI. Table 1.

BS 5268: Part 2: 1991. *Structural Use of Timber*. BSI. Appendix D.

BS 5268: Part 2: 2002. *Structural Use of Timber*. BSI. Tables 8, 17, 19, 20, 21, 24 and extracts from Tables 28, 31, 33 and 34.

BS 5606: 1990. *Guide to accuracy in building*. Figure 4.

BS 5628: Part 1: 2005. *Structural use of unreinforced masonry*. BSI. Tables 1, 5 and 9, extracts from Tables 2 and 3, Figure 3.

BS 5628: Part 3: 2005. *Materials and components, design and workmanship*. BSI. Table 9 and Figure 6.

BS 5950 Part 1: 2000. *Structural Use of Steelwork in Buildings. Code of Practice for Design – Rolled and Welded Sections*. BSI. Tables 2, 9, 14 and 22.

BS 5977: Part 1: 1981. *Lintels. Methods for assessment of load*. BSI. Figures 1 and 4.

BS 6180: 1999. *Code of Practice for Protective Barriers In and About Buildings*. BSI. Table 1.

BS 6399 *Loading for Buildings*. Part 1: 1996. *Code of Practice for Dead and Imposed Loads*. BSI. Tables 1 and 4.

BS 8004: 1986. *Code of Practice for Foundations*. BSI. Table 1 adapted.

BS 8102: 1990. *Protection of structures against water from the ground*. BSI. Table 1 adapted to CIRIA Report 139 suggestions.

BS 8110: Part 1: 1997. *Structural Use of Concrete. Code of Practice for Design and Construction*. BSI. Tables 2.1, 3.3, 3.4, 3.9, 3.19, 3.20 and Figure 3.2.

BS 8118: Part 1:1991. *Structural Use of Aluminium. Code of Practice for Design*. BSI. Tables 3.1, 3.2 and 3.3 and small extracts from Tables 2.1, 2.2, 4.1 and 4.2.

BS 8666: 2005. *Specification for scheduling, dimensioning, bending and cutting of steel reinforcement for concrete*. BSI. Table 3.

Building Regulations, Part B (1991). HMSO. Table A2, Appendix A.

Building Regulations, Part A3 (2004). MSO. Table 11.

CIBSE (2006). Guide A: Environmental Design. Table 3.37 and 3.38 extracts.

Corus Construction (2007). *Structural Sections to BS4*. Corus.

CP111: 1970. *Code of Practice for the Design of Masonry in Building Structures*. BSI. Tables 3a and 4.

CPIC (1998). *Arrangement of Work Sections*, 2nd Edition. CPIC. Selected headings.

DETR (1997). *The Party Wall etc. Act: explanatory booklet*. HMSO.

Dow Corning (2002). *Strength values for design of structural silicon joints*.

Finnforest (2002). *LVL section sizes and grade stresses*.

Hansen, J. Brinch (1961). A General Formula for Bearing Capacity. *Danish Geotechnical Institute Bulletin*. **No. 11**. Also Hansen, J. Brinch (1968). A Revised Extended Formula for Bearing Capacity. *Danish Geotechnical Institute Bulletin*. **No. 28**. Also Code of Practice for Foundation Engineering (1978), *Danish Geotechnical Institute Bulletin*. **No. 32**.

Highways Agency, Design Note HD25/Interim Advice Note 73/06, HMSO. Table 3.1, Chapter 3.

Howe, J. A. (1910). *The Geology of Building Stones*. Edward Arnold. Reprinted Donhead Publishing (2000). Table XXVI.

IG Limited (2008). *Loading data for steel lintels*.

Kulhawy, F. H. (1984). Limiting Tip and Side Resistance. *Proceedings of Symposium Analysis and Design of Piled Foundations*, edited by Meyer, J. R. California, 80–98. Tables 1 and 2.

Manual for the Design of Reinforced Concrete Building Structures (2002). ICE/IStructE. Reinforced Concrete Column Design Charts Appendix.

Manual for the Design of Steelwork Building Structures (2002). ICE/IStructE. Section 11.3, Figures 12, 13, 14 and 15.

NHBC (2007). *National House Building Council Standards*. Appendix 4.2B and 4.2C.

Nickel Development Institute (1994). *Design Manual for Structural Stainless Steel*. NIDI, 12011. Tables 3.1, 3.12, 3.5, 3.6 and A.1.

Pilkington (2000). *Glass and Mechanical Strength Technical Bulletin*. Tables 3, 4, 5, 6, 7 and 8.

Richardson, C. (2000). The Dating Game. *Architect's Journal*. **23/3/00**, 56–59. **30/3/00**, 36–39. **6/4/00**, 30–31.

RMD Kwikform (2002). *Loading data and charts for Super Slim Soldiers*.

Supreme Concrete (2004). Loading data for precast prestressed concrete lintels.

University of Bath (2006). Inventory of Carbon and Energy (ICE). Version 1.5 Beta. Prof Geoffrey Hammond and Craig Jones. Sustainable Energy Research Team (SERT), Department of Mechanical Engineering. www.bath.ac.uk/mech-eng/sert/embodied.

See the 'Useful Addresses' section for contact details of advisory organizations and manufacturers.

Index